Computational Methods in Applied Sciences

Volume 59

This series publishes monographs and carefully edited books inspired by the thematic conferences of ECCOMAS, the European Committee on Computational Methods in Applied Sciences. As a consequence, these volumes cover the fields of Mathematical and Computational Methods and Modelling and their applications to major areas such as Fluid Dynamics, Structural Mechanics, Semiconductor Modelling, Electromagnetics and CAD/CAM. Multidisciplinary applications of these fields to critical societal and technological problems encountered in sectors like Aerospace, Car and Ship Industry, Electronics, Energy, Finance, Chemistry, Medicine, Biosciences, Environmental sciences are of particular interest. The intent is to exchange information and to promote the transfer between the research community and industry consistent with the development and applications of computational methods in science and technology.

Book proposals are welcome at
Eugenio Oñate
International Center for Numerical Methods in Engineering (CIMNE)
Technical University of Catalunya (UPC)
Edificio C-1, Campus Norte UPC Gran Capitán
s/n08034 Barcelona, Spain
onate@cimne.upc.edu
www.cimne.com
or contact the publisher, Dr. Mayra Castro, mayra.castro@springer.com

Indexed in SCOPUS, Google Scholar and SpringerLink.

Tero Tuovinen · Jacques Periaux ·
Dietrich Knoerzer · Gabriel Bugeda ·
Jordi Pons-Prats
Editors

Advanced Computational Methods and Design for Greener Aviation

Springer

Editors
Tero Tuovinen
University of Jyväskylä of Applied Sciences
Jyväskylä, Finland

Dietrich Knoerzer
DK Aeronautics Consultant
Brussels, Belgium

Jordi Pons-Prats
Technical University of Catalonia (UPC)
International Centre for Numerical
Methods in Engineering (CIMNE)
Barcelona, Spain

Jacques Periaux
International Centre for Numerical
Methods in Engineering (CIMNE)
Barcelona, Spain

Gabriel Bugeda
Technical University of Catalonia (UPC)
International Centre for Numerical
Methods in Engineering (CIMNE)
Barcelona, Spain

ISSN 1871-3033 ISSN 2543-0203 (electronic)
Computational Methods in Applied Sciences
ISBN 978-3-031-61108-7 ISBN 978-3-031-61109-4 (eBook)
https://doi.org/10.1007/978-3-031-61109-4

This Springer imprint is published by the registered company Springer Nature Switzerland AG
The registered company address is: Gewerbestrasse 11, 6330 Cham, Switzerland

Preface

The Special Technology Sessions (STS) in ECCOMAS events have a long tradition, as the industry of Europe has a need for advanced methods and tools, especially in Computational Mechanics. Among others, Aeronautics is a discipline progressing specially in design with advanced numerical methods. Special Technology Sessions are mainly focused in aeronautics research and industry topics. Since 1996 in Paris, ECCOMAS Congresses have presented during almost the last three decades aeronautics Special Technology Sessions (STS), dealing with the progress of industrial technologies and research. During these sessions, Senior experts in aeronautics technology areas invite knowledgeable specialists from industry and research institutions ensuring a high scientific and technical quality level of STS.

This book is composed of invited articles collected from the STSs in the CM3 2021 Conference (Computational Multiphysics, Multi scales and Multi Big Data) in Barcelona and the ECCOMAS Congress 2022 in Oslo. These series are launched by the Industry Interest Group (IIG) of ECCOMAS, in charge of the industrial liaison and contacts with the different DGs of the European Commission (EC).

During the CM3 Transport Thematic Workshop 2021 in Barcelone, Spain and the ECCOMAS Congress 2022 in Oslo, Norway eight (8) Mini Symposia (MS) in Aeronautics and seven (7) Special Technology Sessions were addressed and discussed respectively on a relevant technological theme (see below). In these STS (ECCOMAS Congress) or MS sessions (CM3 Transport) several technological achievements of projects from the Research and Innovation Framework Programme Horizon 2020 of the European Union were presented.

Both events included the following themes: aero elastic design, transonic buffet simulation, aerodynamic shape optimization, optimization of aeroengines, additive manufacturing, morphing wing design, transition prediction with suction, refrigerated transport, shock oscillation in transonic compressors, high lift systems, Reduced Order Models (ROM) for supersonic propulsion, GPU software, sparse Linear Systems software, multi-fidelity Metamodels, validation... Among the disciplines discussed in this volume, contributors presented their views on new societal and environmental challenges in four main sections targeting a greener, safer and digitalized climate neutral Transport.

The STSs and MSs are classified into four major components:

- Advanced Computational Methods for Aviation Applications,
- Advanced Computational Methods for Aeronautics Propulsion,
- Aeronautics High-Lift Application and Wing Design,
- Simulation and Application for Transport and Manufacturing.

We are thankful to the organizers of the CM3 Transport Workshop 2021 and the ECCOMAS Congress 2022 for including the STS in the programme of these two events.

Chairs and speakers of STS and MS are also acknowledged as they ensured the STS quality by their effort in setting up interesting sessions and attractive discussions with the audience.

The editors are not forgetting to thank the CIMNE Congress Bureau team, and their appreciated help in integrating STSs and MSs in the two programmes.

Editors are also very grateful to both ECCOMAS and Springer of the series 'Computational Methods in Applied Sciences' for their patience and for disseminating with this volume the scientific computing state of the art in Green Transport digitalized technologies. The STS Organizing team hopes that young and not so young scientists and engineers involved in the computational methods and software tools fields of Transport will enjoy reading this book and benefit from its advanced technical contributions linked to R&D applications.

Jyväskylä, Finland
December 2023

Gabriel Bugeda
Dietrich Knoerzer
Jacques Periaux
Jordi Pons-Prats
Tero Tuovinen

Contents

Part I Advanced Computational Methods for Aviation Applications

1 **Evaluating Performance and Scalability of the Sparse Linear Systems Solver Spliss** 3
Michael Wagner, Jasmin Mohnke, Olaf Krzikalla, and Arne Rempke

2 **Transonic Buffet Simulation Using a Partially-Averaged Navier-Stokes Approach** 15
Andrea Petrocchi, Rene Steijl, and George N. Barakos

3 **Progress of ARI_OPT Software for Aerodynamic Shape Optimization** 29
Chuang Wei, Zhansen Qian, and Long Yang

4 **Verification of Transition Prediction for Flows with Suction Using Linear Stability Theory and e^N-Method** 41
Normann Krimmelbein and Andreas Krumbein

Part II Advanced Computational Methods for Aeronautics Propulsion

5 **Continuous Adjoint-Based Aerothermal and Aeroacoustic Optimization of Aero Engine Components Using PUMA** 59
X. Trompoukis, M. Monfaredi, V. Asouti, K. Tsiakas, and K. C. Giannakoglou

6 **Towards Understanding and Resolving Natural Shock Oscillation in Transonic Compressors** 75
Philipp Leonard Nel, Patrick Grothe, Marius Swoboda, Sergio Pirozzoli, and Julien Weiss

7 **Reduced Order Computational Methods for the Development of Propulsive Technologies for Supersonic Aviation to Achieve Climate Neutrality** ... 95
Karel Van den Borre, Ali C. Ispir, Bora O. Cakir, and Bayindir H. Saracoglu

Part III Aeronautics High-Lift Application and Wing Design

8 **Lattice Boltzmann Simulation of a Deploying Krueger Flap Device** ... 111
Jorge Ponsin and Carlos Lozano

9 **Lessons Learnt from Chimera Method Application to a Deploying Krueger Device** ... 125
Apurva Hasabnis, Hans Maseland, Frédéric Moens, Aleš Prachař, and Jochen Wild

10 **A Validation Program for Dynamic High-Lift System Aerodynamics** ... 145
Jochen Wild, Henning Strüber, Frédéric Moens, Bart van Rooijen, and Hans Maseland

11 **Morphing Effects on the Aerodynamic Performances of a Supercritical Wing's Prototype in Transonic Flow Conditions** ... 159
C. Jimenez-Navarro, J. B. Tô, C. Rouaix, P. Doerffer, A. Marouf, R. El Akoury, and M. Braza

12 **Numerical Simulation of an A320 Morphing Wing Through Frequency Modulation in the Vicinity of the Trailing Edge in Subsonic Regimes** ... 175
C. Rouaix, C. Jiménez-Navarro, M. Carvalho, A. Marouf, Y. Hoarau, R. El Akoury, C. Raibaudo, G. Harran, H. Hangan, J. F. Rouchon, and M. Braza

Part IV Simulation and Application for Transport and Manufacturing

13 **Design Optimization Based on Multi-fidelity Metamodels** ... 191
Alberto Clarich, Luca Battaglia, Carlo Poloni, Lucia Parussini, Haysam Telib, and Angela Scardigli

14 **Krueger High-Lift System Design Optimization** ... 205
Emiliano Iuliano, Domenico Quagliarella, and Jochen Wild

15 **Toward Integrated Aircraft Design with Carbon Fiber Reinforced Thermoplastic (CFRTP)** ... 229
Yoshiaki Abe, Keiichi Shirasu, Shugo Date, Tomonaga Okabe, and Shigeru Obayashi

16 Challenges and Current Solutions of Refrigerated Transportation 241
Pekka Neittaanmäki, Kirill Akimov, Veronika Akimova, and Ronja Tuovinen

17 Effect of Computational Generative Product Design Optimization on Part Mass, Manufacturing Time and Costs: Case of Laser-Based Powder Bed Fusion 257
Patricia Nyamekye, Rohit Lakshmanan, and Heidi Piili

18 A Review of Microscale and Mesoscale Simulation of Laser Powder Bed Fusion 275
Aditya Gopaluni, Heidi Piili, Ashish Ganvir, and Antti Salminen

Part I
Advanced Computational Methods for Aviation Applications

Chapter 1
Evaluating Performance and Scalability of the Sparse Linear Systems Solver Spliss

Michael Wagner, Jasmin Mohnke, Olaf Krzikalla, and Arne Rempke

Abstract Solving linear equation systems is an integral part of implicit methods in computational fluid dynamics (CFD). The sparse linear system solver Spliss aims to provide a linear solver library that, on the one hand, is tailored to requirements of CFD applications but, on the other hand, independent of the particular CFD solver. Focusing on the specific task of solving linear systems allows for integrating more advanced, but also more complex, hardware-specific optimizations, while at the same time hiding this complexity from the CFD solver. Spliss enables the execution of the computationally intensive linear solver on GPUs without the necessity of any code adaption. This work evaluates performance and scalability of Spliss using CODA as an example. CODA is the CFD software being developed as part of a collaboration between the French Aerospace Lab ONERA, the German Aerospace Center (DLR), Airbus, and their European research partners. CODA is jointly owned by ONERA, DLR and Airbus. The evaluation includes an assessment of the scalability on an HPC system based on the AMD Naples architecture and demonstrates the seamless integration of GPUs on a cluster based on Nvidia V100 GPUs.

Keywords Computational fluid dynamics (CFD) · Sparse linear system solver · Aircraft aerodynamics

1.1 Introduction

For future aircraft design, Computational fluid dynamics (CFD) is a key technology on the road to solving global challenges like climate change or mobility. CFD simulations are indispensable to reach European goals of drastic reductions in aviation emissions such as a reduction of 75% of CO_2 emissions, 90% of NOx emission and 65% of perceived aircraft noise by the middle of the century while at the same

M. Wagner (✉) · J. Mohnke · O. Krzikalla · A. Rempke
German Aerospace Center (DLR), Institute of Software Methods for Product Virtualization, Dresden, Germany
e-mail: m.wagner@dlr.de

T. Tuovinen et al. (eds.), *Advanced Computational Methods and Design for Greener Aviation*, Computational Methods in Applied Sciences 59,
https://doi.org/10.1007/978-3-031-61109-4_1

time increasing affordable and reliable connectivity within the European Union, its neighbors and partners and ensuring the competitiveness of European industry [1]. For this, newly developed aircraft have to become significantly lighter and more aerodynamically efficient, in combination with the introduction of innovative flight control and an intelligent mix of alternative propulsion system concepts.

Computational fluid dynamics simulations for aircraft aerodynamics are already today imperative in the aircraft design process. They allow to reduce cost and time of aircraft development by omitting unnecessary prototyping, wind tunnel experiments and real flight tests and, thus, accelerate the introduction of progressive technology and dynamic improvements. In particular, the progress towards the virtual product, i.e., high-precision numerical representation of an aircraft and all its characteristics and components, leads to faster development cycles; starting from product development up to approval, production, maintenance and decommissioning [2].

Moreover, high-precision numerical simulations are inevitable for the assessment of new aircraft designs to provide reliable insight into new aircraft technologies and reach best overall aircraft performance through integrating aerodynamics, structural mechanics and systems design. Such simulations require vast computational resources that are met partly by today's high performance computing (HPC) systems. This year marks the advent of the first true exascale HPC system, i.e. a system capable of calculating more than 10^{18} double precision operations per second, consisting of over 8.7 million cores [3]. CFD simulations aiming to utilize these powerful systems need to be, first, highly parallel to take advantage of the increasing number of compute cores and, second, support the usage of heterogeneous systems consisting of new processor architectures and hardware accelerators such as GPUs.

The sparse linear systems solver Spliss [4] is designed to assist with this challenge by offering the efficient solving of large linear equation systems that result from the discretization of the Reynolds-averaged Navier-Stokes (RANS) equations in CFD methods. Since Spliss focuses on the specific task of solving linear equation systems, it integrates more advanced, but also more complex, hardware-specific optimizations, while at the same time hiding this complexity from the CFD solver. Spliss enables the efficient and transparent execution of the computationally intensive linear solver on new architectures and hardware accelerators such as GPUs. This way, the CFD solver can leverage new architectures and hardware accelerators without the necessity of any code adaptation in the CFD solver.

This work evaluates performance and scalability of Spliss using the CODA CFD software as an example. CODA is the CFD software being developed as part of a collaboration between the French Aerospace Lab ONERA, the German Aerospace Center (DLR), Airbus, and their European research partners. CODA is jointly owned by ONERA, DLR and Airbus. It is one of the key next-generation engineering applications represented in the European Centre of Excellence for Engineering Applications (EXCELLERAT) [5].

The contribution of this work is, first, an assessment of the performance and scalability of Spliss with CODA on the largest available partition of the German Aerospace Center's CARA HPC production system based on the AMD Naples architecture using the the NASA common research model in a strong scaling scenario.

Second, a demonstration of the seamless integration of hardware accelerators such as GPUs into CODA via Spliss on a cluster based on Nvidia V100 GPUs.

The following sections provide background on the sparse linear systems solver Spliss (Sect. 1.2) and the CFD software CODA (Sect. 1.3). Section 1.4 introduces the CARA HPC system, the used test case and presents the results of the scalability assessment. Similarly, Sect. 1.5 introduces the Nvidia V100 test cluster, the used test case and presents first results of Spliss executing the linear solver of CODA on GPUs. Finally, Sect. 1.6 summarizes the presented work and draws conclusions.

1.2 The Sparse Linear Systems Solver

The sparse linear systems solver Spliss is a library that enables a CFD software like CODA to solve large distributed linear equation systems originating from the discretization RANS equations. Such linear equation systems are sparse, since each equation only relates a few of the millions of unknowns. The prescribed couplings however have a structure that results in a block-structured sparse matrix for the linear equation system to be solved.

Spliss allows to specify and solve such linear equations with sparse matrices of dense blocks. A single dense matrix block typically corresponds to the couplings between the different physical quantities within a physical element or on the facette between two elements. For a classical finite volume discretization, the blocks have a size of e.g. 7×7 for a two-equation turbulence model, and for higher order Discontinuous Galerkin approaches these dense blocks can become much larger. Spliss supports both fixed block sizes for the whole matrix (same order approach) and different block sizes within the same equation system (different order approach).

Another important feature of Spliss is that matrix entries can be of mixed scalar data type. This is particularly important when solving harmonic balance or time-spectral methods, where part of the matrix (typically the diagonal blocks) contain complex numbers, while other parts remain real-valued, see Fig. 1.1. Also using

Fig. 1.1 Supported matrices in Spliss: sparse block-structure with constant (left) or variable (middle) block sizes, different scalar types (right)

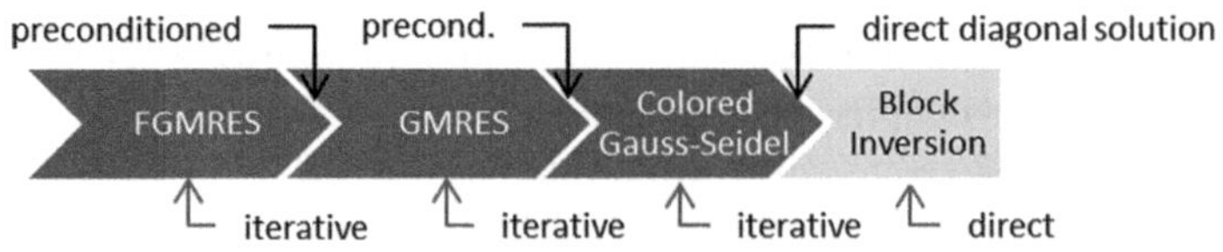

Fig. 1.2 A possible solver chain in Spliss modeling the inner-outer GMRES method preconditioned by a point-implicit multi-color Gauss-Seidel method

different precision scalar types is possible with this approach, which allows saving memory and computational effort, e.g., when evaluating matrix vector products.

In order to solve the sparse linear equation system, Spliss includes a wide range of iterative solvers and preconditioners. These solver components can be stacked together in a flexible way, allowing multiple layers of different solver types operating on potentially different linear operators, see Fig. 1.2. This permits providing fast preconditioners in the inner hot loop while utilizing more exact higher-level operators in the outer loops. Some solver components are specifically tailored for the demands of CFD simulations, e.g., the LinesInversion for handling the high aspect-ratio cells in the boundary layer of a mesh for flows with high Reynolds numbers [4].

Spliss is designed and implemented to efficiently leverage the computational resources of contemporary and emerging HPC platforms. This includes techniques such as one-sided communication, hybrid and heterogeneous parallelization. The use of the C++ template mechanism shifts decisions such as the usage of a GPU and the selection of an appropriate memory layout to the compile time. Thus, a shared library object of Spliss compiled and built on a particular HPC cluster respects the properties of that cluster. On the other hand, Spliss is designed in such a way, that it does not expose these properties to the library interface.

By that means, it is sufficient to link against a GPU-enabled Spliss library in order to benefit from available GPUs. An application is not even required to use the GPU framework during its own compilation process thanks to an automatic explicit template instantiation process. If multiple GPUs are available on a compute node, Spliss distributes those GPUs to the processes running on the node. In addition, Spliss supports CUDA-aware MPI implementations [6]. Those implementations combine MPI and CUDA, so that GPU buffers can be directly passed to MPI calls; omitting a detour via the host buffer.

1.3 The CODA CFD Software

The development of computational fluid dynamics software has a long history at the German Aerospace Center (DLR). Currently, the *TAU* CFD package [7] developed and maintained by DLR is in production in the European aircraft industry, research organizations and academia since more than 20 years. For instance, TAU played a vital role in the Airbus A380 and A350 wing design. As state-of-the-art for its time,

TAU implements a classical MPI parallelization to simulate steady and unsteady external aerodynamic flows using a second order finite-volumes discretization.

To design a modern concept for HPC from scratch, in 2012 DLR initiated the development of a new, flexible, unstructured CFD solver called *Flucs* [8]. Thereby, the focus was set on algorithmic efficiency using strong implicit solvers, higher-order spatial discretization via the Discontinuous Galerkin method featuring hp-adaptation in addition to finite volumes with maximum code share, and seamless integration into Python-based multi-disciplinary process chains via *FlowSimulator* [9, 10]. While the development of Flucs had been started at DLR, it since has become part of a larger cooperation that is driven by Airbus, the French aerospace lab ONERA, and DLR. The joint development of the CFD software based on Flucs was named *CODA* (CFD for ONERA, DLR and Airbus) to honor the new collaboration and the involvement of all three partners pursuing the joint effort and co-development.

The CODA CFD software implements classical domain decomposition to support distributed-memory parallelism via MPI and the GASPI [11] implementation GPI-2. This Partitioned Global Address Space (PGAS) library features efficient one-sided communication to reduce network traffic and latency. Furthermore, CODA improves scalability by allowing the overlap of halo-data communication with computation to hide network latency. In addition to classical domain decomposition, CODA employs a hybrid two-level parallelization to utilize shared-memory parallelism for multi- and many-core architectures [12]. CODA implements sub-domain decomposition, where each domain is further partitioned into sub-domains, each of which being processed by a dedicated software thread that is mapped one-to-one to a hardware thread to maximize data locality. The hybrid approach allows utilizing all parallelism layers and provide a flexible adaption to different hardware architectures [13, 14].

1.4 Evaluation on an AMD Naples HPC System

This section provides an assessment of the scalability of Spliss with CODA on the largest available partition of the German Aerospace Center's CARA HPC production system based on the AMD Naples architecture using the the NASA common research model in a strong scaling scenario. The *Computer for Advanced Research in Aerospace* (CARA) is one of the German Aerospace Center's main HPC systems providing 1.7 TFlop/s peak performance. The system offers 2280 compute nodes, whereas each compute node consists of two AMD EPYC 7601 (32 cores at 2.2 GHz) with four dies of eight cores each. In total, the system offers 145,920 compute cores.

The AMD Naples architecture within this system includes eight NUMA (non-uniform memory access) domains and three NUMA distances: first, to the memory of the seven other cores on the same die, second, to the memory on the three other dies on the same chiplet (socket) and, third, to the memory located on the other chiplet. In addition, only four of the eight cores on each die share a last level cache (L3 cache), which presents an additional difference in memory access latency depending on the locality of the data; whether it is in the shared L3 cache of the according core or in the adjoining L3 cache on the same die.

1.4.1 The Test Case

The test case for the scalability evaluation is based on the NASA Common Research Model (CRM) from the fifth AIAA CFD Drag Prediction Workshop [15]. It simulates the external airflow at subsonic speed and computes typical characteristics like air velocity and direction, pressure and turbulence via a one-equation turbulence model. Figure 1.3 visualizes the computed air pressure and shows aircraft configuration and mesh on the left and the airflow around the wing and fuselage on the right. The CRM test case is well studied and provides experimental data as well as numerical solutions by other CFD applications for comparison.

For the CRM test case, CODA solves the Reynolds-averaged Navier-Stokes equations (RANS) with a Spalart-Allmaras one-equation turbulence model in its negative form (SAneg). It uses a second-order finite-volume spatial discretization with an implicit Euler time integration. For the linear problem, a Block-Jacobi solver with LU decomposition is applied and solved via Spliss. For this case, the vast majority of the iteration phase is spent in the linear solver, i.e. Spliss, thus, measuring the iteration time of CODA provides a very close estimation of the performance and scalability of Spliss within a real-world example. While results may be biased by CODA, the measurements show the minimum performance and scalability of CODA together with Spliss since performance degrading effects accumulate. In this sense, CODA and Spliss may achieve better performance and scalability individually.

The test case operates with a small, unstructured mesh with 5.2 million points and 10.2 million prisms that is obtained by splitting each hexahedron in the original mesh into two prisms such that the geometry's surface mesh is purely triangular. This mesh is about one order of magnitude smaller than typical industrial cases and was chosen to allow a strong-scaling analysis at relatively small core counts, i.e.,

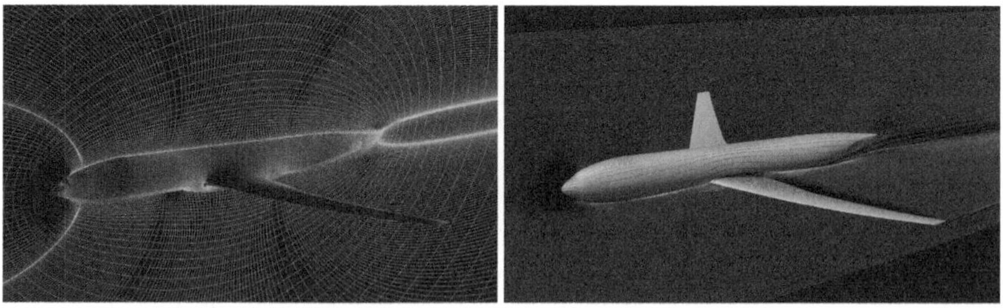

Fig. 1.3 Visualization of the test case simulation: aircraft configuration with mesh (left) and airflow around wing and fuselage (right); both with air pressure as color gradient

neither the purely prismatic volumes nor the small number of cells allow for high CFD accuracy in the boundary layer.

1.4.2 Measurement Setup

For the scalability evaluation all software threads are bound to a hardware thread to ensure thread affinity and three hybrid-parallel setups are evaluated to identify the impact of the memory hierarchy:

- 16 MPI processes per node with 4 OpenMP threads each. This way all four threads are in the same NUMA domain and share the same L3 cache.
- 8 MPI processes per node with 8 OpenMP threads each. This way all eight threads are in the same NUMA domain but are split across two L3 caches.
- 4 MPI processes per node with 16 OpenMP threads each. This way the 16 threads are split across two NUMA domains.

In addition to the above setups using one hardware thread per core, the according setups with two-way simultaneous multi-threading are recorded, too. For these setups the number of OpenMP threads per MPI process is doubled, e.g. the version with 16 MPI processes and 4 OpenMP threads each is also measured with 16 MPI processes and 8 OpenMP threads each, whereas the 8 OpenMP threads run on the same four cores as the 4 OpenMP threads.

All measurements were executed only a single time due to the large core counts, according costs and wait times in the queue. This must be kept in mind when discussing the significance of individual data points. In general, the recorded runtimes are consistent in themselves; nonetheless, data points should be considered as general trend rather than exact values. Parallel runtimes, in particular, can be affected by the specific scheduling to nodes and the overall load on the system, among others. In that sense, the recorded runtimes reflect typical behavior that users can expect in production mode; not isolated benchmark runs in a close-to-perfect environment.

1.4.3 Results

Figure 1.4 shows the general scaling behavior for the different setups of MPI processes to OpenMP threads without and with enabled two-way simultaneous multithreading for 1 to 512 nodes, i.e. 64 to 32.768 cores. On the CARA system 512 nodes was the largest partition that could be reasonably used during normal operation.

Without simultaneous multi-threading CODA with Spliss achieves about 90% parallel efficiency at 4096 cores and 59% parallel efficiency at 32,768 cores within the iterative phase. This represents very good strong-scaling behavior for such a small mesh, where at 32,768 cores on average only 312 elements are assigned to

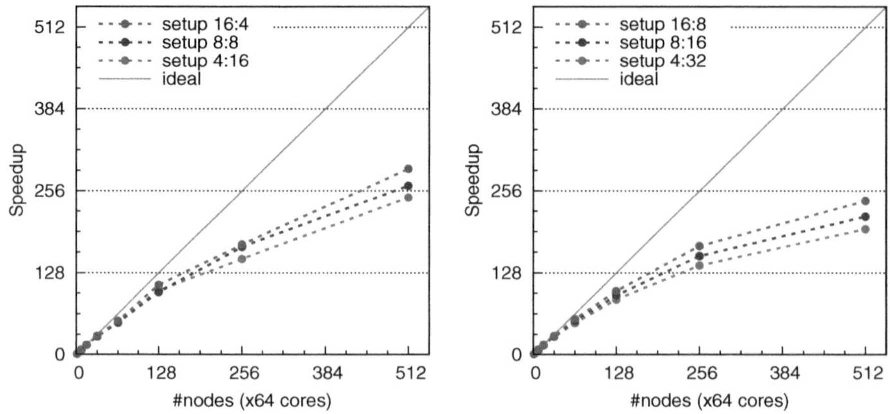

Fig. 1.4 Speedup for 1 to 512 nodes (64 to 32.768 cores) for different MPI rank to OpenMP thread ratios: without (left) and with simultaneous multi-threading (right)

each software thread. As expected for the AMD Naples architecture, the best setup is with four threads per MPI process, so that all four threads are executed on the four cores that share the same last-level cache. The second-best setup is with eight threads per MPI process, so that all eight threads are executed within a single NUMA domain. The execution of threads across NUMA domains (16 threads per MPI process) leads to further reduced performance.

With enabled simultaneous multi-threading CODA with Spliss achieves about 88% parallel efficiency at 4096 cores and 47% parallel efficiency at 32,768 cores in the iterative phase. Again, this represents very good strong-scaling behavior for such a small mesh, where on average only 150 elements are assigned to each software thread at 32,768 cores. Consequently, the scalability is slightly reduced since each thread has only half the computational load. In that sense, computing a test case with 10.2 million prisms across 65,536 threads sets an extreme case and highlights the excellent scaling behavior of CODA even on very little computational load per thread. In comparison, typical workloads used in production have at least one or two orders of magnitude more elements per thread.

Although the setups with enabled simultaneous multi-threading show slightly lower parallel efficiency at scale, they provide significantly better compute performance. Comparing the individual simultaneous multi-threading setups with their non-simultaneous multi-threading counterparts, the setups with enabled simultaneous multi-threading have a 15–20% reduced runtime, which might also be a factor in the slightly reduced scalability.

1.5 Evaluation on an Nvidia V100 GPU System

This section highlights the benefits of using Spliss to incorporate accelerators, in this case GPUs, for solving large, sparse linear equations systems while no changes need to be implemented in the CFD software, in this case CODA. For this assessment, a test case is run on an Nvidia V100 GPU system, which has compute nodes consisting of two Intel Xeon 6230 CPUs with 20 cores each and four Nvidia Tesla V100 GPUs.

1.5.1 The Test Case

The test case for the evaluation is based on the NASA 3D Onera M6 wing test case in version 2308 [16]. It is widely used for research purposes and well-studied. Both experimental data and numerical solutions are available for validation of results.

The test case solves the Reynolds-averaged Navier-Stokes equations (RANS) with a Spalart-Allmaras one-equation turbulence model in its negative form (SAneg). A Spliss Block Inversion preconditioned Block-Jacobi Solver is applied to solve the linear system. The mesh used for the test case consists of about 8.65 M volume elements which is large enough to achieve peak performance per single V100 GPU with the chosen linear solver components for the scale of multi-GPU results presented.

1.5.2 Measurement Setup

The evaluation compares a single- and multi-GPU setup with the Spliss linear solver offloaded to GPUs on one and multiple compute nodes to a CPU-only setup on the same system. To compare one GPU against one CPU only two of the four GPUs per compute node are used. For the parallel setups on CPU, two MPI proecesses per node, i.e. one MPI processes per socket, with 20 OpenMP threads each are launched and bound to a core. For the GPU accelerated execution the CPU setup is identical and additionally each process offloads to one GPU. Switching from the CPU-only setup to taking advantage of Spliss GPU acceleration merely requires linking to a Spliss variant that is compiled for GPU, no changes on the CODA side are necessary.

Both the runtime for the entire time integration iteration phase with a fixed number of 100 iterations and solely the time spent in the linear solver are measured. The latter includes any host-to-device or device-to-host memory transfers that are necessary when offloading this section of computation to GPUs.

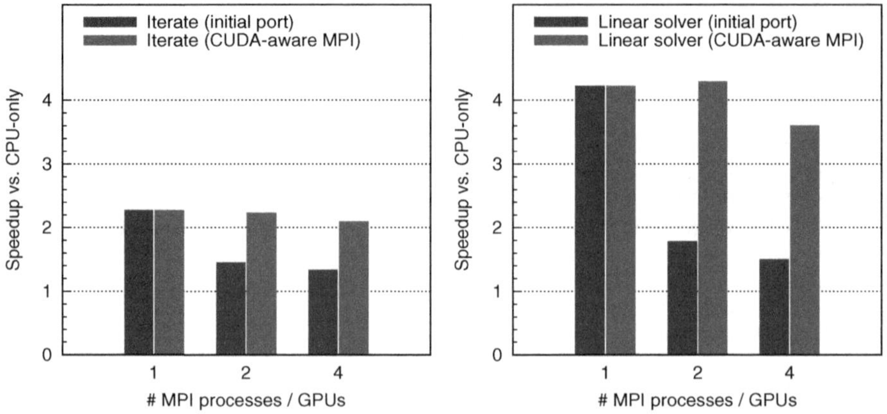

Fig. 1.5 GPU speedup of the entire iteration phase (left) and solely the linear solver (right) with MPI communication via host buffers (initial port) and using CUDA-aware MPI

1.5.3 Results

Acceleration of the Spliss linear solver by offloading computations provides performance gain with a single GPU compared to a single 20-core CPU on the test system compute node. With the use of multiple GPUs both intra- and inter-node MPI communication required for halo updates can prove to be a bottleneck when being carried out via host memory despite all data is residing in the GPU memory during the computation. In order for a gain seen with GPU acceleration to translate to a multi-GPU case, CUDA-aware MPI with GPUDirect acceleration and, thus, optimized transfers from host to device buffers and vice versa need to be taken advantage of (currently work in progress). This gain can be observed both for the intra-node case as depicted in Fig. 1.5 with two MPI processes as well as inter-node communication with four MPI processes distributed on two compute nodes.

For the parallel configuration, the number of available GPUs predetermines the number of used MPI processes that are used. Though this is generally the preferred setup from a GPU performance perspective it is not a fixed constraint given by Spliss and can be circumvented by using Nvidia MPS (Multi Process Service) [17], which enables the efficient offloading of work from multiple processes to a single GPU. Therefore, the seamless integration of GPU usage in application codes via Spliss also extends to the parallel configuration setup.

1.6 Conclusion

This work presents an evaluation of performance and scalability of the sparse linear systems solver Spliss with the CODA CFD software for aircraft aerodynamics. The test case based on the NASA common research model achieves 90% parallel efficiency at 4096 cores and 59% parallel efficiency at 32,768 cores in a strong scaling scenario despite running on a very small mesh with very little computational load per thread; an extreme case rarely approached in production simulations. Furthermore, the assessment highlights that best hybrid-parallel performance is reached when using only four threads per MPI process, so that these threads share the same last level cache, which is specific to the memory layout of the AMD Naples architecture and stands in contrast to other architectures. A second test case based on the 3D ONERA M6 wing shows that Spliss allows leveraging GPUs without the necessity of any code adaption in the CFD solver. By providing the seamless integration of GPUs, Spliss accelerates the test case by a factor of 2.2 in a node-wise comparison.

Acknowledgements This work has been supported by the EXCELLERAT project, which has received funding from the European Union's Horizon 2020 research and innovation programme under grant agreement No. 823691.

References

1. Directorate-General for Mobility and Transport (European Commission), Directorate-General for Research and Innovation (European Commission) (2012) Flightpath 2050: Europe's vision for aviation: maintaining global leadership and serving society's needs. https://doi.org/10.2777/15458
2. Guiding concepts for DLR aeronautics research. https://www.dlr.de/EN/research/aeronautics/guiding-concepts.html. Accessed 29 July 2022
3. Strohmaier E, Dongarra J, Simon H, Meuer M (2022) The 59th top 500 list. https://www.top500.org/lists/top500/2022/06/. Accessed 29 July 2022
4. Krzikalla O, Rempke A, Bleh A, Wagner M, Gerhold T (2021) Spliss: a sparse linear system solver for transparent integration of emerging HPC technologies into CFD solvers and applications. In: New results in numerical and experimental fluid mechanics XIII, pp 635–645. https://doi.org/10.1007/978-3-030-79561-0_60
5. The European Centre of Excellence for Engineering Applications (EXCELLERAT). http://www.excellerat.eu. Accessed 29 July 2022
6. Kraus J (2013) An introduction to CUDA-Aware MPI. https://developer.nvidia.com/blog/introduction-cuda-aware-mpi. Accessed 29 July 2022
7. Schwamborn D, Gerhold T, Heinrich R (2006) The DLR TAU code: recent applications in research and industry. In: Proceedings of the European conference on computational fluid dynamics, ECCOMAS CFD (2006)
8. Leicht T, Vollmer D, Jägersküpper J, Schwöppe A, Hartmann R, Fiedler J, Schlauch T (2016) DLR-Project Digital-X—next generation CFD solver 'Flucs'. Deutscher Luft-und Raumfahrtkongress 2016
9. Meinel M, Einarsson G (2010) The FlowSimulator framework for massively parallel CFD applications. In: PARA
10. Huismann I, Reimer L, Strobl S, Eichstädt J, TschÜter R, Rempke A, Einarsson G (2021) Accelerating the flowsimulator: profiling and scalability analysis of an industrial-grade CFD-CSM toolchain. In: 9th Edition of the international conference on computational methods for coupled problems in science and engineering (COUPLED PROBLEMS 2021). https://doi.org/10.23967/coupled.2021.008

11. Alrutz T, Backhaus J, Brandes T, End V, Gerhold T, Geiger A, Grünewald D, Heuveline V, Jägerskùpper J, Knüpfer A, Krzikalla O, Kügeler E, Lojewski C, Lonsdale G, Müller-Pfefferkorn R, Nagel WE, Oden L, Pfreundt FJ, Rahn M, Sattler M, Schmidtobreick M, Schiller A, Simmendinger C, Soddemann T, Sutmann G, Weber H, Weiss JP (2013) GASPI—a partitioned global address space programming interface. In Facing the multicore-challenge III, LNCS 7686, pp 135–136
12. Jägerskùpper J, Vollmer D (2022) On highly scalable 2-level-parallel unstructured CFD. In: 8th European congress on computational methods in applied sciences and engineering (ECCOMAS 2022). https://doi.org/10.23967/eccomas.2022.208
13. Wagner M, Jägerskùpper J, Molka D, Gerhold T (2021) Performance analysis of complex engineering frameworks. In: Tools for high performance computing, pp 123–138. https://doi.org/10.1007/978-3-030-66057-4_6
14. Wagner M (2023) Scalability evaluation of the CFD solver CODA on the AMD naples architecture. In: Sustained simulation performance 2021, pp 95–106. https://doi.org/10.1007/978-3-031-18046-0_7
15. Vassberg J (2011) A unified baseline grid about the common research model wing/body for the fifth AIAA CFD drag prediction workshop. In: 29th AIAA applied aerodynamics conference. https://doi.org/10.2514/6.2011-3509
16. Destarac D, Dumont A (2016) ONERA M6 Wing Test-Case, Original and TMR. https://turbmodels.larc.nasa.gov/onerawingnumerics_val.html. Accessed 29 July 2022
17. Nvidia: Multi-Process Service (2021) https://docs.nvidia.com/deploy/pdf/CUDA_Multi_Process_Service_Overview.pdf

Chapter 2
Transonic Buffet Simulation Using a Partially-Averaged Navier-Stokes Approach

Andrea Petrocchi, Rene Steijl, and George N. Barakos

Abstract This work investigates the ability of the partially averaged Navier-Stokes (PANS) method to accurately predict self-sustained shock oscillations, also known as transonic buffet, on a supercritical aerofoil at high Reynolds numbers. Attention is paid to the comparison with unsteady Reynolds-averaged Navier Stokes (URANS) results to show the benefits of PANS, in resolving flow unsteadiness on affordable CFD grids. The role of the mesh metrics in the formulation of the PANS model is emphasized, as well as the relation of the mesh metrics with the spatiotemporal discretisation used for the numerical simulations. The aim is to extend the use of PANS to flow cases involving shock-wave boundary layer interactions to obtain accurate predictions without the need for very expensive computations.

Keywords Computational fluid dynamics (CFD) · Shock oscillations PANS · URANS

2.1 Introduction

Computational fluid dynamics (CFD) has been extensively used to simulate Shock wave/boundary layer interaction (SBLI), especially at transonic conditions. Reynolds-averaged Navier Stokes (RANS) simulations have become the most popular tool in CFD because of their reduced CPU cost and wide range of applicability. Nevertheless, the prediction of SBLI phenomena with RANS simulations is not always accurate, especially when the flow is unsteady or separated. Following the increase in the availability of computational resources, direct numerical simulations

A. Petrocchi · R. Steijl · G. N. Barakos (✉)
CFD Lab, School of Engineering, University of Glasgow, Glasgow G128QQ, Scotland, UK
e-mail: george.barakos@glasgow.ac.uk

A. Petrocchi
e-mail: andrea.petrocchi@glasgow.ac.uk

R. Steijl
e-mail: rene.steijl@glasgow.ac.uk

T. Tuovinen et al. (eds.), *Advanced Computational Methods and Design for Greener Aviation*, Computational Methods in Applied Sciences 59,
https://doi.org/10.1007/978-3-031-61109-4_2

(DNS) and large eddy simulations (LES) are no longer out of reach for canonical flows at moderate Reynolds numbers. Nevertheless, cases of aeronautical interest, e.g. flows around transonic wings, cannot yet be studied with such expensive methods. Therefore, in recent years, hybrid RANS/LES approaches have been introduced as a more viable option even for high Reynolds number flows.

In this context, the partially-averaged Navier-Stokes (PANS) [1] method was introduced as a bridging model between RANS and DNS. The ratios of modelled-to-total turbulent kinetic energy and dissipation, f_k and f_ϵ, respectively, are introduced to control the filter width of the turbulence simulation. When both values are equal to unity the PANS equations are identical to RANS, whereas a DNS solution is obtained when these coefficients tend to zero. The main advantage over URANS is that the resolution of a part of turbulent fluctuations is enabled, leading to an improvement in terms of predictions and greater detail of the flow structures. This feature is particularly useful for flows with separation, whether shallow or fixed-point. Buffet flows are in the former category since they exhibit an alternation of boundary layer separation and reattachment. An advantage of PANS over other hybrid RANS/LES models is the potential to deliver improved results even adopting a RANS-like grid, as long as the values of the two parameters f_k and f_ϵ remain reasonably high [2]. Conversely, the switch to LES-mode of techniques like the DES family [3] in separated flow regions, reintroduces the requirements in terms of the spatiotemporal discretisation present in LES computations, leading to a noticeable increase in CPU costs. So far, PANS simulations have been mainly tested on canonical incompressible flows. Flows around circular [4, 5] and square [6] cylinders, backward-facing steps [7], turbulent channels [8] or humps and hills [9, 10] were mostly investigated. Rare exceptions were found for transonic and supersonic flows [11–13].

Over the last two decades, there has been substantial research on transonic buffet flows, and the prediction of this phenomenon revealed to be challenging and not always accurate. For extensive coverage of the topic, the reader is referred to the review papers of Lee [14] and Giannelis et al. [15]. A number of authors [16–18] investigated the ability of URANS to capture buffet. Still, there is no consensus on the topic, and hybrid RANS/LES approaches became more popular in the analysis of the flow around two-dimensional [18, 19] and three-dimensional configurations [20]. Scale-resolving simulations were found to represent the buffet phenomenon with higher accuracy and to show significant improvements in the prediction of buffet. The main drawback of these methods is the considerably higher CPU costs stemming from the fine spatiotemporal discretisation required. Regardless, even if computations of this type can be affordable for a simplified geometry, the cost becomes prohibitive for complex 3D cases. Here rises the need for a computational approach with is accurate for the prediction of the buffet without introducing the costs of the aforementioned approaches.

In this paper, we aim at investigating the capability of PANS to predict transonic buffet around the OAT15A aerofoil [21]. Attention is paid to the efficiency of the computational method over URANS, and the capability of PANS to capture the correct flow physics at a reasonable CPU cost. The article is structured as follows: in

Sect. 2.2 the compressible PANS formulation is presented; in Sect. 2.3 the results for the flow around the OAT15A aerofoil at both pre- and post- buffet onset are presented before drawing some conclusions in Sect. 2.4.

2.2 Mathematical Model

2.2.1 PANS Formulation

The partially-averaged Navier-Stokes (PANS) formulation was introduced by Girimaji et al. [1] as a bridging model between RANS and DNS. This method is based on a RANS paradigm, where the blending is obtained by means of the user-prescribed unresolved-to-total ratios of turbulent kinetic energy f_k and dissipation f_ϵ, bounded between 0 and 1, acting on the turbulence closure equations. They read:

$$f_k = \frac{k_u}{k}, \quad f_\epsilon = \frac{\epsilon_u}{\epsilon}, \tag{2.1}$$

where u subscripts stand for unresolved and the quantities at the denominator are the total ones. The PANS method was initially derived for k-ϵ closures and then extended to the k-ω and the SST models, where f_ϵ is replaced by the unresolved-to-total turbulence frequency $f_\omega = f_\epsilon / f_k$. These formulations inherit from the parent RANS models an eddy viscosity based on a Boussinesq approximation, that is reduced with respect to the RANS case because of the effect of the f_k parameter: since only a fraction of the turbulent kinetic energy is modelled, the corresponding value of the eddy viscosity is reduced.

In this work, the SST-PANS formulation is adopted. It reads

$$\frac{\partial(\rho k)}{\partial t} + \frac{\partial(\rho U_j k)}{\partial x_j} = P_k - \beta^* \rho k \omega + \frac{\partial}{\partial x_j}\left[\left(\mu + \mu_t \sigma_k \frac{f_\omega}{f_k}\right)\frac{\partial k}{\partial x_j}\right], \tag{2.2}$$

$$\frac{\partial(\rho \omega)}{\partial t} + \frac{\partial(\rho U_j \omega)}{\partial x_j} = \frac{\gamma}{\nu_t} P_k - \beta' \rho \omega^2 + \frac{\partial}{\partial x_j}\left[\left(\mu + \mu_t \sigma_\omega \frac{f_\omega}{f_k}\right)\frac{\partial \omega}{\partial x_j}\right] \tag{2.3}$$

$$+ 2\frac{f_\omega}{f_k}(1 - F_1)\frac{\rho \sigma_{\omega 2}}{\omega}\frac{\partial k}{\partial x_j}\frac{\partial \omega}{\partial x_j}, \tag{2.4}$$

where ρ is the density, U_j is the flow velocity, μ is the dynamic molecular viscosity and μ_t is the turbulent viscosity. Here, the turbulent kinetic energy k and frequency ω are the modelled fractions and the u subscripts were dropped for sake of simplicity. In the ω-equation, $\beta' = \left(\gamma\beta^* - \frac{\gamma\beta^*}{f_\omega} + \frac{\beta}{f_\omega}\right)$; F_1 is the SST blending function while

γ, β, $\beta*$, σ_k, σ_ω are the model coefficients [22]. The turbulent viscosity is calculated as in the reference [22].

A debated point is how to prescribe the value of f_k. Three main approaches can be distinguished: *Constant*, f_k is fixed in space and time; *Static*, f_k follows an optimal spatial distribution found after a preliminary RANS simulation; *Dynamic*, f_k is a function of space and time. There is no consensus on which of these three approaches should be used; the constant approach is preferred by its advocates because no commutation errors are introduced adopting a spatially and temporally constant filter, although an *a priori* calibration is required. A variable approach is preferred in flows where confined regions of high turbulent content and anisotropies in the grid are present since a variable filter allows an optimal usage of resources by adapting to both the local turbulent content and the grid size. The dynamic approach introduces some additional CPU costs over the static and constant approaches due to the dynamic evaluation of f_k. Therefore, a static approach is more suitable to problems with localised statistically steady turbulence, whereas the dynamic approach is required where coming to intrinsically unsteady flows like for the case of buffet flows and hence is adopted in this work. The estimate adopted in this work is:

$$f_k = C_{\text{PANS}} \left(\frac{\Delta}{L_t} \right)^{2/3}, \tag{2.5}$$

where $L_t = \sqrt{k}/(C_\mu \omega)$ it the local turbulent length scale, Δ is the local grid size and C_{PANS} is a model constant. The f_ϵ parameter is usually set to one, under the assumption that all dissipative scales are not resolved in the computation. This approach is suitable for high Reynolds numbers for which there is a net separation between energy-containing and dissipative scales [1].

2.2.2 HMB3 Solver

The PANS method was implemented in the helicopter multi-block (HMB) solver of the University of Glasgow [23], a three-dimensional, fully implicit, structured, multi-block code for the solution of the Navier-Stokes equations. The Navier-Stokes equations are discretised using a cell-centred finite volume approach. The computational domain is divided into a finite number of non-overlapping control volumes, and the governing equations are applied to each cell. Also, the Navier- Stokes equations are re-written in a curvilinear co-ordinate system which simplifies the formulation of the discretised terms since body-conforming grids are adopted here. The spatial discretisation of the equations leads to a set of ordinary differential equations in time,

$$\frac{d}{dt}(\mathbf{W}_{ijk} V_{ijk}) = -\mathbf{R}_{ijk}(\mathbf{W}), \tag{2.6}$$

where **W** and **R** are the vectors of cell conserved variables and residuals respectively. The convective terms are discretised using the Osher's upwind scheme. MUSCL variable extrapolation is used to provide second-order accuracy with the Van Albada limiter to prevent spurious oscillations around shock waves. The integration in time of the previous equation to a steady-state solution is performed using an implicit time-marching scheme. The linearized system of equations is solved using the generalized conjugate gradient method with a block incomplete lower-upper (BILU) factorization as a pre-conditioner. The solver offers several one-equation, two-equation, three-equation, and four-equation turbulence models. LES, DES, delayed DES (DDES), improved DDES (IDDES), SAS and PANS methods are also available.

2.3 Application of PANS to Buffet

2.3.1 Description of the Employed Test Case

In the present study, the flow around the supercritical aerofoil OAT15A is investigated. This configuration was studied [21] in the S3Ch wind tunnel of ONERA. The wing section had a chord of $c = 0.23$ m and a span of 0.78 m, a thickness-to-chord ratio of $t/c = 0.123$, and a trailing edge thickness-to-chord of 0.005. The wing was mounted in a squared section wind tunnel having nominal dimensions of $0.78 \times 0.78 \times 2.2$ m. An adaptation technique was used at the lower and upper tunnel walls to reproduce free-stream conditions. Measurements were collected at free-stream Mach numbers of 0.7–0.75, chord-based Reynolds number of $\mathrm{Re}_c = 3 \times 10^6$, and angles of attack of 1.36–3.9 deg. Static and dynamic pressure measurements in the vicinity of the mid-span section allowed the detection of the flow unsteadiness at an angle of attack of 3.1 deg, culminating in a self-sustained shock-induced oscillation at higher angles. At $\mathrm{M} = 0.73$ and $\alpha = 3.5$ deg, a laser Doppler velocimetry (LDV) system was also employed. These measurements were complemented with oil-flow and schlieren visualisations.

Here, the study is focused on two different flow conditions, at $\mathrm{M} = 0.73$ and angles of attack of $\alpha = 2.5$, 3.5 deg, representative of a statistically steady flow, a fully-established buffet flow, respectively.

2.3.2 CFD Grids and Numerical Setup

The three-dimensional flow domain (3D in Table 2.2) consists of half of the wind tunnel used in the experimental campaign and it is sketched in Fig. 2.1. The grids adopted for this configuration are indicated 3D in Table 2.1 At the inflow, free stream values of pressure, density and velocity were imposed, while adiabatic wall conditions were imposed on both the aerofoil and the tunnel sidewall, since no detail

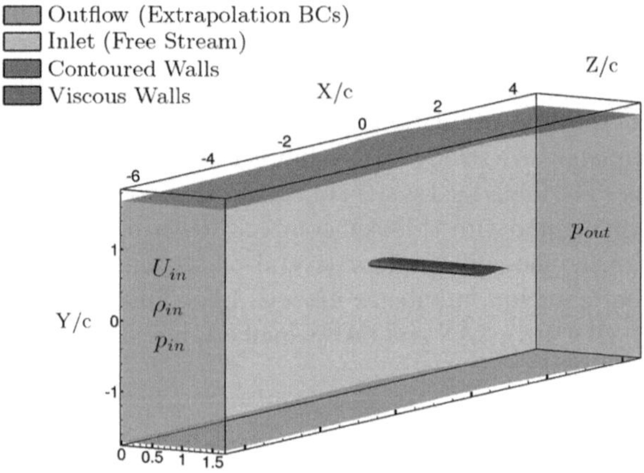

Fig. 2.1 Computational domain with coloured by boundary conditions. The symmetry plane at $z/c = 1.7$ is not coloured

Table 2.1 Main features of the different grids used for computations

Grid #	N_{airf}	N_z	$\Delta z_{max}/c$	N_y	$\Delta n_{airf}/c$	N_{wake}	$\Delta x_{TE}/c$	N_{tot}
2D	710	60	0.0042	115	5×10^{-6}	220	5×10^{-6}	9.07×10^6
3D	405	76	0.033	102	5×10^{-6}	87	5×10^{-6}	$\simeq 5.03 \times 10^6$

Table 2.2 Computations performed at different angles of attack, on different grids, and using different timesteps. All computations were run at $M_\infty = 0.73$. In the run ID, U stands for URANS while P for PANS

Run #	α	2D/3D	C_{PANS}	Δt	$\langle C_l \rangle$	St
Exp.	–	–	–	–	$\simeq 0.875$	–
U1	2.5	3D	–	0.1	0.874	
P1	2.5	3D	3.3	0.1	0.879	
P1.1	2.5	3D	1.0	0.1	0.841	
P1.2	2.5	3D	0.5	0.1	0.733	
Exp.	–	–	–	–	$\simeq 0.875$	0.066
U4	3.5	3D	–	0.01	0.82	–
P4	3.5	3D	1.0	0.01	0.833	0.042
P4.1	3.5	3D	0.5 and clip 0.2	0.01	0.746	0.043
P4.2	3.5	3D	0.5 and clip 0.4	0.01	0.739	0.046
P4.4	3.5	3D	0.5 and clip 0.6	0.01	0.835	–
P5	3.5	2D	0.5	0.005	0.748	0.065

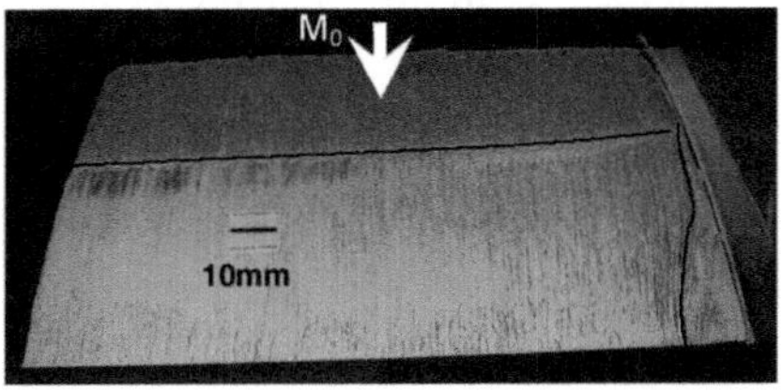

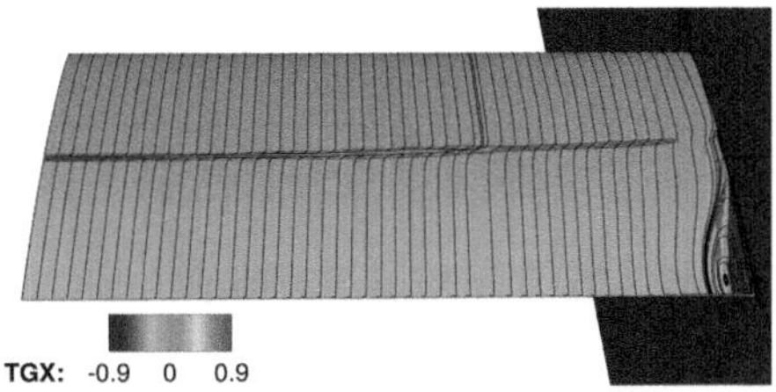

Fig. 2.2 Left: oil flow visualization around the OAT15A [21] at $\alpha = 2.5$ deg (shock position and separation line marked with a solid, black line); right: friction lines from CFD. The surface is coloured by the streamwise component of $\frac{\partial \mathbf{u}}{\partial \mathbf{n}}$, i.e. the spatial derivative of the velocity field in the direction normal to the wall

was given on the sidewall treatment of the boundary layer. The symmetry boundary conditions imposed at the centre plane allowed simulating just half of the domain. The upper and lower walls were modelled as slip-walls, and the shape was extracted from a preliminary 2D RANS simulation to replicate the results of the adaptive technique used in the experimental campaign. At the outlet, the tunnel back pressure was imposed in the subsonic region. Computations on an unconfined configuration with reduced spanwise extent ($L_z = 0.25c$) were also performed (2D in Tables 2.1 and 2.2).

2.3.3 *Pre-buffet Flow*

Figure 2.2 shows the effect of the wind tunnel walls at pre-buffet condition. The shock position can be detected by the negative values of the streamwise component of $\frac{\partial \mathbf{u}}{\partial \mathbf{n}}$ at the surface and, going towards the sidewall, from a sudden deflection of the friction lines. The shock curves when approaching the sidewalls, and a region of separated flow is confined within the 10% of the span, in agreement with the experimental results.

Moreover, because of the viscous sidewall, the turbulence level in the boundary layer grows as it approaches the aerofoil. This coincides with a growth in the turbulence length scales and a following reduction in f_k at the wall. Although the behaviour of the method is correct, this introduces the need for an overly fine spatial and temporal distribution in a region of limited interest for the development of transonic buffet. Therefore, in this work, an f_k value of 1 was imposed within a distance of 0.15c from the wall. This URANS region encompasses the separation region at the wing-wall junction throughout the buffet period. Since the vortices in the detached region at the walls were deemed to be steady in the experiments, this approach has been used.

Practically identical C_p distributions were obtained with different PANS coefficients and in good agreement with the experimental results.

2.3.4 Buffet Simulation

Here the results from the simulations performed at $\alpha = 3.5$ deg are shown. Figure 2.3 shows the lift coefficient history for the simulations (U4, P4-P4.2) on the coarser grid C3. The grid under analysis (details are given in Table 2.1) is rather coarse in the spanwise direction presenting the features of usual RANS grids. Even at this condition, the usage of PANS led to benefits in the prediction of transonic buffet. The URANS simulation converged to a steady-state after about ten travel times, while the PANS ones are able to predict the buffet. In the former case, the corner separation induced a spanwise flow deviation at the trailing edge preventing the flow from separating. Even if the flow separated at the shock foot, the separation region did not merge with the one at the trailing edge and the buffet was inhibited. The main reasons can be found in an over-estimated flow separation at the corner, together with a high level of eddy viscosity predicted from the URANS solution in the post-shock region. Therefore, the presence of the tunnel walls does not help the developing of the self-sustained shock oscillation. Conversely, the PANS simulation predicts the buffet developing according to a precise pattern, shown in Fig. 2.4. Approaching the most downstream position (a), the flow is attached and the shock strength increases until the flow separates underneath; at the same time, trailing edge separation occurs (b). The two regions merge around the centerline inducing a flow acceleration between the corner and central separated regions. This results in a decrease in the size of corner separation and the creation of vortical structures at the boundary with the central separated flow. When the flow is fully separated after the shock, this is affected by the disturbances coming from the trailing edge and the shock begins to move upstream. In parallel, the aforementioned vortical structures propagate in the separated region and extend to the entire wingspan giving rise to a region characterized by large cells (c), in agreement with what observed in the experiments [21]. A comparison between

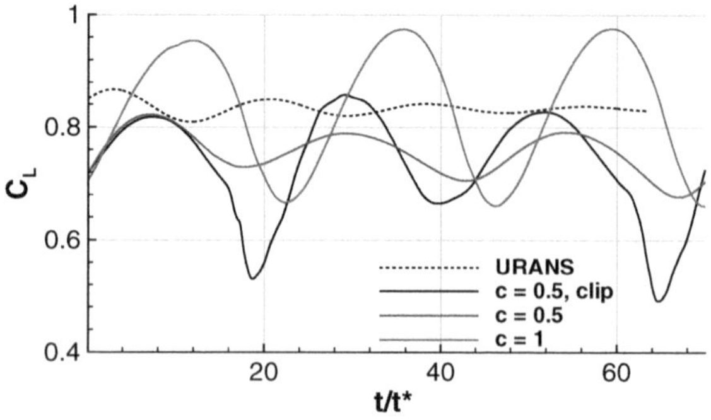

Fig. 2.3 Temporal evolution of the lift coefficient around the OAT15A at $\alpha = 3.5$

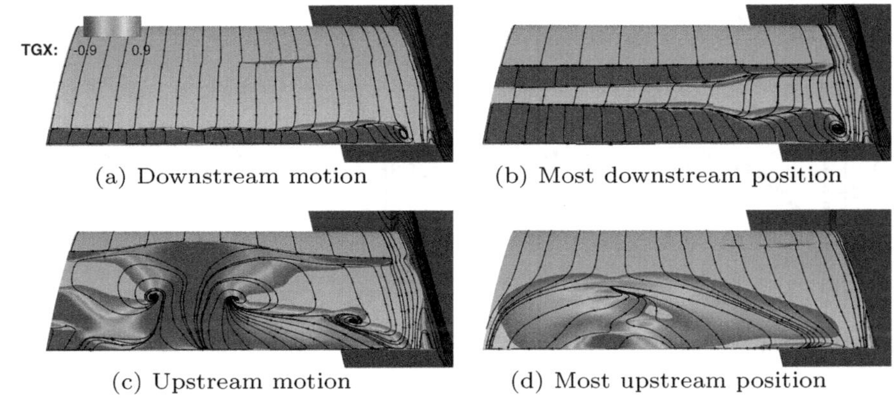

(a) Downstream motion (b) Most downstream position

(c) Upstream motion (d) Most upstream position

Fig. 2.4 Surface friction lines at different phases of the buffet period. Results from simulation P4.2 in Table 2.2

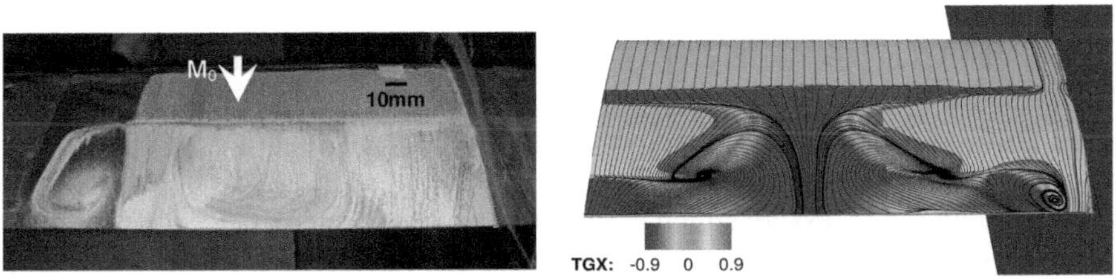

Fig. 2.5 Left: oil flow visualization around the OAT15A (picture taken from [21]) at $\alpha = 3.5$ deg; right: friction lines from CFD. Results from simulation P4.2 in Table 2.2

the computational and experimental results is given in Fig. 2.5. Approaching the most upstream position, the shock strength decreases, the flow reattaches completely (d), and a new oscillation begins with the shock moving downstream.

The computations on the unconfined configuration, not shown here, did not present such a three-dimensional development of buffet. Indeed, the reduced spanwise extent and the absence of the viscous walls resulted in 2D buffet behaviour, only displaying the alternated boundary-layer separation and reattachment and a straight shock front.

2.3.4.1 Influence of the C_{PANS} Coefficient

Here, the influence of the C_{PANS} coefficient and of f_k are investigated. Adopting the estimate of Eq. 2.5, the value of f_k tends toward the zero in the separated region, and, therefore, the parameter must be clipped to a suitable value to prevent an excessive reduction of the eddy viscosity. Figure 2.3 shows the time evolution of the aerodynamic coefficients for two values of C_{PANS}. For $C_{\text{PANS}} = 0.5$, two different values of the clip are used. In the case of $C_{\text{PANS}} = 1$ the oscillations are regular, while in the case of $C_{\text{PANS}} = 0.5$ the oscillations become irregular. In the latter case, the mean

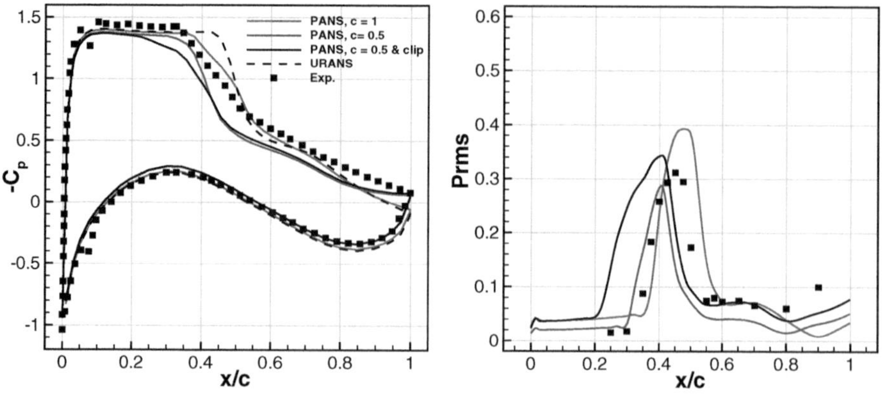

Fig. 2.6 Left: time-averaged pressure coefficient at $\alpha = 3.5$ deg; right: root means square of the pressure. Experiments from [21]

value of the lift coefficient decreases as well as the mean amplitude of oscillation. The same behaviour was found for the other aerodynamic coefficients, not shown here. This is reflected in the pressure coefficient and fluctuations on the aerofoil, shown in Fig. 2.6. The pressure coefficient distribution and RMS values were evaluated, averaging the pressure signal extracted at probes located at the same locations of the pressure sensors used in the experiments [21]. A wider oscillation coincides with a higher pressure RMS, while in the case of $C_{\text{PANS}} = 0.5$ the shock excursion was smaller, as well as the peak RMS. Conversely, the higher the C_{PANS} the lower C_p is. In both cases, the RMS at the trailing edge was underestimated. Figure 2.7 complements the previous figure adding some more insight. In the case of $C_{\text{PANS}} = 1$ the separation was incipient at both the trailing edge and at the shock foot. The two recirculation regions were generated at the most downstream shock position, but they did not merge in a unique separation bubble. Conversely, the lower eddy viscosity obtained by lowering the PANS parameter allows for the flow separation at the trailing edge during a great part of the shock oscillation. The separated region at the shock foot is generated when the shock approaches its most downstream position and merges with the other separation region, giving rise to a single separation bubble. Nevertheless, the formation of the two recirculation zones should occur almost at the same time, as shown in [18], e.g., suggesting that the trailing edge eddy viscosity level is too low. This in turn reflects in a too upstream average shock position. From Fig. 2.7, a different behaviour at the sidewall is found to influence the buffet dynamics. In the first case ($C_{\text{PANS}} = 1$), a much wider separation region is present at the corner. The blockage effect delays the trailing edge separation, that cannot catch up with the shock foot separation. Moreover, the corner separation changes in size over the buffet period and influences the shock position at the centre plane. The opposite happens when $C_{\text{PANS}} = 0.5$ is adopted. The excessive separation at the centre plane, where PANS is active, forces the flow to bend towards the wall, causing a decrease in the corner separation size. This reflects in an almost straight shock front. In addition,

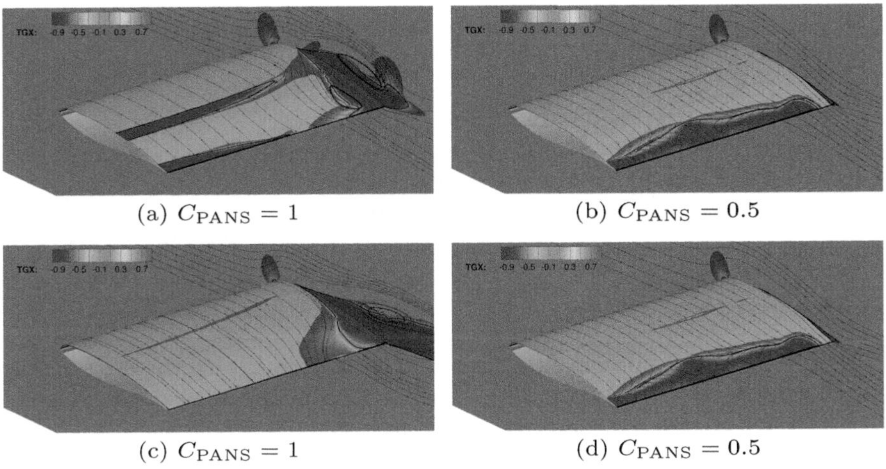

Fig. 2.7 Friction lines over the aerofoil surface. The black line marks the reversed flow region

the vortical structures generated in the recirculation region on the wall can propagate towards the symmetry plane in the separated boundary layer extending throughout the wingspan. When clipping f_k to a higher value (P4.1 of Table 2.2), the eddy viscosity level at the trailing edge increased and the flow was less prone to separation over the buffet period. This reflected in an effective alternation of attached and separated boundary layer that, in turn, caused the shock oscillation to be wider and the RMS value at the trailing edge to improve. Nevertheless, an average upstream position is still predicted. Increasing again the value of the clip ($f_{k,inf} = 0.6$) led to a failure in predicting the self-sustained oscillations. Therefore, the adopted estimate for the f_k parameter was found to be sensitive to both parameters investigated. Nonetheless, when set properly, the computations were able to predict the observed flow patterns, and provided a good comparison with the experiments.

2.3.5 Computational Performance and Cost

Table 2.3 presents the CPU cost associated with PANS and URANS simulations. The comparison was over one characteristic time, i.e. over 100 computational timesteps. In the analysis, the same grid and the same number of processors were used. Since URANS tends to a steady solution while the PANS describes an intrinsically unsteady flow, the number of inner iterations required to meet the convergence criterion on the flow residual is different; therefore, the averages were taken over the number of inner iterations of the implicit dual-time stepping scheme. This was done to identify the net additional cost of the PANS routine in the CFD solver. From the table, it can be seen that the higher increase is that on the evaluation of the residual Jacobian, where the

Table 2.3 Comparison between computational times of PANS and URANS. ABAH: apply boundary and halo; IATS: initialize and time step; CHOG: calculate high order gradients; CRJ: calculate residual Jacobian; CP: calculate preconditioner; SLS: solve linear system

	GCG iter.	ABAH [s]	IATS [s]	CHOG [s]	CRJ [s]	CP [s]	SLS [s]	Total
URANS	3.86	0.31	0.055	0.00024	0.65	0.09	0.83	**1.95**
PANS	4.59	0.31	0.055	0.00025	0.70	0.09	0.85	**2.00 (+3.1%)**

PANS routines are modifying the source terms of the k and ω equations. Moreover, the number of iterations of the GCG scheme for the solution of the linear scheme is slightly increased. This reflects in an increase in the total computational time of about 3% when compared to URANS. The ability of PANS to work in URANS-mode in agreement with the grid capability to resolve structures was shown.

2.4 Conclusions

In this work, the application of PANS for the flow around the OAT15A section was presented. Computations on confined and unconfined configurations were carried out since the wind tunnel presence influenced the buffet dynamics by introducing a strong three-dimensionality to the flow. At pre-buffet conditions, the presence of the wind tunnel walls resulted in an irregular shock front and a region of separated flow at the wing/tunnel wall junction. At pre-buffet condition, the PANS approach recovered the URANS behaviour, given a reasonable distribution of the f_k parameter. To avoid the costs associated with the PANS approach at the wall, where a high level of turbulence develops in the boundary layer, the transition from RANS to PANS was fixed at $0.15c$ from the sidewall. After the onset, although the buffet mechanism maintained its two-dimensional nature, the interaction with the separated region at the wall-wing junction was shown to be crucial in the generation of a pair of cells at the mid-span of the wing. Conversely, the unconfined configuration showed a typical two-dimensional behaviour with an alternated separation and reattachment of the flow on the suction side of the aerofoil. In both cases, the k-ω SST model was not able to predict the buffet although the angle of attack was well beyond the buffet onset, while the use of PANS unlocked the shock oscillations.

The role of f_k in the PANS formulation was crucial. High values of f_k did not reduce enough the eddy viscosity to allow for trailing edge separation and an increased size of the separation region at the corner was found. The bubble pulsation induced a shock oscillation all along the wingspan. In that case, there was no presence of 3D vortical structures. A reduced value of f_k, corresponding to $C_{\text{PANS}} = 0.5$ in this case, allowed to predict the correct flow topology, with the trailing edge separation that resulted in a reduced size of the corner recirculation. The formation of a separation region, spanning from the shock foot to the trailing edge, was detected in this case. Within that region, large three-dimensional structures were developed and propagated from the sidewall to the symmetry plane. Additional care must be taken in clipping the f_k to a minimum value according to the grid resolution. Although the agreement with the experiments [21] was not perfect (buffet frequency and shock position were slightly mispredicted), the flow displayed the same three-dimensional development, as shown in Figs. 2.2 and 2.5. The average quantities were in fair agreement with experiments. Moreover, the PANS approach proved to be helpful even at URANS resolution, at little additional expense ($\simeq$ 3%). The approach provided enhanced results with respect to URANS and looks promising for large-scale simulations.

Future efforts will be devoted on the application of this technique to flows with SBLI, paying attention to the corner flow behaviour, which in this work proved to be troublesome. Moreover, a different estimate of the f_k parameter accounting for the local grid size, and flow features will be pursued to make the preliminary calibration of the method unnecessary.

Acknowledgements The present work is supported by the TEAMAero European project No. 860909—Towards Effective Flow Control and Mitigation of Shock Effects In Aeronautical Applications. The use of the Cirrus system of EPCC for HPC is gratefully acknowledged.

References

1. Girimaji SS, Abdol-Hamid KS (2005) Stokes partially-averaged Navier, model for turbulence: implementation and validation. AIAA Aerospace Sciences Meeting and Exhibit, Reno. NE, USA
2. Luo D (2019) Numerical simulation of turbulent flow over a backward facing step using partially averaged Navier-Stokes method. J Mech Sci Technol 33:2137–2148
3. Spalart PR, Jou WH, Strelets M, Allmaras SR (1997) Comments on the feasibility of LES for wings, and on a hybrid RANS, LES approach. Advances in DNS, LES: Direct numerical simulation and large eddy simulation, Reno. NE, USA
4. Abdol-Hamid KS, Girimaji SS (2004) A two-stage procedure toward the efficient implementation of PANS and other hybrid turbulence models. National Aeronautics and Space Administration, NASA/TM-2004-213260
5. Pereira FS, Vaz G, Eca L, Girimaji SS (2018) Simulation of the flow around a circular cylinder at Re=3900 with partially-averaged Navier-Stokes equations. Int J Hea Fluid Flow 69:234–246
6. Basara B, Pavlovic Z, Girimaji SS (2018) A new approach for the calculation of the cut-off resolution parameter in bridging methods for turbulent flow simulation. Int J Heat Fluid Flow 74:76–88
7. Frendi A, Tosh A, Girimaji SS (2007) Flow past a backward-facing step: comparison of PANS, DES and URANS results with experiments. Int J Comput Methods Eng Sci Mech 8:23–28
8. Basara B, Krajnovic S, Girimaji S, Pavlovic Z (2011) Near-Wall formulation of the partially averaged Navier-Stokes turbulence model-layer interaction. AIAA J 49:2627–2636
9. Davidson L, Friess C (2019) A new formulation of f_k for the PANS model. J Turbulence 20:322–336
10. Friess C, Davidson L (2020) A formulation of PANS capable of mimicking IDDES. Int J Heat Fluid Flow 86:1–25

11. Basu D, Hamed A, Dias K (2007) Assessment of partially averaged Navier/Stokes (PANS) multiscale model in transonic turbulent separated flows. In: ASME/AJME (2007) 5th joint fluids engineering conference. San Diego, CA, USA
12. Luo D, Yan C, Wang X (2015) Computational study of supersonic turbulent-separated flows using partially averaged Navier-stokes method. Acta Astronautica 107:234–246
13. Bonnifet V, Gerolymos GA, Vallet I (2017) Prediction transonic buffet, using partially averaged Navier-Stokes. In: 23rd AIAA computational fluid dynamics conference, Denver. CO, USA
14. Lee BHK (2001) Self-sustained shock oscillations on airfoils at transonic speeds. Progr Aerosp Sci 37:147–196
15. Giannelis NF, Vio GA, Levinski O (2017) A review of recent developments in the understanding of transonic shock buffet. Progr Aerosp Sci 92:39–84
16. Barakos G, Drikakis D (2000) Numerical simulation of transonic buffet fow using various turbulence closures. Int J Heat Fluid Flow 21:620–626
17. Goncalves E, Houdeville R (2004) Turbulence model and numerical scheme assessment for buffet computations. Int J Numer Methods Fluids 46:1127–1152
18. Grossi F, Braza M, Hoarau Y (2014) Prediction of transonic buffet by delayed detached-eddy simulation. AIAA J 52:2300–2312
19. Deck S (2005) Numerical simulation of transonic buffet over a supercritical airfoil. AIAA J 43:1556–1566
20. Masini L, Timme S, Pace AJ (2020) Scale-resolving simulations of a civil aircraft wing transonic shock-buffet experiment. AIAA J 58:4322–4338
21. Jacquin L, Molton P, Deck S, Maury B, Soulevant D (2009) Experimental study of shock oscillation over a transonic supercritical profile. AIAA J 47:1985–1994
22. Menter FR (1994) Two-equation eddy-viscosity turbulence models for engineering applications. AIAA J 32:1598–1605
23. Steijl R, Barakos G, Badcock K (2006) A framework for CFD analysis of helicopter rotors in hover and forward flight. Int J Numer Methods Fluids 51:819–847

Chapter 3
Progress of ARI_OPT Software for Aerodynamic Shape Optimization

Chuang Wei, Zhansen Qian, and Long Yang

Abstract In order to improve the design efficiency and design effect of the aerodynamic shape of the aircraft, the aerodynamic optimization design tool ARI_OPT was developed based on the numerical optimization design technology. ARI_OPT includes modules such as aerodynamic shape parameterization, grid automatic generation, surrogate model optimization, multi-level numerical simulation, and high-efficiency optimization algorithm. The basic principles, methods and verification results of each module are briefly introduced. The functional verification results of numerical function cases and the application results of typical single-objective and multi-objective aerodynamic shape optimization problems such as airfoil, multi-element airfoil and wing-body by ARI_OPT, which strongly prove its effectiveness.

Keywords Aerodynamic shape · Numerical optimization · ARI_OPT · Surrogate model

3.1 Introduction

The traditional cut and try based manual modification aerodynamic design method, which relies on experience, has low design efficiency and is difficult to obtain the best aerodynamic shape. With the development and maturity of the techniques of Computational Fluid Dynamics (CFD), Computer Aided Geometric Design (CAD) and numerical optimization, aerodynamic analysis and optimization design technology

---

C. Wei · Z. Qian (✉) · L. Yang
AVIC Aerodynamics Research Institute, No. 1, Yangshan Road, Shenyang 110034, China
e-mail: qianzs@avic.com

C. Wei
e-mail: agangood@sina.com

L. Yang
e-mail: jerryyl216@163.com

Aero Science Key Lab of High Reynolds Aerodynamic Force at High Speed, No. 1, Yangshan Road, Shenyang 110034, China

T. Tuovinen et al. (eds.), *Advanced Computational Methods and Design for Greener Aviation*, Computational Methods in Applied Sciences 59,
https://doi.org/10.1007/978-3-031-61109-4_3

based on highly fidelity CFD has been widely used in the design and development of aviation and aerospace aircraft, and is playing an increasingly important role in improving the aerodynamic and comprehensive performance of aircraft [1, 2].

The complete aerodynamic shape optimization process includes aerodynamic shape parameterization, automatic mesh generation, aerodynamic performance evaluation and efficient optimization algorithm, etc. According to category of optimization algorithms, the existing optimization design methods are mainly divided into three categories [3]: genetic algorithm, particle swarm optimization algorithm and other heuristic optimization algorithms; Gradient optimization algorithm; Proxy optimization algorithm. The three methods have their own advantages and disadvantages, and researchers domestic and abroad have carried out extensive research. For example, Jameson of Stanford University carried out fruitful pioneering research on adjoint method [4], and then Alonso team of Stanford University developed SU2 [5] open source aerodynamic optimization program based on adjoint method, and Boeing company developed MDOPT [6] optimization software for aerodynamic shape optimization. The CFD solver FUN3D [7, 8] of NASA Ames Research Center, the unstructured solver TAU [9, 10] of German Space Agency and the solver e1sA [11, 12] of France Aerospace Have all integrated and developed the aerodynamic shape optimization design function. China Aerodynamics Research and Development Center [13], Northwestern Polytechnical University [14–16], Tsinghua University [17], Nanjing University of Aeronautics and Astronautics [18, 19], AVIC Aerodynamics Research Institute [20] and other research institutes have also carried out extensive and in-depth research, and established their own aerodynamic optimization design tools.

In this paper, the Aerodynamics Research Institute Optimization Code (ARI_OPT) are briefly introduced, and the verification examples based on numerical functions are given. In order to provide a reference for promoting the exchange of aerodynamic shape optimization design, the application examples of the method to typical single-objective and multi-objective aerodynamic shape optimization problems such as airfoil, multi-segment airfoil and flying wing layout are described in detail.

3.2 ARI_OPT Aerodynamic Optimization Software

ARI_OPT is an in-house code designed by AVIC Aerodynamic Research Institute (ARI) for aerodynamic shape optimization of aircraft. We chosed the fortran and C/C++ rogramming languages to develop ARI_OPT in Windows and Linux system. It can solve any constraints of single objective and multi-objective optimization problems, mainly applied to the aerodynamic optimization design based on high-accuracy CFD code. The optimization software is integrated into the large computing cluster of ARI, which can carry out massively parallel optimization by MPI. The software includes aerodynamic shape parameterization, grid deformation, high fidelity CFD and optimization algorithm modules. ARI_OPT contains a variety of aerodynamic shape parameterization methods, such as Class-Shape Transformation (CST)

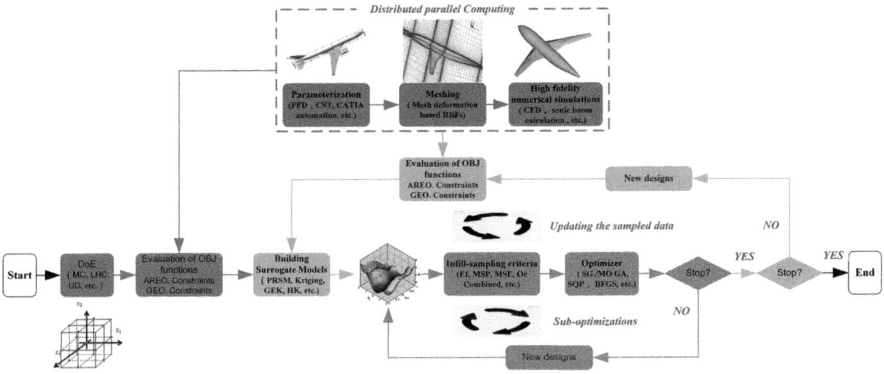

Fig. 3.1 Flow chart of surrogate-based optimization of ARI_OPT

[21], hicks-henne [22], Free-form deformation (FFD) [23] used in airfoil and wing parameterization, elliptic equation used in multi-element airfoil parameterization. The Radial Basis Function (RBF) mesh deformation technique [24] is used to realize automatic mesh generation. Aerodynamic performance evaluation in the optimization process relies structured-grid Reynolds-averaged Navier-Stokes (RANS) code named ARI_ENSMB. The code mainly adopts optimization strategy based on the surrogate-based optimization (SBO), the main ingredients of such a SBO process are: design of experiment (DoE) method, surrogate modeling, infill-sampling criteria and sub-optimization. The DoE methods implemented in ARI_OPT are the Latin hypercube (LHS), uniform design (UD), Monte Carlo sampling (MC), etc. Quadratic response surface model (RSM) and Kriging surrogate are available in ARI_OPT. Several infilling-sampling criteria is available in ARI_OPT, such as maximizing the probability of improvement (PI), maximizing the expected improvement (EI), minimizing the surrogate prediction (MSP), etc.

The flow chart of a typical SBO for solving the optimization problem is shown in Fig. 3.1. The typical optimization process is as follows: (1) Initial sample points are chosen by DoE method, response values are calculated by CFD solver and build the initial surrogate model. (2) Based on the surrogate model, the traditional optimization algorithm is used to solve the sub-optimization, and the new sample points are obtained according to infill-sampling criteria. (3) New sample points are evaluated by CFD added to the existing dataset, the surrogate model was constantly updated until global optimal solution is available.

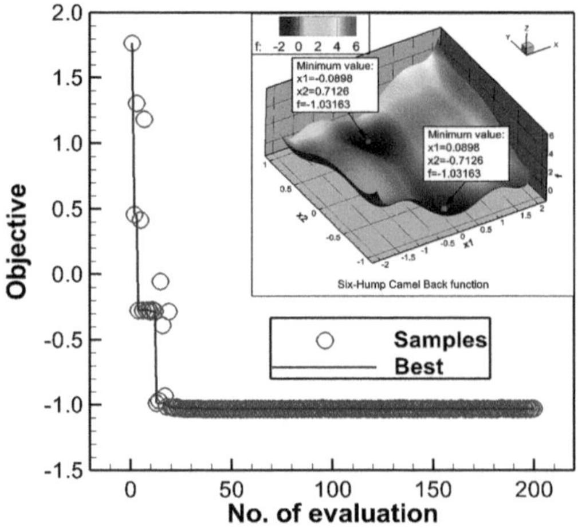

Fig. 3.2 Convergence histories of SBO for six-hump camel function

3.3 Optimization Test of Numerical Functions

3.3.1 Six-Hump Camel Function

The Six-hump camel back function is a global optimization test function. Within the bounded region are six local minima, two of them are global minima.The optimization problem is described as follows:

$$\begin{aligned} &Min: f(x_1, x_2) = (4 - 2.1x_1^2 + \frac{x_1^3}{3})x_1^2 + x_1x_2 + 4(x_2^2 - 1)x_2^2 \\ &S.T.: -3 < x_1 < 3, -2 < x_2 < 2 \end{aligned} \tag{3.1}$$

The multi-island genetic algorithm (MIGA) and the surrogate-based optimization method were tested respectively. For the MIGA, the initial population is divided into 4 islands, the population number of each island is set to 32, the number of generations is 30, and called 1600 samples evaluation in total. For the SBO, the initial sample point is set to 4, the DOE method is LHS, the surrogate model is Kriging, the combined infill-sampling criteria of EI+MSP+LCB+PI were used to select 4 samples at each each updating cycle and call 200 samples evaluation in total. Convergence histories of SBO for Six-hump camel function shown in Fig. 3.2 and Table 3.1 shows the average results of 30 tests. It can be seen that the surrogate-based optimization method is superior to the multi-island genetic algorithm in both efficiency and optimization results.

Table 3.1 Comparison of test results of SBO with MIGA

Method	x_1	x_2	Result	Error/(%)
True	−0.0898	0.7126	1.03163	0
MIGA	−0.0816	0.71025	1.03133	0.029
SBO	−0.0898	0.7126	1.03163	0

3.3.2 ZDT3 Multi-objectives Function

ZDT3 is a standard test function, commonly used in the field of multi-objective optimization. The true Pareto solutions are convex and discontinuous. The optimization problem is described as follows:

$$Min: \begin{cases} f_1(x) = x_1 \\ f_2(x) = g(x)[1 - \sqrt{\frac{x_1}{g(x)}} - \frac{x_1}{g(x)} sin(10\pi x_1)], g(x) = 1 + \frac{9}{n-1} \sum_{i=1}^{n} x_i \end{cases}$$
$$S.T. : 0 < x_i < 1, i = 1, 2, \cdots, n \tag{3.2}$$

The fast and elitist non-dominated sorting in genetic algorithms (NSGA-2) and SBO were tested respectively. Figure 3.3 shows the optimization results of the two methods. For NSGA-2, the population number is set to 100, the Pareto solutions of 50 generations cycle is significantly different from true Pareto front, good Pareto solutions were obtained until 100 generations which called 10000 samples evaluation. For the SBO, the initial sample point is set to 60, the DoE method is LHS, the surrogate model is Kriging, the infill-sampling criteria of MSP for Pareto optimization was used to select 100 samples at each each updating cycle and call 960 samples evaluation in total. We can be seen that the number of function evaluation of using SBO is much less than that of using NSGA-2, when reaching a better solution quality.

3.4 Typical Applications

3.4.1 Single Objective Optimization of Airfoil Drag Reduction

This is a benchmark test case (case 2) defined by AIAA aerodynamic design optimization discussion group (ADODG) [25]. The design sate is Ma=0.734 and Re=6.5×10^6. The optimization problem is described below:

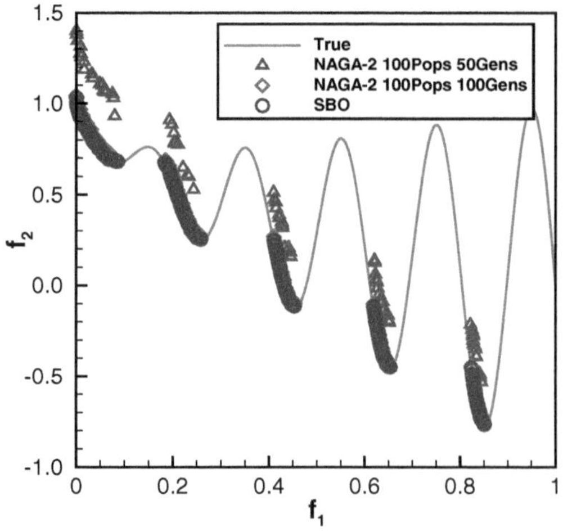

Fig. 3.3 Test results of SBO and NSGA-2 for ZDT3 function

$$
\begin{aligned}
&Min: \ C_D \\
&S.T.:(1)C_L = 0.824 \\
&\qquad (2)C_m \geq -0.092 \\
&\qquad (3)Area \geq Area_{initial}
\end{aligned}
\tag{3.3}
$$

CST parameterization method is used with 18 design variables along the upper and lower airfoil surface. The aerodynamic performance of the airfoil during optimization was obtained by solving the compressible RANS equation, and the turbulence model was κ-ω SST two-equation model. The initial sample point is set to 36, the DoE method is LHS, the surrogate model is Kriging, the combined infill-sampling criteria of EI+MSP+LCB+PI were used to select 4 samples at each each updating cycle and call 600 samples CFD calculation, and the convergence history of drag coefficient is plotted in Fig. 3.4. The aerodynamic performances comparison between the optimum airfoil and RAE2822 is shown in Table 3.2, which shows that drag coefficient is reduced by 31.74%.

Figure 3.5 show the geometric shape and pressure distribution between the optimum airfoil and RAE2822. It can be seen that the leading edges of optimized airfoil gets sharper which is good for the shock wave strength reduction on the upper surfaces, and thus reducing wave drag.

3.4.2 Multi-objective Optimization of Multi-element Airfoil

A multi-objective optimization problem is presented dealing with a multi-element airfoil used in high-lift devices. The objectives considered are to maximize lift coefficient at angles of attack of 6 deg and 20 deg at landing operations by searching for

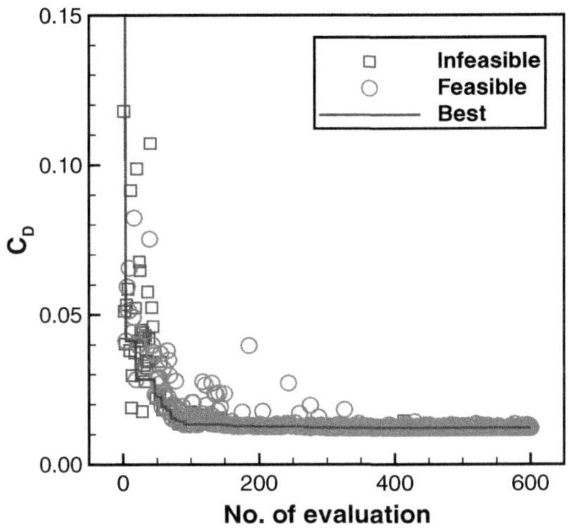

Fig. 3.4 Convergence history of drag coefficient

Table 3.2 Comparison of performance of RAE2822 and optimum airfoil

Airfoil	C_L	C_D	C_m
RAE2822	0.82398	0.018977	−0.097898
Optimum	0.82398	0.012151	−0.091617
Δ	0	35.97%	–

optimal shape and setting parameters for both slat and flap. The design sate is Ma = 0.2 and Re = 45.6×10^6. The optimization problems were described as follows:

$$\begin{aligned} &Max: \begin{cases} C_{L-DP1} = C_{L,\alpha=6^\circ} \\ C_{L-DP2} = C_{L,\alpha=20^\circ} \end{cases} \\ &S.T.:(1)\ C_{L,\alpha=8^\circ} \geq C_{L,\alpha=6^\circ @baseline} \\ &\quad (2)\ C_{L,\alpha=20^\circ} \geq C_{L,\alpha=20^\circ @baseline} \\ &\quad (3)\ C_{m,\alpha=8^\circ} \geq C_{m,\alpha=6^\circ @baseline} \\ &\quad (4)\ C_{m,\alpha=20^\circ} \geq C_{m,\alpha=20^\circ @baseline} \end{aligned} \tag{3.4}$$

Elliptical equations are used to automatically generate the slat and flap shape, 7 Control point coordinates are only necessary, as well as gap overlap and deflection angle were selected as the design variables for setting optimization,there are 14 design variables. The initial sample point is set to 28, the DoE method is LHS, the surrogate model is Kriging, the infill-sampling criteria of MSP for Pareto optimization was used to select 56 samples at each each updating cycle and call 949 samples CFD evaluation in total. Figure 3.6 shows the final Pareto front, solutions named OPT1 and OPT3 are of primary importance, since they qualify the improvements in one objective while the other is kept fixed. From OPT1 to OPT3, the lift coefficient

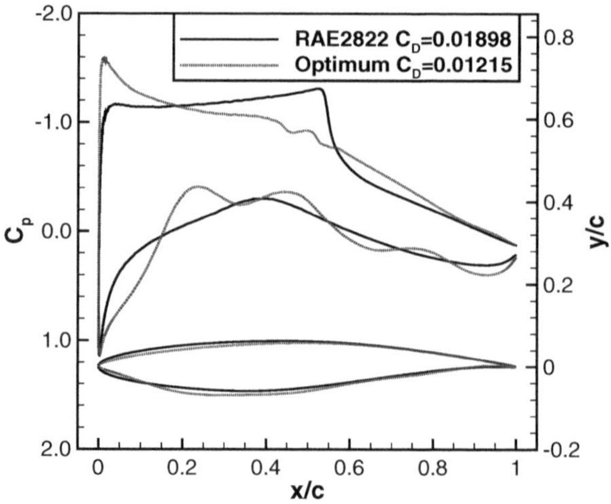

Fig. 3.5 Geometric shapes and pressure distributions for RAE2822 and optimum airfoil

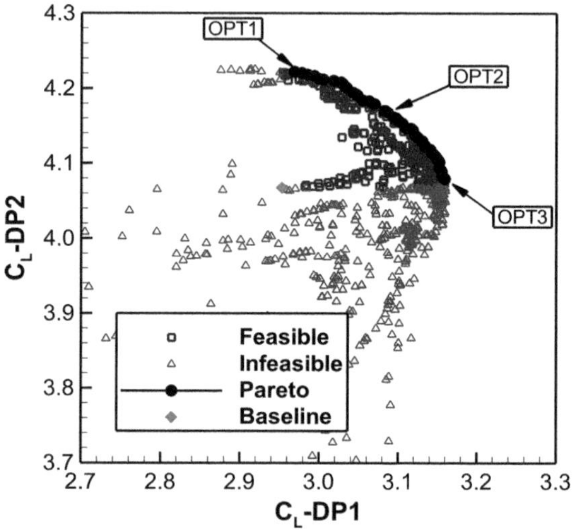

Fig. 3.6 Final Pareto front of multi-element airfoil optimization

becomes larger and larger at α =6°, On the other hand, the lift coefficient becomes smaller and smaller at the α =20°. OPT2 has larger lift both at α=6° and α =20°.

The aerodynamic performances of the baseline and typical optimized configurations are listed in Table 3.3, A graphical representation of the three optimized geometries is given in Fig. 3.7, which shows that the maximum increase of lift coefficient at α =6° is 6.73% (OPT3), and at α =20° is 3.78% (OPT1). The lift coefficient curve among OPT1-3 and baseline is showed in Fig. 3.8. The $C_L - \alpha$ curve of OPT1-3 is above the baseline. Compare to baseline, OPT1 configuration has the largest maximum lift coefficient with a smaller increase at linear segment of $C_L - \alpha$ curve, while OPT3 configuration has the largest lift coefficient at linear segment of $C_L - \alpha$ curve

Table 3.3 Comparison of lift coefficient of baseline and optimal multi-element airfoil

Design point	6°	$\Delta\ C_{L,\alpha=6^\circ}$	20°	$\Delta\ C_{L,\alpha=20^\circ}$
Baseline	2,9585	–	4.0741	–
OPT1	2.9685	0.34%	4.2282	3.78%
OPT2	3.0458	2.95%	4.1737	2.45%
OPT3	3.1575	6.73%	4.0867	0.31%

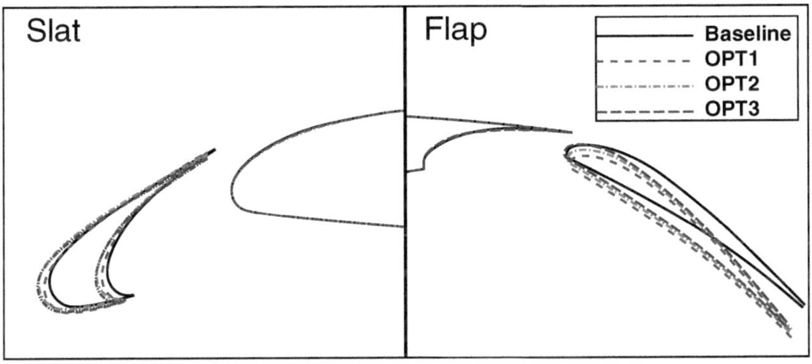

Fig. 3.7 Comparison of shapes for baseline and optimum multi-element airfoil

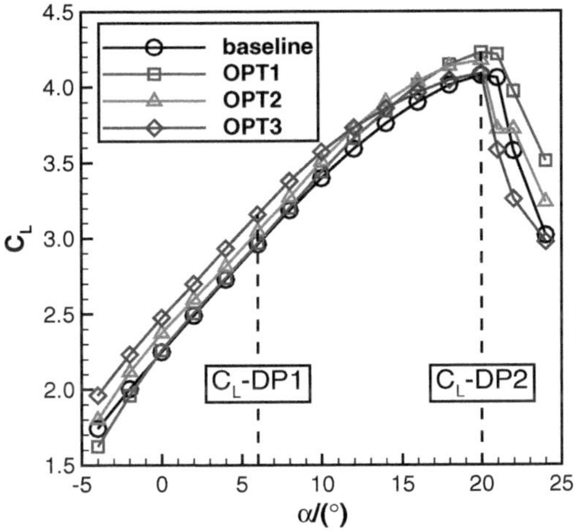

Fig. 3.8 Comparison of lift curve of baseline and optimum multi-element airfoil

with a smaller increase of maximum lift coefficient. OPT2 configuration has larger lift both at 6°(2.95%) and 20°(2.45%).

3.5 Conclusion

This paper summarizes the development of a the aerodynamic optimization design tool called ARI_OPT and its recent Progress and applications for aerodynamic shape design optimization.

The results of numerical function indicates that surrogate-based optimization method can gain better optimization results by less sampling points compared with genetic algorithm, which is suited for engineering designs driven by expensive numerical analysis.

The ARI_OPT is an effective aerodynamic optimization design tool with high efficiency, which is validated by numerical function and other aerodynamic optimization problems such as airfoil and multi-element airfoil.

In the future, We will continue to improve the optimization efficiency, universal and scalability of ARI_OPT, MDO applications will be important concerns for future development.

References

1. Peigin S, Zhu ZQ, Epstein B (2014) Applicable numerical optimization methods for aerodynamic design of civil aircraft. Acta Aeronautica et Astronautica Sinica 35(1):58–69
2. Zhou Z, Huang JT, Gao ZH, Huang Y, Chen ZB, Yu J (2019) Challenges and prospects of numerical optimization design for large civil aircraft aerodynamic shape. Acta Aeronautica et Astronautica Sinica 40(1):69–90
3. Han ZH, Zhang Y, Xu CZ, Wang K, Wu MM, Zhu Z, Song WP (2019) Aerodynamic optimization design of large civil aircraft wings using surrogate-based model. Acta Aeronautica et Astronautica Sinica 40(1):155–170
4. Shankaran S, Jameson A, Martinelli I (2008) Continuous adjoint method for unstructured grids. AIAA J 46(5):226–239
5. Palacios F, Economon TD, Aranake AC (2014) Stanford University Unstructured (SU2): open-source analysis and design technology for turbulent flows. In: AIAA science and technology forum and exposition 2014: 52nd aerospace sciences meeting, National Harbor, MD, 13–17 Jan 2014
6. LeDoux ST, Herling WW, Fatta GJ (2004) MDOPT-A multidisciplinary design optimization system using higher order analysis. In: 10th AIAA/ISSMO multidisciplinary analysis and optimization conference
7. Nielsen EJ, Diskin B, Yamaleev NK (2010) Discrete adjoint-based design optimization of unsteady turbulent flows on dynamic unstructured grids. AIAA J 48(6):1195–1206
8. Park M (2011) Low boom configuration analysis with FUN3D adjoint simulation framework. In: 29th AIAA applied aerodynamics conference
9. Brezillon J, Dwight RP (2009) Aerodynamic shape optimization using the discrete adjoint of the Navier-Stokes equations: applications towards complex 3D configurations. In: CEAS/KATnet II conference on key aerodynamic technologies, Bremen Germany. machtWissen
10. Widhalm M, Ronzheimer A, Hepperle M (2009) Comparison between gradient-free and adjoint based aerodynamic optimization of a flying wing transport aircraft in the preliminary design. In: 25th AIAA applied aerodynamics conference
11. Carrier G, Destarac D, Dumont A, Meheut M, Salah El Din I, Peter J, Saloua I, Ben Khelil S, Brezillon J, Pestana M (2014) Gradient-based aerodynamic optimization with the elsA software. In: 52th Aerospace sciences meeting

12. Dumont A, Le Pape A, Peter J, Huberson S (2011) Aerodynamic shape optimization of hovering rotors using a discrete adjoint of the Reynolds-averaged Navier-Stokes equations. J Am Helicopter Soc 56(3):1–11
13. Huang JT, Zhou Z, Gao ZH (2017) Aerodynamic multi-objective integrated optimization based on principal component analysis. Chin J Aeronaut 30(4):1336–1348
14. Li J, Gao ZH, Zhao K (2015) Robust design of transonic laminar wing-body configuration based on direct manipulated FFD technique. Scientia Sinica Technologica 46(9):964–974
15. Bai JQ, Lei RW, Yang TH, Wang H, He XL, Qiu YS (2019) Progress of adjoint-based aerodynamic optimization design for large civil aircraft. Acta Aeronautica et Astronautica Sinica 40(1):103–120
16. Han ZH (2016) Kriging surrogate model and its application to design optimization: a review of recent progress. Acta Aeronautics et Astronautica Sinica 37(11):3197–3225
17. Zhao T, Zhang YF, Chen HX, Chen YC, Zhang M (2015) Aerodynamics optimization method of supercritical airfoil geared to the performance of swept and tapered wing. Scientia Sinica Technologica 45(10):89–101
18. Zhao QJ, Zhang W, Yuan X, Zhu Z, Wang B (2019) Optimization design of coaxial rotor aerodynamic planform. J Nanjing Univ Aeronaut Astronaut 51(2):160–165
19. Tang Z, Huang M (2001) Control theory based airfoil design using Euler equations. Acta Aerodynamica Sinica 19(3):262–270
20. Wei C, Zhang TJ, Liu TZ (2017) Multi-objective optimization of fowler flap on general aircraft. Acta Aerodynamica Sinica 35(4):572–578
21. Kulfan BM, Bussoletti JE (2006) Low boom configuration analysis with FUN3D adjoint simulation framework. In: AIAA/ISSMO multidisciplinary analysis and optimization conference, Portsmouth, Virginia, AIAA paper 2006-6948
22. Hicks RM, Henne PA (1978) Wing design by numerical optimization. J Aircr 15(7):407–412
23. Sederberg TW, Parry SR (1986) Free-form deformation of solid geometric models. In: Proceedings of the 13th annual conference on computer graphics and interactive techniques
24. De Boer A, Van der Schoot MS, Bijl H (2007) Mesh deformation based on radial basis function interpolation. Comput Struct 85(11–13):784–795
25. Nadarajah S (2013) Fundamental parametric geometry representations for aircraft component shapes. https://drive.google.com/file/d/13uJ_KxpPqGuCBs5KxCy08EUcz71wqXWE/view

Chapter 4
Verification of Transition Prediction for Flows with Suction Using Linear Stability Theory and e^N-Method

Normann Krimmelbein and Andreas Krumbein

Abstract A transition prediction approach using linear stability theory and e^N-method is used for the evaluation of the stability of flows with suction. The input data for the stability analysis are directly taken from the solution of a RANS computation, where the suction velocity at the surface is introduced by a mass flux boundary condition. The prediction capability of the e^N-method is validated using different two– and three–dimensional experimental test cases, such as the flow over a NACA 64_2-A-215 and a NACA 64-A-010 airfoil and the flow over swept wings with ONERA D and ONERA DTP-B airfoil sections normal to the leading edge. A reduced critical N-factor for flows with suction, compared to a critical N-factor for the unaffected flow, is intended to account for disturbances possibly introduced to the flow due to the suction process itself or due to the surface roughness.

Keywords Stability · Transition prediction · e^N method

4.1 Introduction

An important practical problem in the design of aerodynamic wings is the possibility to delay the onset of laminar-turbulent transition and thus to maximize the extent of laminar flow to reduce the overall friction drag. Different concepts exist to affect the transitional behaviour of flows around general wing geometries in a possibly favourable way: (a) natural laminar flow (NLF), (b) laminar flow control (LFC), and (c) hybrid laminar flow control (HLFC). Natural laminar flow is generally realized

---

N. Krimmelbein (✉)
DLR - Deutsches Zentrum für Luft- und Raumfahrt, Institute of Aerodynamics and Flow Technology, Lilienthalplatz 7, 38118 Braunschweig, Germany
e-mail: normann.krimmelbein@dlr.de

A. Krumbein
DLR - Deutsches Zentrum für Luft- und Raumfahrt, Institute of Aerodynamics and Flow Technology, Bunsenstraße 10, 37073 Göttingen, Germany
e-mail: andreas.krumbein@dlr.de

T. Tuovinen et al. (eds.), *Advanced Computational Methods and Design for Greener Aviation*, Computational Methods in Applied Sciences 59,
https://doi.org/10.1007/978-3-031-61109-4_4

through an optimization of the airfoil shape to generate large portions of the flow which exhibit a favourable pressure gradient and thus potentially retain laminar flow over a large distance. Laminar flow control usually requires a certain kind of active control of the boundary layer to maintain laminar flow over significant portions of a wing. Finally, hybrid laminar flow control is a combination of both, natural laminar flow and laminar flow control. For the laminar flow control aspect of the approaches to extent the laminar flow over airfoils and wings, applying suction to stabilize the laminar boundary layer is an effective and established method.

Previous work on applying linear stability theory and the e^N-method to flows with suction indicates that the suction process itself or the alteration of the surface (e.g. roughness due to suction holes) may have a negative influence on the stability of the flow, compensating some of the benefit of the stabilizating effect of suction. Van Ingen in his early work [13] had to lower the critical Tollmien-Schlichting N-factor from $N_{TS} = 11.2$ for flows with no or little suction to $N_{TS} = 4$ for flows with stronger suction in order to accurately match the experimental transition locations. Arnal et al. [2] also noted much lower computed cross flow N-factors for a 3D suction case compared to the no-suction equivalent. They attributed this effect to local distortions of the laminar base flow due to the suction through discrete holes which are essentially comparable to the destabilizing effects of surface roughness. More recently Schrauf et al. analysed flight tests using a simple HLFC system installed on the vertical tail plane of a passenger aircraft [10]. In order to better match the experimental data with suction they proposed a limiting N_{CF}-N_{TS}-correlation which has somewhat lower values compared to a previously published correlation for natural laminar flow without suction.

The present work revisits the aforementioned observation of lower critical N-factors for flows with suction. It addresses the prediction of laminar-turbulent transition with linear stability theory and the e^N-method for two–dimensional and three–dimensional flows. In an attempt to eventually derive a formulation for a variable critical N-factor for streamwise and cross–flow instabilities for flows with and without suction an adaptation or even a new calibration of the critical N-factors based on suction strength and size of suction area is applied. With these reduced critical N-factors, that account for a possible negative effect on the flow quality due to the suction process itself or due to the surface roughness caused by the suction holes, transition is predicted for a range of different two–dimensional and swept wing configurations. The objective is to further validate the general applicability of linear stability theory for flows with suction and to give some ideas on a possible dependency of critical N-factors on the implementation of the suction arrangement in experiments.

4.2 Results

For the stability analysis with suction a coupled program system is used, consisting of a transition prediction module [7] connected to the unstructured/hybrid RANS solver TAU of the German Aerospace Center (DLR) [11]. The transition module applies the linear stability theory in conjunction with the e^N-method for Tollmien-Schlichting and cross flow instabilities with amplification rates determined by an automated local, linear stability code.

All computations presented in the following have been carried out utilizing an implicit LU-SGS time integration scheme, a $4w$ multigrid cycle, matrix artificial dissipation and the Spalart-Allmaras turbulence model for the modeling of turbulent flow. The suction velocities at the viscous walls are introduced by a suction mass flux boundary condition in the RANS solver. The unstructured computational grids used all have a structured layer of 128 grid points normal to the wall, covering the entire boundary layer. The first spacing used for the structured layer leads to y^+ values lower than 0.5. Amplification rates and N-factor curves have been computed using incompressible linear stability theory with the boundary layer velocity profiles needed for the stability analysis directly extracted from the RANS solution.

4.2.1 NACA 64_2-A-215

Experiments with and without suction on a NACA 64_2-A-215 airfoil section at different Reynolds numbers and angles of attack were carried out in the low speed wind tunnel of Delft Technological University [13]. The wind tunnel itself has a cross section of 1.8m $\times$ 1.25m and has a nominal turbulence intensity of $Tu = 0.04\%$ at a free stream velocity of 28m/s. The model has a chord length of $c = 1.35$m and suction was applied between 30 and 90% of the length using 40 different suction compartments.

For an angle of attack of $\alpha = 0°$ and a Reynolds number of $Re = 2.75 \times 10^6$ velocity profiles have been measured without and with suction. For both cases, while using the experimental suction distribution (Fig. 4.1) in the computations for the suction case, a very good agreement is achieved for the computed and measured displacement thicknesses (Fig. 4.2). Applying here a critical Tollmien–Schlichting N-factor of $N_{TS} = 10.0$ (corresponding to $Tu = 0.04\%$ and using Mack's correlation [9]), transition is predicted at approximately $x/c = 0.6$ for the case without suction. If suction is applied, the flow stays laminar up to the trailing edge of the airfoil.

For a range of Reynolds numbers between $Re = 2 \times 10^6$ and $Re = 9 \times 10^6$ transition locations have been measured for two angles of attack ($\alpha = 0°$ and $\alpha = 3°$) without suction being applied. Using the critical N-factor of $N_{TS} = 10.0$, for $\alpha = 0°$ the predicted transition locations are in very good agreement with the experiment over the whole Reynolds number range (Fig. 4.3). For $\alpha = 3°$, transition is predicted

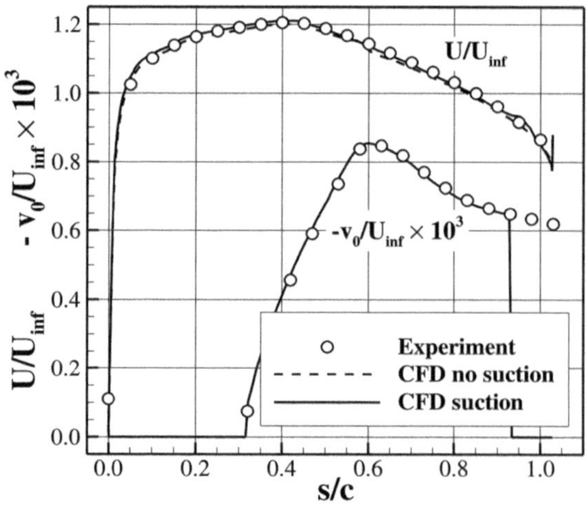

Fig. 4.1 Pressure and suction distribution, NACA 64_2-A-215

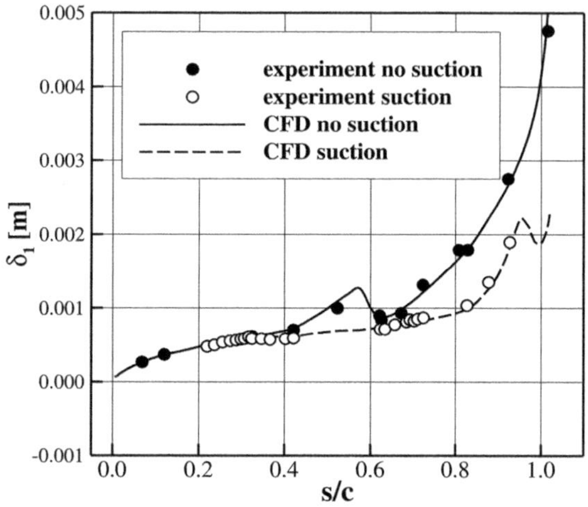

Fig. 4.2 Displacement thickness with and without suction, NACA 64_2-A-215

in the simulations in very good agreement with the experiment for Reynolds numbers larger than $Re = 6 \times 10^6$ as well. However, for lower Reynolds numbers, transition is predicted slightly too far upstream compared to the experiment but is nevertheless still in acceptable agreement with the experiment [8].

At a Reynolds number of $Re = 3.37 \times 10^6$ and an angle of attack of $\alpha = 0°$ suction has been applied in the experiments and transition locations have been measured for different suction strengths. If for these cases a critical N-factor of $N_{TS} = 10.0$ is used for the transition prediction, transition locations are predicted too far downstream compared to the experiment. However, if a noticeably reduced critical Tollmien–Schlichting N-factor of $N_{TS} = 4.0$ is used in the computations a very good agreement with experimental transition locations for a broad range of suction-flow coefficients can be achieved (Fig. 4.4).

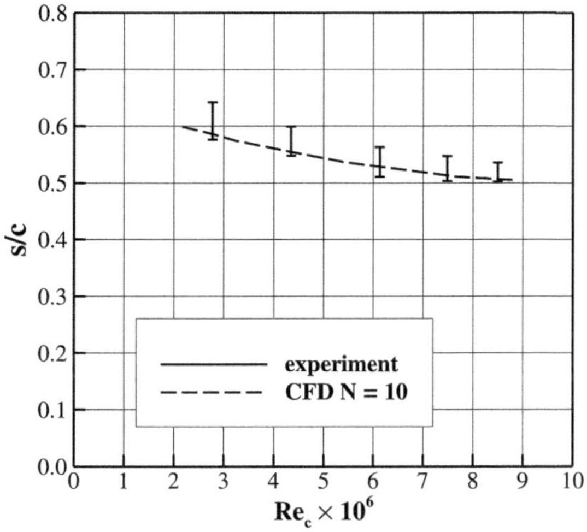

Fig. 4.3 Measured transition regions and predicted transition locations, $\alpha = 0°$, no suction, NACA 64_2-A-215

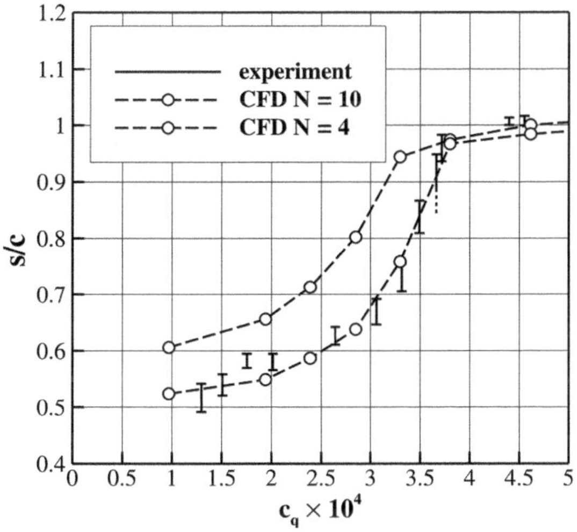

Fig. 4.4 Measured transition regions and predicted transition locations, $\alpha = 0°$, suction, NACA 64_2-A-215

The lowering of the critical N-factor may be attributed to account for destabilizing effects of the surface roughness of the suction panels and is in accordance with findings from the experiments, where due to the surface roughness no significant amount of laminar flow could be achieved for certain flow conditions, even if suction was applied: "*From the experiments (...) it was found to be impossible to keep the boundary layer laminar at values of* Re_c *above* 3.5×10^6. *At higher speeds the inherent surface roughness (...) became supercritical causing transition due to roughness*" (Van Ingen [13], page 164). This is an indication that the surface roughness of the suction panels, and possibly also the suction process itself, had a negative effect on the flow quality in the boundary layer in the experiment and this deterioration of the flow quality can be represented by a somewhat lower critical Tollmien–Schlichting N-factor.

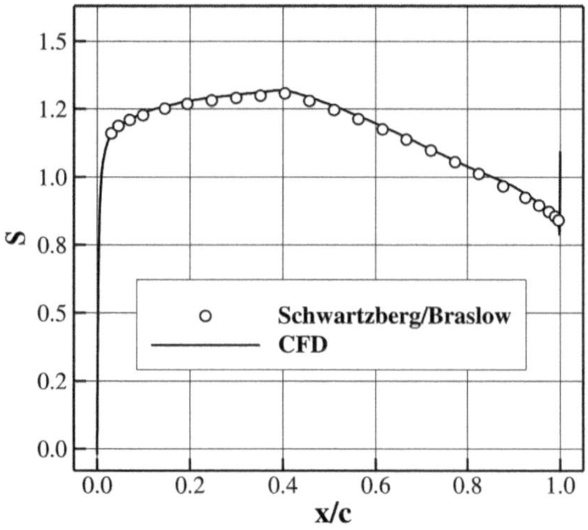

Fig. 4.5 Pressure distribution, NACA 64-A-010

4.2.2 NACA 64-A-010

The effects of suction on the flow around a NACA 64-A-010 airfoil have been experimentally investigated in the Langley Low-Turbulence Pressure Tunnel [12]. The tunnel has a 3ft $\times$ 7.5ft rectangular test section of 7.5ft length. The airfoil used in the experiments has a chord length of $c = 3$ft and spans the complete test section of the wind tunnel. Suction for this model was applied from 5% of the airfoil chord up to the trailing edge through a porous sintered-bronze surface [12]. The experiments where conducted at a Mach number of $Ma = 0.3$, a Reynolds number of $Re = 6 \times 10^6$ and an angle of attack of $\alpha = 0°$ for various suction velocity distributions.

The local suction velocities in the experiment can be derived from the external pressure distribution S (Fig. 4.5) and the suction-air pressure-loss coefficient C_p by the following formula (see Ref. [4]):

$$- v_0/U_\infty = \left(c_{v_0}/2ct\right) \cdot Re \cdot \left(C_p - S\right) \tag{4.1}$$

Here, c is the airfoil chord length, t is the thickness of the porous material and c_{v_0} is the porosity factor. For different suction-air pressure-loss coefficients C_p, the local suction velocities can then be integrated to give the suction-flow coefficient C_Q [12].

For transition prediction three different critical Tollmien-Schlichting N-factor values have been used ($N_{TS} = 10.0$, $N_{TS} = 7.0$ and $N_{TS} = 4.0$). $N_{TS} = 10.0$ corresponds to the measured turbulence level of the test–section of approximately $Tu = 0.4\%$ for a Mach number of $Ma = 0.3$ and a unit Reynolds number of $Re_1 = 2 \times 10^6/\text{ft}$ [14]. However, for a somewhat lower critical N-factor of $N_{TS} = 7.0$ a quite good agreement with the experiment can be achieved (Fig. 4.6, note that the suction coefficient C_Q in the computations was only 0.5 times the values of C_Q from

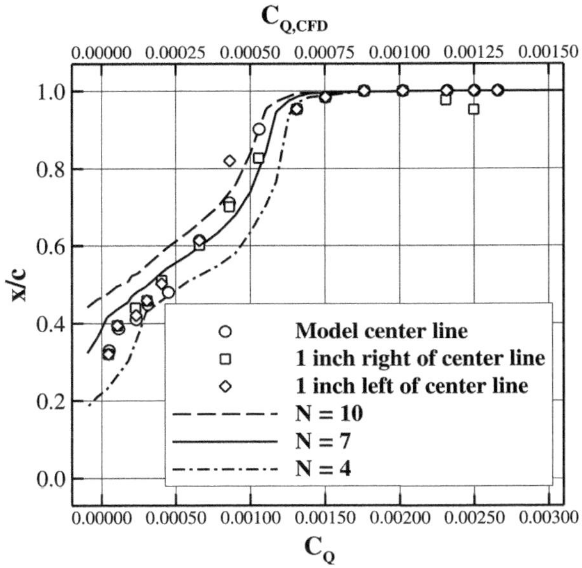

Fig. 4.6 Measured and predicted transition locations, NACA 64-A-010

the experiments [8]). Lowering the critical N-factor to match the experimental data indicates again (cf. Sect. 4.2.1) that the surface roughness of the suction panel and/or the suction process itself seems to have a negative effect on the flow quality.

4.2.3 ONERA D

Experiments on the influence of wall suction on transition in three–dimensional flow for a 60° swept wing model with an ONERA D section normal to the leading edge have been carried out in an Eiffel type wind tunnel with a 0.35m × 0.6m test section of 2.5m length. [1, 3, 5]. The wing model has a chord of $c = 0.35$m normal to the leading edge and is at both ends in direct contact with the wind tunnel walls. On the upper surface the wing has 10 suction ducts which extend from 13% to 68% chord (Fig. 4.7). The experiments were carried out at an angle of attack of $\alpha = -8°$ for two different Reynolds numbers of $Re = 2.56 \times 10^6$ and $Re = 3.2 \times 10^6$.

For the numerical investigations fully three-dimensional computations have been performed which included the tunnel walls taken into account as slip walls. To match the experimental pressure distribution, the angle of attack has been changed to $\alpha = -6.5°$ (Fig. 4.7). Based on the high free stream turbulence intensity of $Tu = 0.15\%$ for the experimental facility [3] travelling cross–flow instabilities have been investigated here.

For the two Reynolds numbers three different configurations with suction have been tested in the experiments: a) suction at only the first of the 10 ducts, b) suction at ducts 1 and 2 and c) suction at ducts 1 to 3. The suction velocity was varied from $v_0 = 0\text{cm/s}$ to $v_0 = -10\text{cm/s}$.

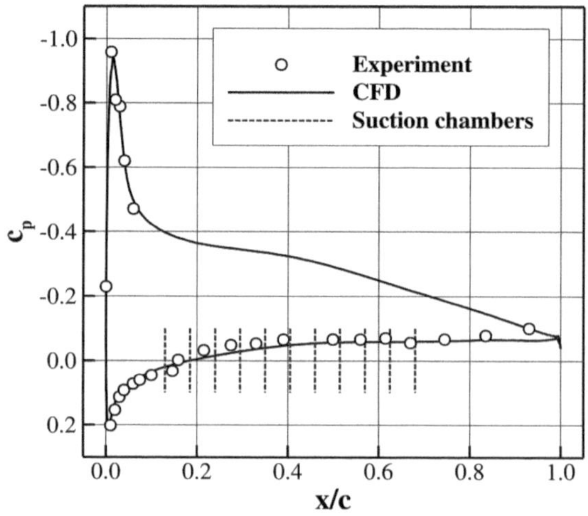

Fig. 4.7 Pressure distribution, ONERA D

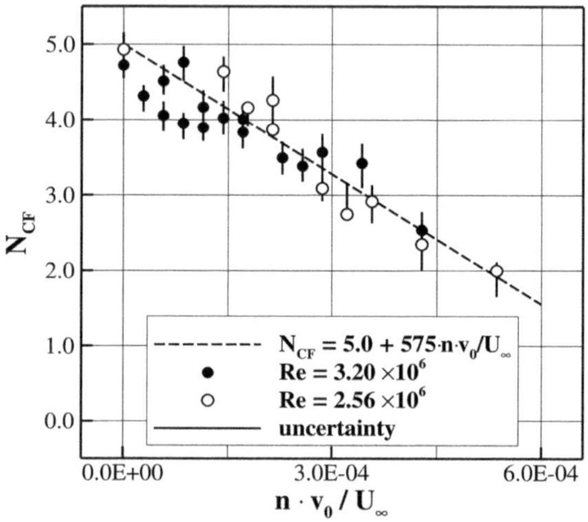

Fig. 4.8 Predicted N_{CF}-values at experimental transition location, ONERA D

An evaluation of the cross–flow N-factors determined at the experimental transition locations for all considered flow conditions (Fig. 4.8) leads to the formulation of a variable critical cross–flow N-factor, where the critical value is linearly dependent on area and strength of the suction, represented by the number n of active suction ducts ($n = 1, 2, 3$) and the non–dimensional suction velocity v_0/U_∞:

$$N_{TCF} = 5 + 575 \cdot n \cdot v_0/U_\infty \tag{4.2}$$

It can be seen, that using a variable N-factor according to Eq. 4.2 for the critical travelling cross–flow N-factor leads to transition locations in the CFD computations which are in very good agreement with the experimental values for the cases with

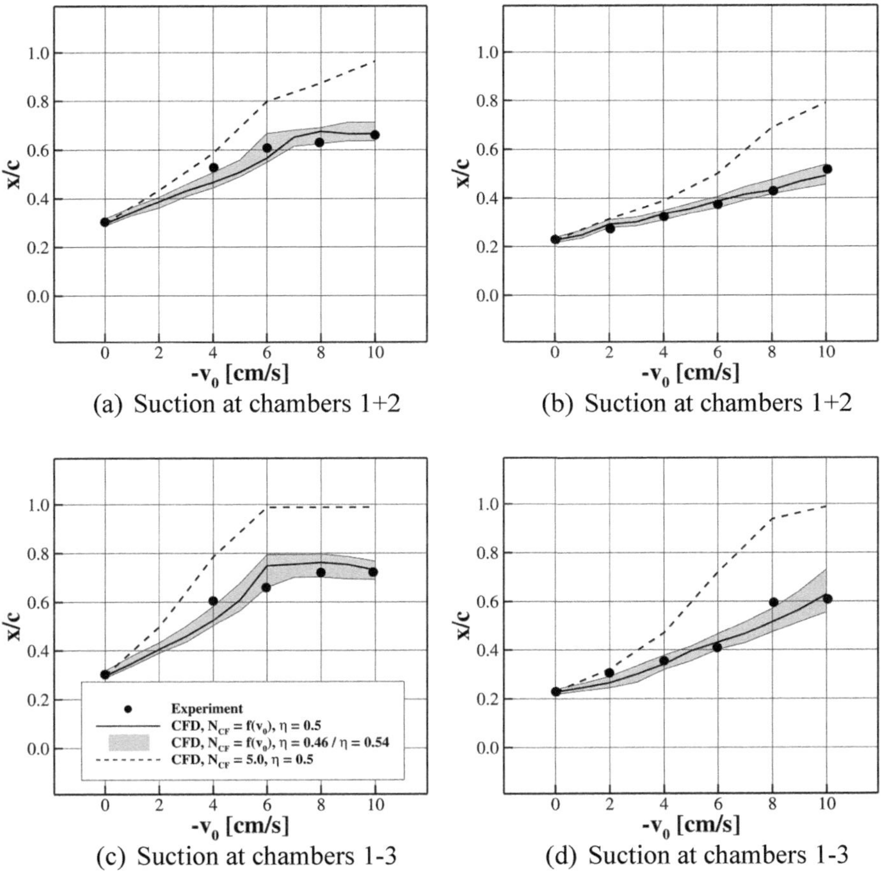

Fig. 4.9 Measured and predicted transition locations, ONERA D. (left: $Re = 2.56 \times 10^6$, right: $Re = 3.20 \times 10^6$)

suction at chamber 1 only (see Ref. [8]), suction at chambers 1+2 (Fig. 4.9a and b) and suction at chambers 1–3 (Fig. 4.9c and d). A constant critical N-factor however, not accounting for a potential influence of the suction process on the stability of the boundary layer, leads to an increased deviation from the experiment regarding predicted transition location with increased amount of suction applied (Fig. 4.9).

4.2.4 ONERA DTP-B

Thorough experimental investigations on the influence of suction on transition in three–dimensional flow have been conducted in the ONERA F2 wind tunnel [6]. The ONERA DTP-B wing model with a span of $s \approx 2.5$m used in the experiments

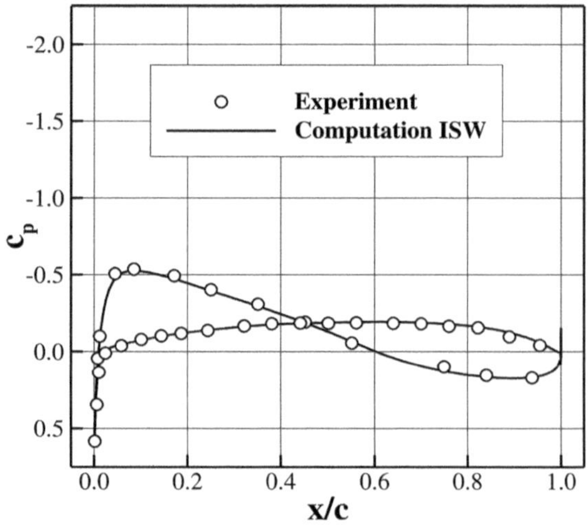

Fig. 4.10 Experimental and computational pressure distributions, $\Lambda = 40°$, $\alpha = 2°$, ONERA DTP-B

is attached on one side to the wind tunnel wall, the other end is positioned freely in the air stream. The chord length normal to the leading edge of the wing is $c = 0.7$m and the lower surface of the geometry is equipped with 7 suction chambers, distributed between 5% and 25% of the chord length. Transition has been detected in the experiment based on infra–red and hot film techniques. The flow conditions investigated in the experiments consider two Reynolds numbers ($Re = 3.2 \times 10^6$, $Re = 4.4 \times 10^6$) and two sweep angels ($\lambda = 40°$, $\lambda = 50°$) at five different suction velocities ($v_0 = 0\text{m/s}$ to $v_0 = -0.1\text{m/s}$, constant over all 7 suction chambers). The angle of attack was varied in the experiment from $\alpha = -1°$ to $\alpha = 8°$.

In the computations infinite swept wing conditions have been assumed, giving a very good agreement for the experimental and numerical pressure distributions (see Fig. 4.10 and Ref. [8]). A thorough evaluation of the predicted N_{TS}-N_{CF}-pairs at the experimental transition locations for the variety of flow conditions lead to the formulation of a variable critical N-factor for stationary cross–flow (Fig. 4.11). From the case-by-case depiction of the the N_{TS}-N_{CF}-pairs in Fig. 4.12 it can be seen, that for the lower Reynolds number case (Fig. 4.12a and b) the influence of the suction strength on the cross–flow N-factor at transition is quite significant, whereas for the higher Reynolds number the variation in critical cross–flow N-factor is relatively small (Fig. 4.12c and d). The final calibrated, variable cross–flow N-factor shows a dependency on the suction strength (v_0/U_∞) and on sweep angle Λ and Reynolds number:

$$N_{SCF} = 7.0 + \Delta N_{SCF} = 7.0 + 5 \cdot 10^4 \cdot \left[v_0/U_\infty \left(10^6 \cdot cos(\Lambda)/Re\right)^2 \right] \quad (4.3)$$

Although there appears to be a larger scatter in the values for the N_{TS}-factor at transition, a constant value of $N_{TS} = 7.5$ has proven to be sufficiently accurate

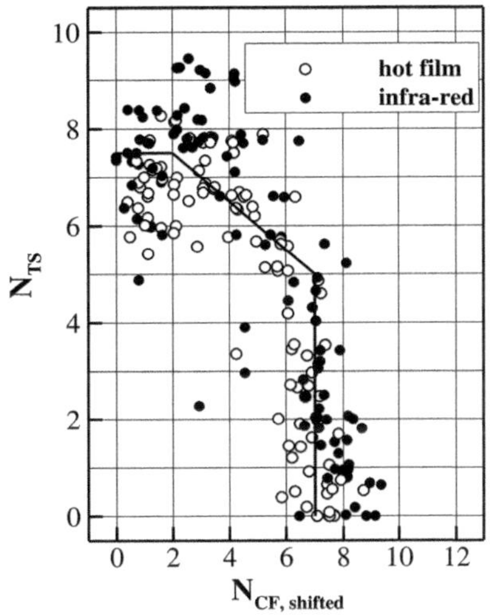

Fig. 4.11 Predicted N_{TS}-N_{CF}-pairs at experimental transition location, ONERA DTP-B

[8] (note: $Tu \approx 0.07\%$ for the wind tunnel would lead to $N_{TS} = 9.0$ using Mack's formula [9]). If Eq. 4.3 for the critical cross–flow N-factor is inversed, that is, the N_{CF}-factor for stationary cross–flow is shifted by ΔN_{SCF}, a constant critical value of $N_{SCF} = 7.0$ is determined, as is evident in Fig. 4.11. The stability limit, also modelling an interaction of Tollmien-Schlichting and cross–flow waves, is then indicated by the solid line in Fig. 4.11.

The predicted transition locations are shown in Figs. 4.13, 4.14, 4.15 and 4.16, using either a constant critical cross–flow N-factor of $N_{SCF} = 7.0$ or the variable critical value according to Eq. 4.3. The most significant improvement in predicted transition location with a variable critical N-factor can be seen for the cases with large suction for the lower Reynolds number of $Re = 3.2 \times 10^6$ (Figs. 4.13c, d and 4.14c and d). For the cases with the higher Reynolds number ($Re = 4.4 \times 10^6$, Figs. 4.15 and 4.16), both approaches for the critical N-factor lead to similar results. Here, the dependency of the critical N-factor on the suction strength seems to be negligible and the predicted transition locations are in very good agreement with the experiment for both considered sweep angles and all investigated suction velocities.

4.3 Conclusion

A verification and validation study of transition prediction using linear stability theory in form of the e^N method in a RANS solver environment for two–dimensional and three–dimensional flows with suction is presented. For all test cases considered, a certain adjustment is needed for the critical N-factors for Tollmien–Schlichting and cross–flow instabilities to account for negative effects on the flow quality due to the suction process itself or due to the surface roughness caused by the suction holes.

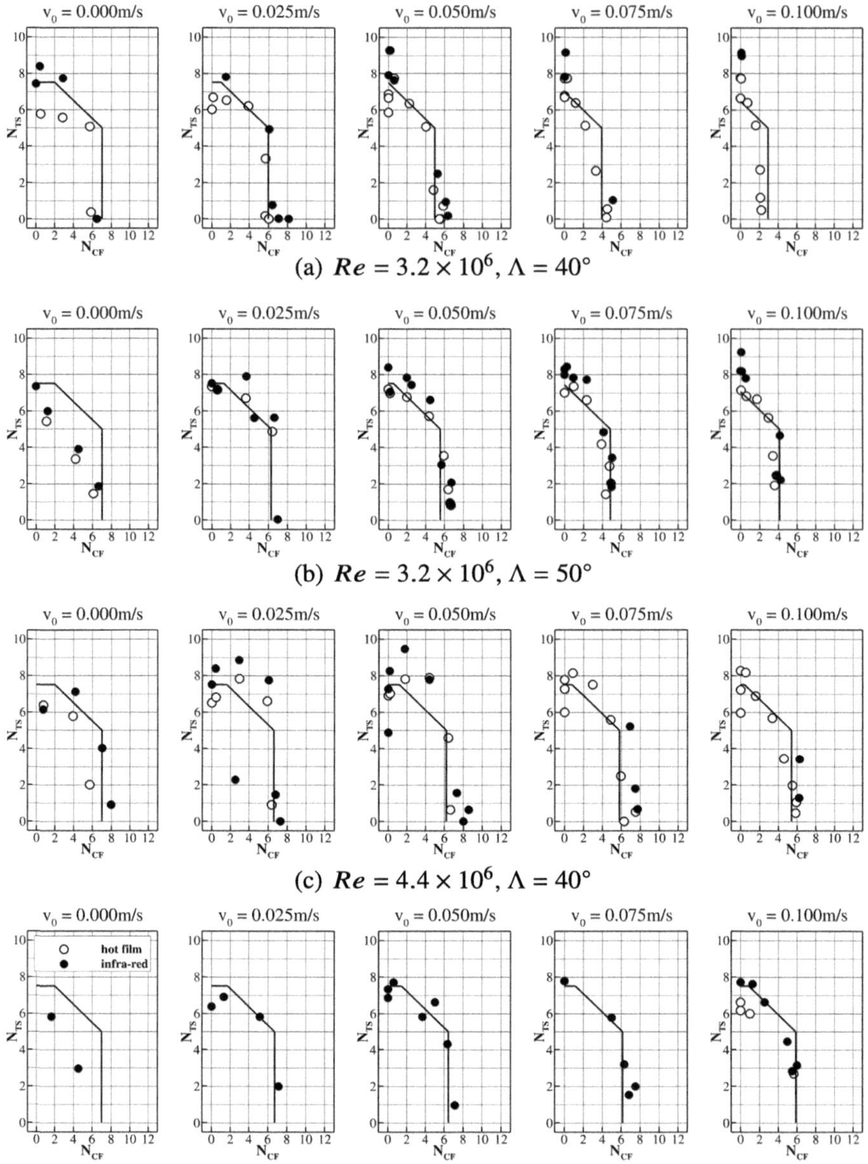

(a) $Re = 3.2 \times 10^6$, $\Lambda = 40°$

(b) $Re = 3.2 \times 10^6$, $\Lambda = 50°$

(c) $Re = 4.4 \times 10^6$, $\Lambda = 40°$

(d) $Re = 4.4 \times 10^6$, $\Lambda = 50°$

Fig. 4.12 Predicted N_{TS}-N_{CF}-pairs at experimental transition location, ONERA DTP-B

For the two–dimensional airfoil test cases, NACA 64_2-A-215 airfoil and NACA 64-A-010 airfoil, a rather arbitrary reduction of the critical N-factor was used which is intended to somehow account for different disturbance sources, such as surface porosity or suction strength. A more thorough derivation of the critical N-factors

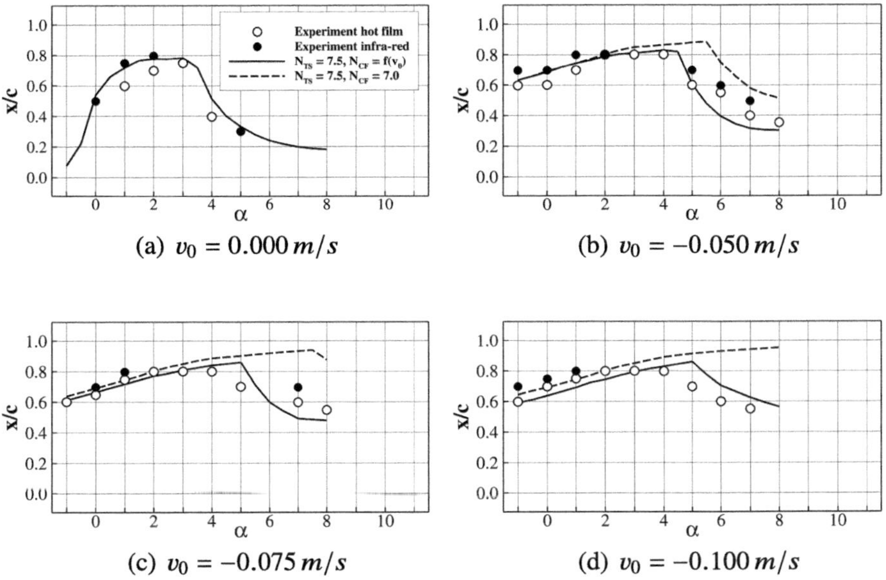

Fig. 4.13 Measured and predicted transition locations, $Re = 3.2 \times 10^6$, $\Lambda = 40°$, ONERA DTP-B

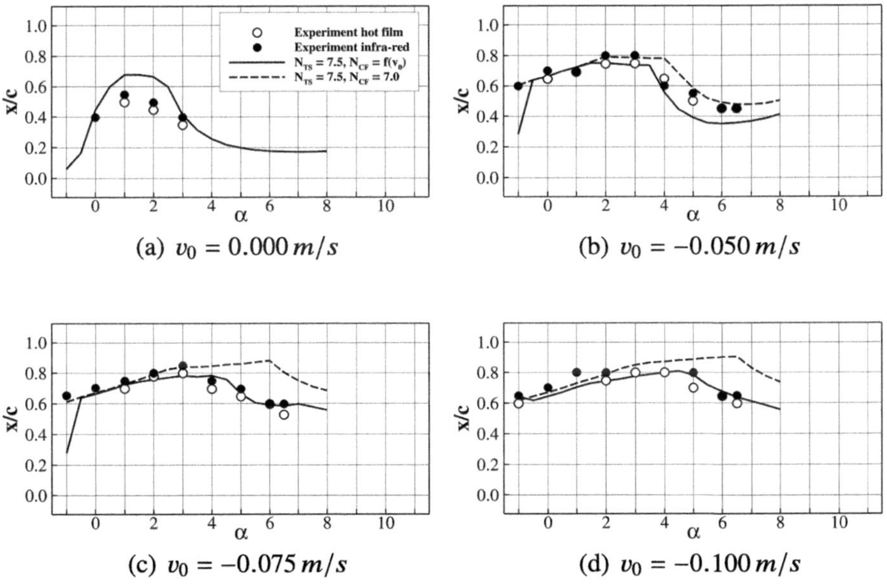

Fig. 4.14 Measured and predicted transition locations, $Re = 3.2 \times 10^6$, $\Lambda = 50°$, ONERA DTP-B

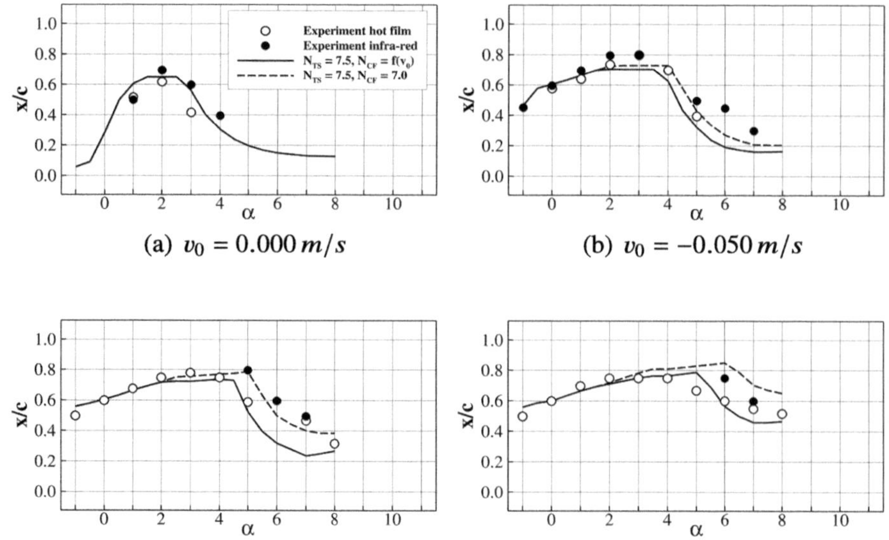

(a) $v_0 = 0.000\,m/s$ (b) $v_0 = -0.050\,m/s$

(c) $v_0 = -0.075\,m/s$ (d) $v_0 = -0.100\,m/s$

Fig. 4.15 Measured and predicted transition locations, $Re = 4.4 \times 10^6$, $\Lambda = 40°$, ONERA DTP-B

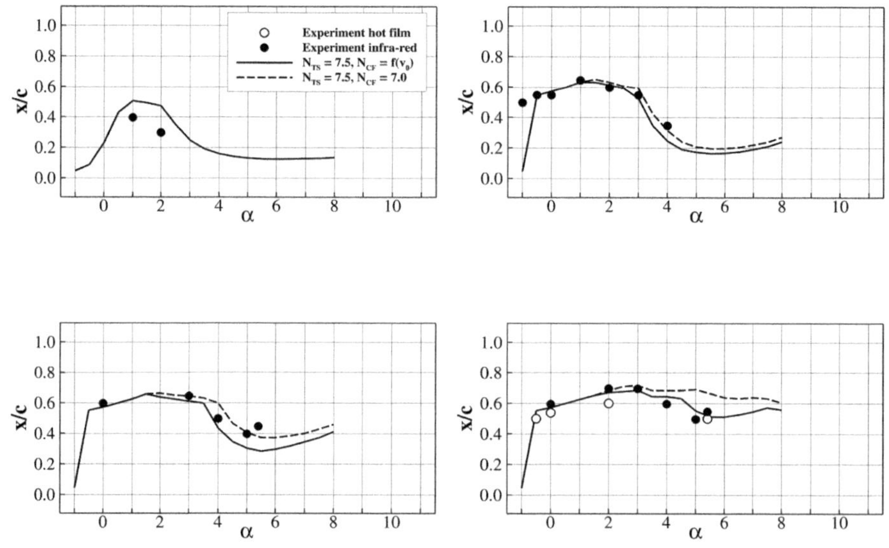

(c) $v_0 = -0.075\,m/s$ (d) $v_0 = -0.100\,m/s$

Fig. 4.16 Measured and predicted transition locations, $Re = 4.4 \times 10^6$, $\Lambda = 50°$, ONERA DTP-B for cross–flow instabilities has been established for the flow over swept wings with ONERA D and ONERA DTP-B airfoil profiles.

By applying modified critical N-factors, compared to the critical N-factors used without suction, for all considered test cases with suction the predicted transition locations are generally in very good agreement with the experimental transition locations for a wide range of flow conditions. However, a final statement on the exact and consistent dependency of the critical N-factor on specific elements of the suction arrangement is yet to be made. For this, a much larger amount of reliable data sets than presented here is needed for further calibration and validation.

References

1. Arnal D, Juillen JC, Laburthe F (1990) Transition de la couche limite. Rapport de synthèse final DERAT 50(5018):37–38
2. Arnal D, de Bruin AC, Cousteix J, Schrauf G, Seitz A, Smith PD (1994) Transition for airfoils and wings. Final report of GARTEUR AD (AG14), GARTEUR TP–081
3. Arnal D, Juillen JC, Casalis G (1991) The effects of wall suction on laminar–turbulent transition in three–dimensional flow. In: Boundary layer stability and transition to turbulence; proceedings of the symposium, ASME and JSME joint fluids engineering conference, vol 114. Portland, pp 155–162
4. Braslow AL, Burrows DL, Tetervin N, Visconti F (1951) Experimental and theoretical studies of area suction for the control of the laminar boundary layer on an NACA 64A010 airfoil. NACA–TR–1025, NACA
5. Cousteix J, Arnal D, Aupoix B, Gleyzes C (1991) Recent studies on transition and turbulence at ONERA-CERT. In: 29th Aerospace sciences meeting, AIAA Paper 91–0332. Reno
6. Gasparian G (1998) Étude expérimentale et modélisation des mécanismes de transition sur aile en flèche, avec application au maintien de la laminarité. Ph.D. thesis, École nationale supérieure de l'aéronautique et de l'espace
7. Krimmelbein N (2021) Transition prediction method for three-dimensional flows using linear stability theory. Ph.D. thesis, Technische Universtät Braunschweig, Institut für Strömungsmechanik, Deutsches Zentrum für Luft- und Raumfahrt, DLR-FB 2021-26. ISSN 1434-8454
8. Krimmelbein N, Krumbein A (2021) Transition prediction for flows with suction using the e^n-method. In: AIAA Scitech 2021 forum, AIAA Paper 2021–0630
9. Mack LM (1977) Transition prediction and linear stability theory. In: Laminar-Turbulent transition. AGARD-CP-224
10. Schrauf G, von Geyr H (2021) Simplified hybrid laminar flow control for the a320 fin-part 2: Evaluation with the e^n-method. In: AIAA Scitech 2021 forum, AIAA Paper 2021–1305
11. Schwamborn D, Gerhold T, Heinrich R (2006) The DLR TAU-code: Recent applications in research and industry. In: Wesseling P, Onate E, Periaux J (eds) European conference on computational fluid dynamics. ECCOMAS CFD 2006 (2006)
12. Schwartzberg MA, Braslow AL (1952) Experimental study of the effects of finite surface disturbances and angle of attack on the laminar boundary layer of an NACA 64A010 airfoil with area suction. NACA–TN–2796, NACA
13. Van Ingen JL (1965) Theoretical and experimental investigations of incompressible laminar boundary layers with and without suction. Technische Hogeschool Delft, Vliegtuigbouwkunde, Rapport VTH-124
14. von Doenhoff AE, Abbott FT (1947) The Langley two–dimensional low turbulence pressure tunnel. NACA–TN–1283, NACA

Part II
Advanced Computational Methods for Aeronautics Propulsion

Chapter 5
Continuous Adjoint-Based Aerothermal and Aeroacoustic Optimization of Aero Engine Components Using PUMA

X. Trompoukis, M. Monfaredi, V. Asouti, K. Tsiakas, and K. C. Giannakoglou

Abstract This article addresses the application of the in-house GPU-enabled CFD analysis and shape optimization tool PUMA in Aeroacoustic and Conjugate Heat Transfer problems. The gradients of the objective functions are obtained using the continuous adjoint method that, also, includes the differentiation of the Spalart-Allmaras turbulence model and the Hamilton-Jacobi equation for distance. The aeroacoustic optimization concern the design of the shape of a 3D aero-engine intake generatrix by minimizing the energy contained in the spectrum of the sound pressure, at prescribed receiver locations; the fan manifests its presence through a separately computed time-dependent pressure distribution at the fan inlet and the noise propagation is based on the Ffowcs Williams and Hawkings (FW-H) acoustic analogy. With regard to the aerothermal application, the optimization of the blade shape of an internally cooled linear turbine cascade and the position of cooling channels aims at min. total pressure drop of the hot gas flow and min. highest temperature within the blade. A free-form parameterization tool, based on volumetric NURBS, controls the blade airfoil contour while the cooling channels can be relocated inside the blade, by changing the coordinates of their centers. Geometric and flow constraints are imposed to keep the cooling channels away from the blade contour and retaining the turbine inlet capacity and flow turning.

Keywords Computational fluid dynamics (CFD) · Aerothermal · Aeroacoustic · Continuous adjoint method

X. Trompoukis · M. Monfaredi (✉) · V. Asouti · K. Tsiakas · K. C. Giannakoglou
Parallel CFD and Optimization Unit, School of Mechanical Engineering, National Technical University of Athens (NTUA), Athens, Greece
e-mail: morteza.monfaredi@gmail.com

V. Asouti
e-mail: vasouti@mail.ntua.gr

K. C. Giannakoglou
e-mail: kgianna@central.ntua.gr

T. Tuovinen et al. (eds.), *Advanced Computational Methods and Design for Greener Aviation*, Computational Methods in Applied Sciences 59,
https://doi.org/10.1007/978-3-031-61109-4_5

5.1 Introduction

Aeroacoustic and aerothermal applications require multi-disciplinary optimization (MDO). They involve different analysis methods for flow simulation and sound propagation (for the former) or the simultaneous solution of the mass and heat transfer equations in the fluid domain as well as the solution of the heat conduction equation in the solid domain (for the latter). To perform gradient-based optimization, the gradients of objective and constraint functions must be computed. Below, the two selected MDO fields are separately overviewed.

It is known that noise emitted by aircrafts affects their public acceptance particularly in areas close to airports [1]. In civil aircrafts powered by high and ultra high by-pass-ratio engines, the blade tips of the large diameter fan are the main source of noise generation. The design of the engine intake is of utmost importance since it is responsible for providing a uniform air supply to the engine while, at the same time, regulates the fan noise propagation towards far-field. Aeroacoustic noise prediction comprises detailed unsteady turbulent flow simulations, acoustic pressure propagation to long distances and evaluation of complex noise metrics. This can be linked with a gradient-based tool supported by the adjoint method in order to perform the optimization with a cost independent of the number of design variables. Adjoint methods are in widespread use in aerodynamic shape optimization [2, 3], while their use in aeroacoustic noise reduction is rather recent [4, 5]. This is mainly due to the unsteady nature of the acoustic problems requiring a time-accurate adjoint solver integrating backwards in time. Compared to its continuous counterpart, the discrete adjoint has received much more attention in aeroacoustic shape optimization [4, 6]. The discrete adjoint to hybrid acoustic models (URANS/FW-H) of [4] used to reduce the turbulent blunt trailing edge noise. A similar method was used for propeller design in [6]. The continuous adjoint for a hybrid URANS/FW-H analogy is developed in [7] for inviscid flows and in [5] for turbulent flows. In [8], the continuous adjoint to the linearized Euler equations is used to optimize the shape of a turbofan inlet duct.

At the same time, the efficient design of aircraft engines aims at improved performance and longer lifetime. Gains are expected from aerodynamically optimized compressors, optimally performing inlet ducts and nozzles, and/or higher turbine inlet temperatures. The latter is limited by the thermal resistance of the first turbine row blades which are able to operate at gas temperatures exceeding the limits set by the materials, thanks to internal and/or external cooling. Relevant studies are based on Conjugate Heat Transfer (CHT) analysis and optimization tools which involve the coupled solution of the governing equations in the adjacent fluid and solid domains. Gradient-based optimization algorithms has been used for the aerothermal design of turbine blades [9] and their internal passages [10, 11]. Verstraete et al. [10] and He et al. [11] used discrete adjoint for the shape optimization of a U-bend channel representing a section of turbine internal cooling system. The objective was the minimization of the weighted sum of total pressure losses and heat transfer. Gkaragkounis et al. [9] developed the continuous adjoint for CHT optimization for incompressible fluid flows in the OpenFOAM environment, and performed the shape optimization of a 2D internally cooled turbine cascade for min. mean temperature over the blade.

This article presents two adjoint-based MDO applications and the relevant computational tools used. The first application deals with the aeroacoustic shape optimization of an aero-engine intake by minimizing the total energy contained in the sound pressure spectrum at a set of receivers. This was based on a hybrid RANS/FW-H noise propagation model in the frequency domain and its adjoint counterpart. The second application targets the aerothermal optimization of the internally cooled C3X blade. Two optimization studies were performed. The first focused on the blade shape optimization while the second controlled both blade shape and its cooling hole positions to optimize its aerothermal performance. In all case studies, the adjoint to the turbulence model is also solved.

5.2 Computational Modules

All studies were performed using the in-house software PUMA [12] which includes a (U)RANS solver, a noise prediction tool based on the FW-H analogy and a heat conduction solver along with their continuous adjoints. All these tools run on GPU clusters employing the MPI protocol for data communication between GPUs on different computing nodes. A distinguishing feature of PUMA is the use of mixed-precision arithmetics (MPA) for high parallel efficiency. According to MPA, the Jacobian of the residuals is computed in double—though stored in single-precision, reducing the required GPU memory space and transactions, without damaging accuracy.

5.2.1 The Flow Analysis Module

The (U)RANS equations are discretized using a vertex-centered finite volume technique on unstructured grids consisting of tetrahedra, pyramids, prisms and/or hexahedra. A multi-stage Runge-Kutta with residual smoothing integrates the governing equations in pseudo-time. The inviscid fluxes are discretized based on a central differences scheme with artificial dissipation consisting of a blend of second- and fourth-order flow solution differences. The Spalart-Allmaras turbulence models is used in the aeroacoustic and the $k-\omega$ SST in the aerothermal application.

5.2.2 The Aeroacoustic Module

To compute the far-field noise and, then, perform the optimization at an affordable cost, a hybrid method is used. The CFD simulation is restricted to a small area close to the solid bodies (near-field flow analysis), and then, noise propagates to the distant receivers, using an acoustic analogy. The noise prediction tool of PUMA is based on the permeable version of the FW-H analogy [13], which provides the sound pressure at the receivers' locations $\mathbf{x}_r$ by computing the FW-H integral, performed on the FW-H surface, located in the part of the grid which is not displaced during the optimization. Since the aeroacoustic objective function is defined in the frequency domain, the FW-H integral should be expressed in the same domain. For each frequency (ω), this yields

$$H(f)\hat{p}'(\mathbf{x}_r,\omega) = -\int_{f=0} \hat{\mathcal{F}}_i(\mathbf{x}_s,\omega)\frac{\partial \hat{G}(\mathbf{x}_r,\mathbf{x}_s,\omega)}{\partial x_{s_i}}ds - \int_{f=0} \mathrm{i}\omega\hat{\mathcal{Q}}(\mathbf{x}_s,\omega)\hat{G}(\mathbf{x}_r,\mathbf{x}_s,\omega)ds \tag{5.1}$$

where quantities marked with a hat ($\hat{}$) have been transformed into the frequency domain using Fourier transform, i is the unit complex number and $\mathbf{x}_s$ denotes position vectors of points (sources) located on the FW-H surface. f is a signed distance from the FW-H surface so that $f = 0$ corresponds to this surface which should include all acoustic sources, and $H(f)$ is the Heaviside function. $\mathcal{Q}$ and $\mathcal{F}_i$ are the monopole and dipole source terms, reading

$$\mathcal{Q}(\mathbf{x},t) = \left(\rho u_i - \rho^\infty u_i^\infty\right) n_i^{\text{FWH}}, \ \mathcal{F}_i(\mathbf{x},t) = \left[\rho\left(u_i - 2u_i^\infty\right)u_j + \rho^\infty u_i^\infty u_j^\infty + p\delta_{ij} - \tau_{ij}\right] n_j^{\text{FWH}} \tag{5.2}$$

where c^∞ and $\mathbf{u}^\infty$ are the far-field's sound speed and velocity. In the current implementation, quadrupole terms are omitted. $p = p^\infty + p'$, $\rho = \rho^\infty + \rho'$ and $u_i = u_i^\infty + u_i'$ are the pressure, density and velocity components, respectively, with far-field quantities indicated by the superscript ∞. $\mathbf{n}^{\text{FWH}}$ is the outward unit vector normal to the FW-H surface. In subsonic flows, the Green function for 3D problems in the frequency domain is $\hat{G}(\mathbf{x}_r,\mathbf{x}_s,\omega) = -\frac{\exp(-ikr^+)}{4\pi r^*}$ where $k = \omega/c_\infty$, $r^+ = \frac{1}{\beta^2}\left(r^* - \frac{\mathbf{u}^\infty\cdot\mathbf{r}}{c^\infty}\right)$, $r^* = \left((\frac{\mathbf{u}^\infty\cdot\mathbf{r}}{c^\infty})^2 + |\mathbf{r}|^2\beta^2\right)^{0.5}$, $\mathbf{r} = \mathbf{x}_r - \mathbf{x}_s$ and $\beta = \sqrt{1-M^2}$ with M the Mach number.

In the aeroacoustic application this article is dealing with, the steady flow equations are solved in the rotating frame, and the unsteady data required to compute Eq. 5.2 result by properly rotating steady fields.

5.2.3 *The Aerothermal Module*

The heat conduction equation is integrated over the solid domain in vertex-centered finite volumes (similarly to the flow solver) and solved using a Krylov based solver. This part is loosely coupled with the flow solver via a fixed point iteration scheme with dynamic relaxation, see [14]. Upon convergence, the temperature and heat flux fields on the fluid-solid interface (FSI) become available. The temperature across the FSI is the boundary condition imposed to the flow solver which after a user-defined number of pseudo-iterations computes the heat flux crossing the FSI and communicates this to the heat conduction solver which updates the temperature along the FSI and so on and so forth, till convergence.

5.2.4 Continuous Adjoint Module

The adjoint solver of PUMA is based on the continuous adjoint approach. Different objective and constraint functions have been used in the two optimizations presented here. The aeroacoustic objective function J^{CAA} is defined in the frequency domain as the energy contained in the sound spectrum at the blade passing frequency ($\omega =$ BPF) averaged over the N_r receivers' locations and reads

$$J^{CAA} = \frac{1}{N_r} \sum_{a=1}^{N_r} \left| \hat{p}'(\mathbf{x}_{r_a}, \omega) \right| \tag{5.3}$$

where $\hat{p}'(\mathbf{x}_{r_a}, \omega)$ results from Eq. 5.1 and $\left|\hat{p}'\right| = \sqrt{\hat{p}'^2_{Re} + \hat{p}'^2_{Im}}$; Re and Im denote the real and imaginary parts of complex variables.

In the aerothermal optimizations, two objective functions and another two constraints are considered, namely, the hot gas total pressure drop (J_1^{CHT}), the blade's highest temperature (J_2^{CHT}), the inlet capacity (J_3^{CHT}) and the hot gas exit angle (J_4^{CHT}). These read,

$$J_1^{CHT} = \bar{p}_t^I - \bar{p}_t^O, \quad J_2^{CHT} = \frac{\sum_{Solid} T^S e^{\alpha T^S}}{\sum_{Solid} e^{\alpha T^S}}, \quad J_3^{CHT} = \dot{m}^I \frac{\sqrt{\bar{T}_t^I}}{\bar{p}_t^I}, \quad J_4^{CHT} = \tan^{-1}\left(\frac{\bar{u}_2^O}{\bar{u}_1^O}\right) \tag{5.4}$$

where overbars represent massflow-averaged flow quantities in the hot gas inlet (^{I}) and outlet (^{O}) boundaries. $\dot{m}^I$ stands for the inlet massflow and p_t, T_t are total pressure and total temperature, respectively. The blade's highest temperature is computed based on a differentiable function which mimics the *max* function, with α being a large constant value and T^S the nodal T values over the solid.

To formulate the adjoint problem, an augmented objective function is defined as

$$J^{aug} = J + \int_{\Omega^F} \left(\Psi_n R_n^{MF} + \tilde{\nu}_a R^{\tilde{\nu}} + d_a R^d \right) d\Omega + \int_{\Omega^S} T_a^S R^S d\Omega \tag{5.5}$$

where J can be any of the aforementioned objective or constraint functions. Ψ_n ($n = 1, ..., 5$), $\tilde{\nu}_a$ and d_a are the adjoint mean-flow (MF), the adjoint turbulence model and the adjoint distance fields, respectively. These are multiplied with the mean-flow (R_n^{MF}), Spalart-Allmaras ($R^{\tilde{\nu}}$) and Hamilton-Jacobi (R^d) equation residuals, respectively, and integrated over the fluid domain Ω^F. The Hamilton-Jacobi equation, [2], computes distances from the wall which are required by the turbulence model. An additional field integral over the blade's solid domain Ω^S is added to J^{aug} and contains the product of the adjoint solid temperature T_a^S with the heat conduction equation residual R^S. In what follows, the development of the continuous adjoint is demonstrated only for the mean-flow equations; the detailed mathematical development for aerodynamic, aeroacoustic and CHT optimization problems can be found in [5, 15, 16].

Next step is to differentiate J^{aug} w.r.t. the design variables b_i giving rise to the derivatives of the primal variables w.r.t. b. To avoid computing these expensive terms, their multipliers are set to zero within the field and along the boundaries that leads to the field adjoint equations and their boundary conditions as

$$-\frac{\partial f_{nk}^{inv}}{\partial U_m}\frac{\partial \Psi_n}{\partial x_k} - \left(\frac{\partial \tau_{qk}^{adj}}{\partial x_k} - \frac{\partial \Psi_5}{\partial x_k}\tau_{kq}\right)\frac{\partial u_q}{\partial U_m} - \frac{\partial q_k^{adj,F}}{\partial x_k}\frac{\partial T^F}{\partial U_m} + S_m^{adj} = 0 \quad (5.6)$$

where U_m and f_{nk}^{inv} stand for the conservative flow variables and inviscid fluxes. τ_{mk}^{adj}, $q_k^{adj,F}$ stand for the adjoint stresses and heat flux, respectively. Source terms S_m^{adj} are non-zero only in the aeroacoustic optimization application and are related with the engine's rotation speed and the derivative of J^{CAA} w.r.t. the conservative flow variables with contribution along the permeable FW-H surface [5]. The adjoint no-slip condition ($\Psi_2 + u_1^f \Psi_5 = \Psi_3 + u_2^f \Psi_5 = \Psi_4 + u_3^f \Psi_5 = 0$), along with the adjoint adiabatic ($q_k^{adj,F} n_k = 0$) or the adjoint FSI ones ($\Psi_5 = T_a$ and $q_k^{adj,F} n_k^F = -q_k^{adj,S} n_k^S$) are imposed over the adiabatic walls and the FSI, respectively, where n_k^F, n_k^S are the unit normals to the wall from the fluid and solid domains, $q^{adj,S}$ is the adjoint solid's heat flux and u_k^f the kth frame velocity Cartesian component. Along the fluid domain inlets and outlets, the following adjoint boundary condition holds

$$\Psi_n A_{nmk} n_k \frac{\partial U_m}{\partial \mathcal{V}_j} + (\tau_{km}^{adj} n_m - \Psi_5 \tau_{km} n_m)\frac{\partial u_k}{\partial \mathcal{V}_j} + q_k^{adj,F} n_k \frac{\partial T}{\partial \mathcal{V}_j} + \frac{\partial J}{\partial \mathcal{V}_j} = 0 \quad (5.7)$$

where $A_{nmk} = \frac{\partial f_{nk}^{inv}}{\partial U_m}$ and $\mathcal{V}_j$ is the jth outgoing Riemann variable. The last term in Eq. 5.7 is non-zero only for the aerothermal objective functions defined along the Ω^F inlet/outlet. The mean-flow adjoint field equation 5.6 is coupled with the Spalart-Allmaras and heat conduction (for CHT applications) adjoint field equations, and, for the aeroacoustic optimization, it is solved in the (relative) frame of reference rotating with the engine's speed. The adjoint distance equation do not contribute to the rest adjoint equations, and is, thus, solved only once before computing the sensitivity derivatives of J w.r.t. b_i, which yield

$$\frac{\delta J}{\delta b_i} = -\int_{\Omega^F} \Big[\Psi_n\Big(\frac{\partial f_{nk}^{inv}}{\partial x_i} - \frac{\partial f_{nk}^{vis}}{\partial x_i}\Big) + \tau_{km}^{adj}\frac{\partial u_m}{\partial x_i} + q_k^{adj}\frac{\partial T}{\partial x_i}\Big]\frac{\partial}{\partial x_k}\Big(\frac{\delta x_i}{\delta b_i}\Big) d\Omega$$
$$- \int_{\Omega^F} \frac{\partial \Psi_n}{\partial x_k} U_n \frac{\delta u_k^f}{\delta b_i} d\Omega + \int_{S_W^F} \Psi_5 \left(q_k \frac{\delta n_k}{\delta b_i} + \frac{\delta (u_k^f n_K)}{\delta b_i} - \tau_{km} n_m \frac{\delta u_m^f}{\delta b_i} \right) dS + \mathcal{I}_i^{SA} + \mathcal{I}_i^d + \mathcal{I}_i^S \tag{5.8}$$

where f_{nk}^{vis}, τ_{mk} are the viscous fluxes and stresses, $\mathcal{I}^{SA}$, $\mathcal{I}^d$, $\mathcal{I}^S$ summarize the contribution from the adjoint Spalart-Allmaras, distance and heat conduction equations. Grid sensitivities on the parameterized surfaces are computed by additionally differentiating the parameterization tool. These are propagated into the computational domain(s) using a variant of inverse distance weighting morphing method.

5.3 Aeroacoustic Optimization—Aero-engine Intake

Since the intake geometry was axisymmetric, a CFD domain that corresponds to a single blade passage of the fan was used, with appropriate periodicity conditions. The far-field flow was still and the pressure distribution along the fan inlet was used as boundary condition, Fig. 5.1a. The computational domain, shown in Fig. 5.1b, had about $3.7M$ nodes arranged on 100 meridional planes.

The outcome of the FW-H analogy was verified by comparing it with pure CFD results. The FW-H surface, shown in Fig. 5.1b, is axisymmetric and has 16,000 nodes. To perform the integration of Eq. 5.1, the full circumference of the FW-H surface was needed. However, instead of rotating the FW-H surface, the receivers were rotated and the acoustic pressure is recovered by superimposing pressure signals from each receiver. The comparison between the hybrid model and pure CFD was made at 3

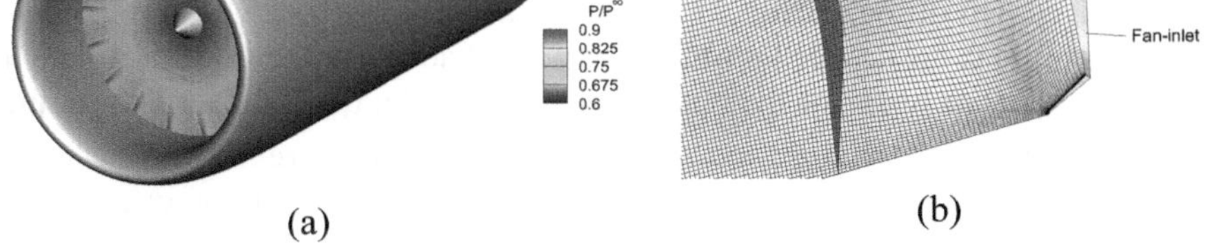

Fig. 5.1 Aeroacoustic optimization. **a** Perspective view of the aero-engine with a snapshot of the rotating static pressure at the engine inlet. **b** Close-up views of the CFD domain. The parameterized part of the nacelle and the FW-H surface are colored in red and blue, respectively

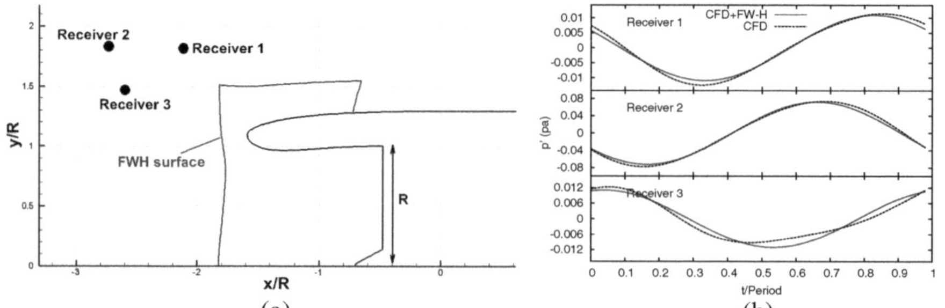

(a) (b)

Fig. 5.2 Aeroacoustic optimization. **a** Location of the FW-H surface, in blue, and 3 receivers within the CFD domain. **b** Pressure fluctuations computed by CFD+FW-H and plain CFD, at these receivers

receiver locations, shown in Fig. 5.2a. As illustrated in Fig. 5.2b, the agreement is good and some discrepancies are attributed to the omission of quadrupole terms, representing noise due to viscous effects and turbulence.

The generatrix of the engine intake was parameterized by a NURBS curve, giving rise to 15 control points controlling the shape of the nacelle lips and the throat area, leaving its outer shape and the part close to the fan intact, Fig. 5.3a. The first and last control points (CPs) were fixed while the other 13 were allowed to move in both the axial and radial direction leading to $2 \times 13 = 26$ design variables.

The aeroacoustic objective function (Eq. 5.3) was averaged over 1800 receivers equally distributed over the entire circumference, located at angle $\theta = 90°$ and radius $r = 8.8R$ (Fig. 5.3b). The convergence history of the J^{CAA} is presented in Fig. 5.4a showing a reduction by 70% after 12 optimization cycles. As a result, the optimized geometry has a much lower amplitude of pressure fluctuations as shown in Fig. 5.4b. Note that the sound pressure signal at all receivers was almost identical, though different in phase; thus, it was sufficient to compare the amplitude only at a single receiver location. Comparison of the directivity plot of the sound pressure level (SPL), normalized by its initial value at $\theta = 90°$, Fig. 5.5a, shows that the reduction is directional; as expected, maximum reduction occurs in the vicinity of the receivers contributing to J^{CAA} while, in some other angles, noise is increased. This was also expected as all receivers were located at the same (r, θ) location, while incorporating more receivers could reduce the noise in other directions as well. Changes in the parameterized part of the nacelle are illustrated in Figs. 5.3a and 5.5b. The biggest change occurred at the nacelle lip which was pushed downward towards the rotation axis. Total pressure losses of the optimized shape are by 0.6% higher than in the initial geometry.

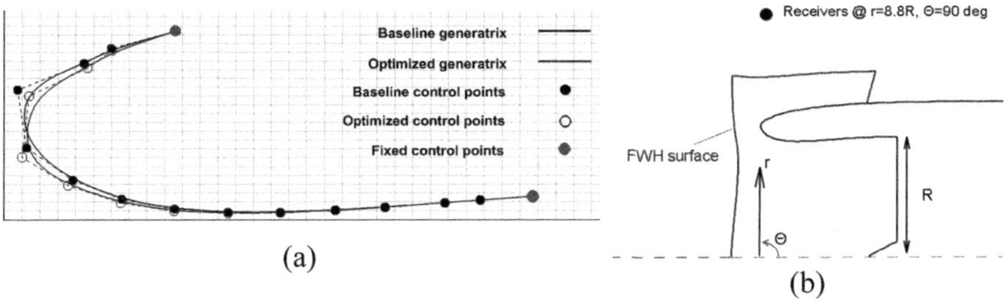

Fig. 5.3 Aeroacoustic optimization. **a** NURBS parameterization of the nacelle generatrix. **b** 2D view of the intake geometry, the FW-H surface at plane $z = 0$, and the receivers' location

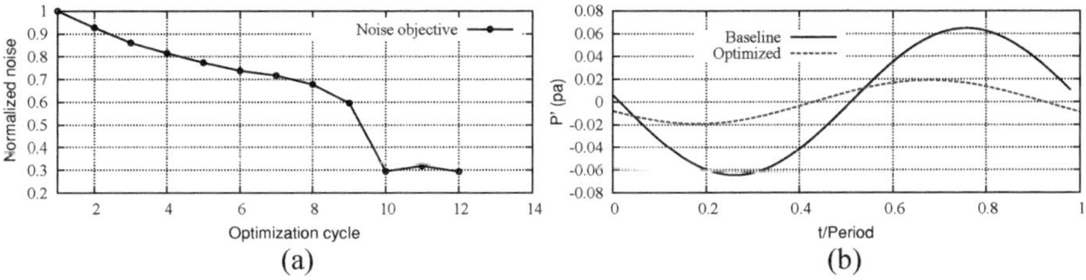

Fig. 5.4 Aeroacoustic optimization. **a** Convergence of J^{CAA}, normalized by its initial value. **b** Pressure fluctuations time-series, within a period, at one receiver

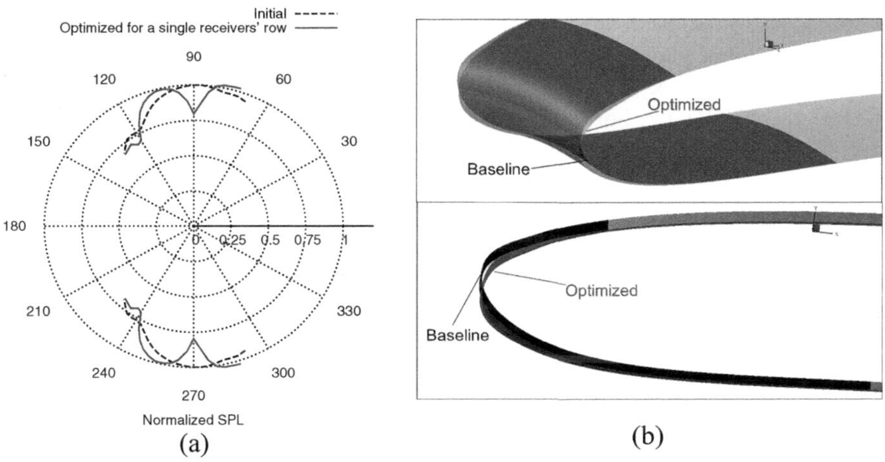

Fig. 5.5 Aeroacoustic optimization. **a** Optimization results. Comparison of the directivity plot of the sound pressure level. Values are normalized based on the initial value at $\theta = 90°$. **b** 3D views of the baseline and optimized geometries

Table 5.1 Average temperatures, flow rates and heat transfer coefficients at the cooling channels (numbered from the leading to trailing edge)

# Channel	T_{avg} (K)	$\dot{m}$ (kg/s)	$h\ (\text{W/m}^2\text{K})$
1	358.14	0.0167	1420
2	359.37	0.0174	1468
3	349.97	0.0148	1284
4	351.51	0.0165	1402
5	342.56	0.0175	1464
6	371.85	0.0165	1413
7	351.85	0.0161	1374
8	385.96	0.0055	2000
9	413.22	0.0035	1403
10	454.87	0.0017	1745

5.4 Aerothermal Optimization of an Internally Cooled Turbine Blade

In this section, the aerothermal optimization of the C3X internally cooled turbine linear cascade is presented. The blade, the geometry of which can be found in [17], is cooled by 10 radial channels. Its span is 7.62 cm long, with an axial chord of 7.816 cm. It is made of stainless steel with density $\rho^S = 7900\,\text{kg/m}^3$ and heat capacity $C^S = 586.15\,\text{J kg}^{-1}\,\text{K}^{-1}$. Its thermal conductivity is a function of temperature, $k^S = 6.8110 + 0.020716T^S$. For the hot gas, at the inlet, $p_t^I = 243{,}700\,\text{Pa}$, $T_t^I = 808\,\text{K}$ and $\text{Tu}^I = 8.3\%$ (turbulence intensity), whereas at the outlet $p^O = 142{,}530\,\text{Pa}$. For each cooling channel, the heat transfer coefficient was computed based on the measured channel's average temperature and coolant flow rate, [17], see Table 5.1.

For the analysis, both Spalart-Allmaras and Menter's SST turbulence models were used. The computed pressure and temperature distributions on the blade sides were compared with measurements in Fig. 5.6. A good agreement was observed, though the temperature along the suction side (close to the leading edge) was overestimated due to the absence of a transition model.

Two optimization studies were performed. In both, the 7×3 volumetric NURBS control grid of Fig. 5.7a was used to control the blade shape. The red control points at the leading and trailing edges were fixed while the rest were free to move in the axial and pitchwise directions, resulting in 38 design variables. The CFD grid of Ω^F is adapted to the changing boundaries using an inverse distance weighting morphing method. The grid inside the blade was generated from scratch each time the blade shape and/or cooling channel locations change. The optimization was based on the method of moving asymptotes [18].

The first optimization study focused solely on the aerothermal optimization of the blade shape without affecting the cooling holes (i.e. their position and shape remained fixed throughout the optimization). The minimization of the weighted sum

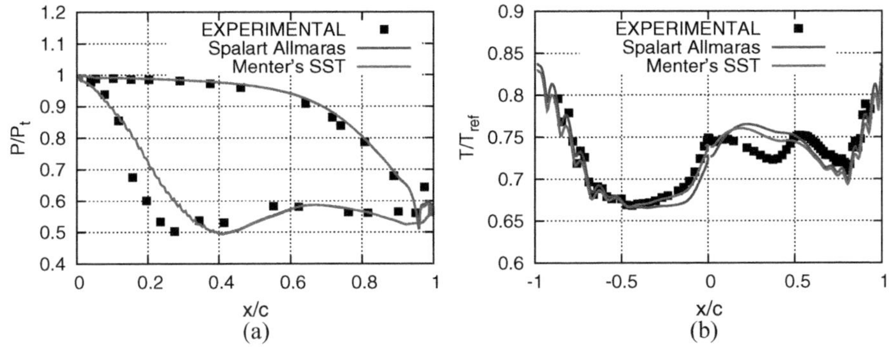

Fig. 5.6 Aerothermal analysis of the baseline geometry. Computed **a** pressure and **b** temperature distributions, non-dimensionalized by the inlet values (243,700 Pa and 811 K)

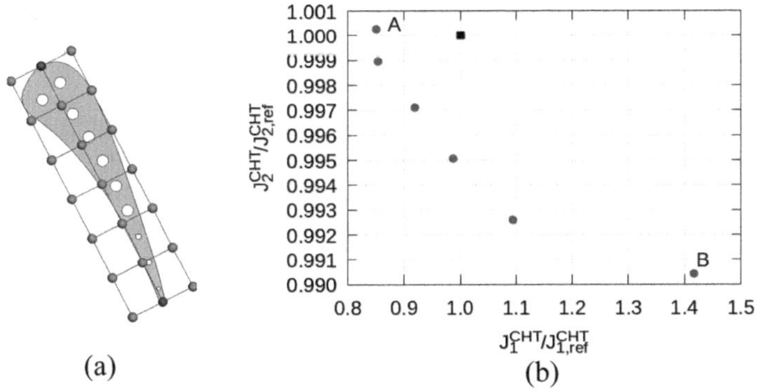

Fig. 5.7 Aerothermal optimization. **a** Parameterization of the C3X blade airfoil shape. **b** Computed front of non-dominated solutions (red circles) along with the baseline geometry (black square)

of J_1^{CHT} and J_2^{CHT} with six different value-sets of the weight factors (w_1, w_2) resulted to the front of six non-dominated solutions in Fig. 5.7b, where J_1^{CHT}, J_2^{CHT} are normalized with the respective values of the baseline geometry. The left edge (**A** in Fig. 5.7b) of the front corresponds to $(w_1, w_2) = (1, 0)$ and has ~14.5% less total pressure drop and slightly higher blade's T_{max}. The right edge (**B**) resulted from $(w_1, w_2) = (0, 1)$ and corresponds to a blade shape with ~6 K lower blade's T_{max} but highly increased losses. Figure 5.8 presents the computed Mach number fields for the baseline and the two front edge configurations. Solution **A** has reduced flow velocity at the throat region and thus, reduced total pressure drop compared to the baseline configuration. Solution **B** reduced the T^F close to the blade's trailing edge where T^S_{max} takes place. This reduced T^S_{max} by ~6 K. However, the passage throat region area became smaller, two strong shock waves appeared in the throat region and close to the trailing edge, and these increased the hot gas total pressure drop, compared to the baseline configuration.

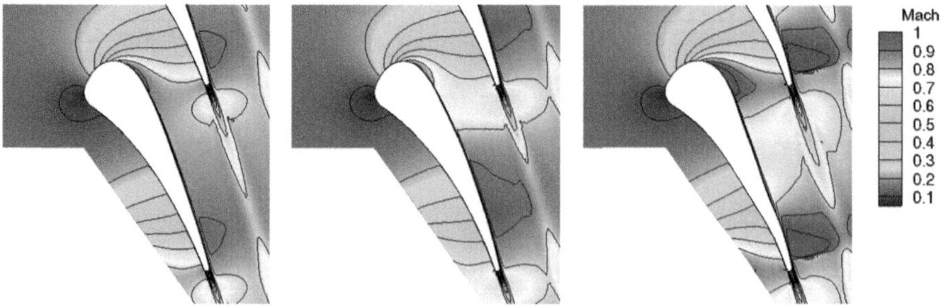

(a) (b) (c)

Fig. 5.8 Aerothermal optimization. Computed Mach number fields for the baseline configuration (left) and the two front edges **A** (middle) and **B** (right)

The target of the second optimization study was to reduce both the total pressure drop and T^S_{max} while constraining the inlet capacity and the hot gas exit flow angle to change no more than 0.1% from their datum values. This was carried out in two optimization steps. The first step dealt with the constrained minimization of the total pressure drop. This time, the cooling channel centers were controlled by the volumetric NURBS grid of Fig. 5.7a while their radii were fixed. The optimization converged as in Fig. 5.9 (left). The total pressure drop was decreased by ~21% after 25 cycles by satisfying the flow constraints. The fact that the cooling holes could be displaced together with the airfoil contour, enabled larger blade deformations which let the total pressure drop decrease more than in the first optimization study.

The best solution of the first optimization step had almost the same T^S_{max} with the baseline since this is mainly affected by the location of the cooling channels. Thus, the outcome of the first step was further optimized for min. T^S_{max} to further refine the location of the cooling holes. In this (second) optimization step, the blade shape remained fixed while the cooling holes were free to move, leading to 20 design variables (2 per cooling hole) in total. To avoid intersection of the cooling holes with the blade sides, the distance of each hole from the blade sides was constrained to be higher than the 30% of their radius. These 20 geometric constraints (2 per cooling hole) kept the holes away from the blade sides, and avoided extremely high $|\nabla T^S|$ (high thermal stresses). After 25 optimization cycles, T^S_{max} was reduced by ~30 K by still satisfying all geometric constraints. Figure 5.9 (right) shows the convergence of the optimization algorithm, where all T^S values are normalized with T^I_t. For clarity, only the distance of the 10th cooling hole (the closest one to the trailing edge) from the blade's sides is plotted in Fig. 5.9 (right), where distances are normalized with the holes' radii. The best solution of the second optimization step had slightly increased (by less than 0.1%) total pressure drop compared to the solution of the first step. Inlet capacity and hot gas exit angle did not differ by more than 0.1% from their datum values.

Figure 5.10 shows the Mach number fields for the baseline and the optimized configuration (i.e. the outcome of the second optimization step). The reduction in

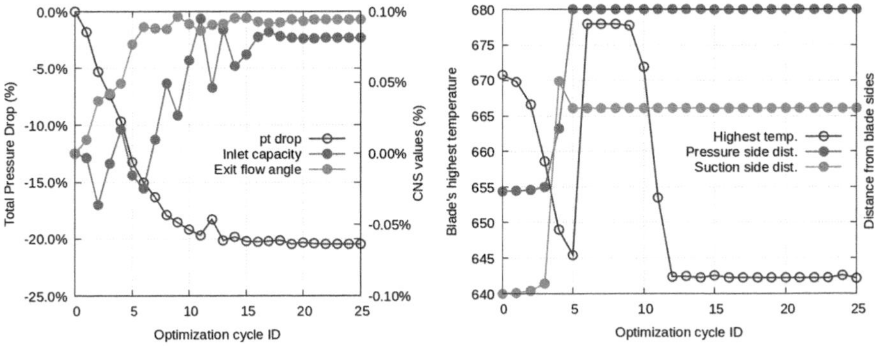

Fig. 5.9 Aerothermal optimization. Convergence history of the two optimization steps

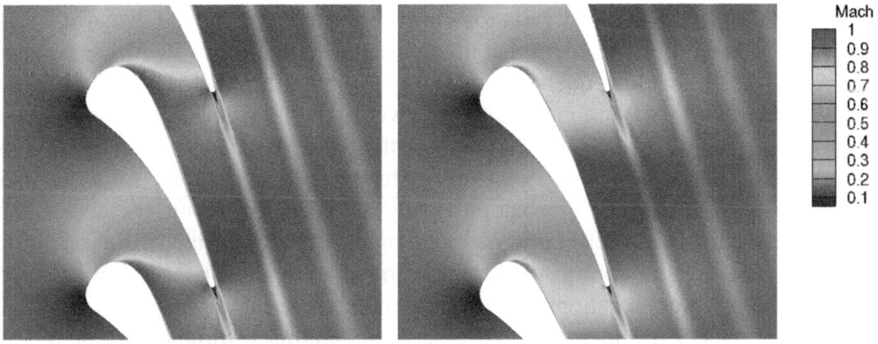

(a) (b)

Fig. 5.10 Aerothermal optimization. Computed Mach number fields for the **a** baseline and **b** optimized (from the two-step procedure) configurations

total pressure drop was mostly achieved by reducing the flow velocity in the passage throat region. Figure 5.11 presents the computed T^S and $|\nabla T^S|$ fields inside the blade for the baseline and the optimized configurations. In the latter, the first two cooling holes were moved towards the leading edge where T^S is high, while the remaining ones were moved towards the overheated trailing edge. Thanks to the geometric constraints, the cooling holes did not come too close to the blade sides and the highest $|\nabla T^S|$ changed by less than ~0.3% compared to the baseline configuration.

5.5 Conclusions

This article demonstrated the capabilities of the in-house CFD software PUMA by presenting its use in the aeroacoustic and Conjugate Heat Transfer optimization of aero-engine components. PUMA employs gradient-based optimization algorithms,

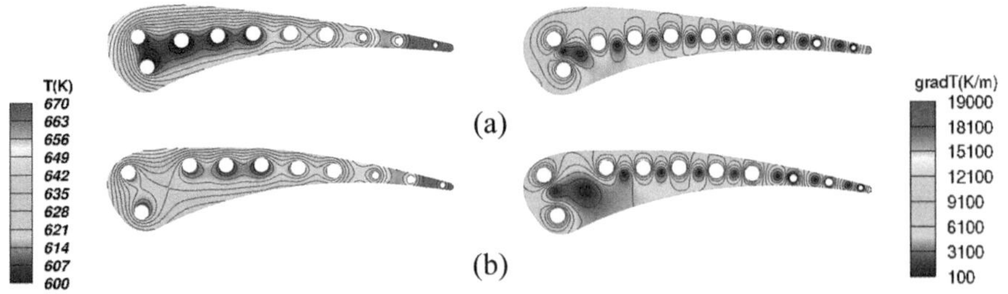

Fig. 5.11 Aerothermal optimization. Computed T^S (left) and $|\nabla T^S|$ (right) fields for the **a** baseline and **b** optimized (from the two-step procedure) configurations

assisted by the continuous adjoint method for computing the gradients of the objective and constraint functions, with variations in turbulence viscosity and distance fields taken into account. It was shown that the aeroacoustic module of PUMA, based on the FW-H analogy in the frequency domain along with its differentiation for this to be used together with the continuous adjoint solver to the flow equations, can be used in air-intake design applications. Results of the 3D aeroacoustic noise prediction and shape optimization showed that using steady flow and adjoint solutions and building the unsteady field by rotating steady fields can overcome the downsides of unsteady adjoint, namely large memory requirements and long solution time. Also, the aeroacoustic optimization demonstrated a significant reduction in noise levels at the receivers locations. Concerning the aerothermal optimization, PUMA was used in the optimization of the internally cooled, linear turbine cascade, C3X, targeting minimum hot gas total pressure drop along with minimum blade's highest temperature, by also imposing flow and geometric constraints. The blade's shape was parameterized by the volumetric NURBS-based parameterization tool of PUMA, while the cooling channels were free to change their location within the blade. A two-step optimization scheme allowed a $\sim$21% reduction in the hot gas total pressure drop simultaneously with a $\sim$30 K reduction in blade's highest temperature.

Acknowledgements This work is part of the MADELEINE project which received funding from the EU's Horizon 2020 research and innovation programme under GA No. 769025. The authors would like to thank Rolls-Royce plc. (Dr. S. Shahpar) and the Institute of Sound and Vibration Research of the University of Southampton (Dr. L. Wu) for providing the initial intake geometry and the boundary conditions of the fan inlet used herein.

References

1. Hansell AL, Blangiardo M, Fortunato L, Floud S, de Hoogh K, Fecht D, Ghosh RE, Laszlo HE, Pearson C, Beale L et al. Aircraft noise and cardiovascular disease near Heathrow airport in London: small area study. BMJ
2. Papoutsis-Kiachagias EM, Giannakoglou KC (2016) Continuous adjoint methods for turbulent flows, applied to shape and topology optimization: industrial applications. Arch Comput Methods Eng 23(2):255–299
3. Papadimitriou DI, Giannakoglou KC (2007) A continuous adjoint method with objective function derivatives based on boundary integrals for inviscid and viscous flows. Comput Fluids 36(2):325–341
4. Rumpfkeil MP, Zingg DW (2010) A hybrid algorithm for far-field noise minimization. Comput Fluids 39(9):1516–1528
5. Monfaredi M, Trompoukis X, Tsiakas K, Giannakoglou KC (2021) Unsteady continuous adjoint to URANS coupled with FW-H analogy for aeroacoustic shape optimization. Comput Fluids 230:105136
6. Içke R, Baysal O, Lopes LV, Zhou BY, Diskin B, Moy A (2020) Toward adjoint-based aeroacoustic optimization for propeller and rotorcraft applications. In: AIAA aviation 2020 forum. AIAA 2020-3140
7. Monfaredi M, Trompoukis XS, Tsiakas KT, Giannakoglou KC (2021) An unsteady aerodynamic/aeroacoustic optimization framework using continuous adjoint. In: Advances in evolutionary and deterministic methods for design, optimization and control in engineering and sciences, pp 147–162
8. Qiu Sh (2018) A continuous adjoint-based aeroacoustic shape optimization for multi-mode duct acoustics. Proc Inst Mech Eng Part C: J Mech Eng Sci 232(21):3897–3914
9. Gkaragkounis KT, Papoutsis-Kiachagias EM, Giannakoglou KC (2018) The continuous adjoint method for shape optimization in conjugate heat transfer problems with turbulent incompressible flows. Appl Therm Eng 140:351–362
10. Verstraete T, Müller L, Müller J (2017) Adjoint-based design optimisation of an internal cooling channel u-bend for minimised pressure losses. Int J Turbomach Propuls Power 2(10)
11. He P, Mader C, Martins J, Maki K (2019) Aerothermal optimization of a ribbed u-bend cooling channel using the adjoint method. Int J Heat Mass Transf 140:152–172
12. Asouti VG, Trompoukis XS, Kampolis IC, Giannakoglou KC (2011) Unsteady CFD computations using vertex-centered finite volumes for unstructured grids on graphics processing units. Int J Numer Methods Fluids 67(2):232–246
13. Williams JEF, Hawkings DL (1969) Sound generation by turbulence and surfaces in arbitrary motion. Philos Trans R Soc Lond Ser A Math Phys Sci 321–342
14. Ganine V, Hills N, Lapworth B (2013) Nonlinear acceleration of coupled fluid-structure transient thermal problems by Anderson mixing. Int J Numer Methods Fluids 71(8):939–959
15. Tsiakas K (2019) Development of shape parameterization techniques, a flow solver and its adjoint, for optimization on GPUs. Turbomachinery and external aerodynamics applications. PhD thesis, NTUA
16. Trompoukis X, Tsiakas K, Asouti V, Kontou M, Giannakoglou KC (2021) Optimization of an internally cooled turbine blade—mathematical development and application. Int J Turbomach Propuls Power 6(2):20
17. Hylton LD, Mihelc MS, Turner ER, Nealy DA, York RE (1983) Analytical and experimental evaluation of the heat transfer distribution over the surfaces of turbine vanes. NASA-CR-168015
18. Svanberg K (2002) A class of globally convergent optimization methods based on conservative convex separable approximations. SIAM J Optim 12(2):555–573

Chapter 6
Towards Understanding and Resolving Natural Shock Oscillation in Transonic Compressors

Philipp Leonard Nel, Patrick Grothe, Marius Swoboda, Sergio Pirozzoli, and Julien Weiss

Abstract The mechanisms of shock oscillation in a transonic compressor with a laminar shock-boundary layer interaction are investigated with Large Eddy Simulation using the Rolls-Royce Hydra LES solver. The effect of shock position (bow, passage, swallowed shock) is investigated. It is found that the collapse and recovery of the laminar separation bubble (shock-induced separation) due to a Kelvin-Helmholtz instability is the root of the oscillation, and that this instability most likely occurs when the shock induced separation bubble is large enough (passage shock) and severe density and velocity gradients exist. Therefore, the oscillation does not occur for a bow shock or swallowed passage shock.

Keywords Shock-boundary layer interaction · Laminar separation bubble · Transonic · Compressor · Kelvin-Helmholtz instability · Highly loaded · Altitude excitation · LES

6.1 Introduction

The current work focusses on the effect of the growth, collapse and stability of a laminar separation bubble (LSB) from a laminar shock-wave/boundary-layer interaction (SBLI) in a transonic compressor, and the consequences on the shock oscillation. More specifically, the formation of a Kelvin-Helmholtz instability on the LSB shear

Supported by European Union Horizon 2020.

P. L. Nel (✉) · P. Grothe · M. Swoboda
Rolls-Royce Deutschland Ltd. Co. KG, Blankenfelde-Mahlow, Brandenburg 15827, Germany
e-mail: nel@ist.rwth-aachen.de

S. Pirozzoli
Department of Mechanical and Aerospace Engineering, Sapienza University of Rome, Via Eudossiana 18, 00184 Roma, Italy

J. Weiss
Institute of Aeronautics and Astronautics, Technical University of Berlin, Marchstraße 12-14, 10587 Berlin, Germany

T. Tuovinen et al. (eds.), *Advanced Computational Methods and Design for Greener Aviation*, Computational Methods in Applied Sciences 59,
https://doi.org/10.1007/978-3-031-61109-4_6

layer, its consequent advection, and effect on the time dependent behaviour of the compressor passage is studied.

The origins of the work are explained next. A laminar boundary layer persists on the suction side of transonic fan at high altitudes, leading to laminar SBLI, shock oscillation and the consequent excitation of structural blade modes. Structural excitation of a problematic mode may be mitigated by promoting transition on the suction side boundary layer upstream of the shock. The oscillation mechanism however, is not understood: it is not understood whether when mitigated, the (aerodynamic) oscillation amplitude reduces, or whether a shift toward off-resonant frequencies occurs. Furthermore, the dimensionality of the oscillation mechanism is not clear. This also means that the requirements for resolving the oscillation are uncertain.

It is known experimentally that the (structural) excitation occurs for a specific range of operating conditions at high blade loading. Therefore, a representative operating condition within this range is investigated in both full span and Quasi-2D (Q2D) configuration, using Reynolds Averaged Navier Stokes (RANS), Unsteady Reynolds Averaged Navier-Stokes (URANS), and Large Eddy Simulation (LES) with and without sub-grid model.

So far, LES results suggest a 2D mechanism, with origins at a shear layer instability, more specifically of the Kelvin-Helmholtz (KH) type, forming for specific conditions, and setting off the oscillation cycle, hence the objective of the current work is to investigate this phenomenon using LES.

Work on the stability of separated boundary layers dates back almost 150 years ago. The stability of LSB's has also been the topic of extensive research. The Taylor-Goldstein equation can describe the dynamics of the KH instability, and the onset of instability is governed by the Richardson number. Gropengiesser investigated the stability of the compressible shear layer and showed that a density gradient could increase the amplification rate of instability waves [3]. In a further investigation, Masslowe and Kelly showed that density gradient can increase as well as reduce disturbance amplification, with an increase occurring when density and velocity gradients oppose one another [10]. Diwan et al. showed that these instability modes are convectively unstable and are therefore advected downstream into the separation region [1]. Michelis et al. studied the response of laminar separation bubbles to impulsive forcing, and showed that the separation height/displacement thickness across the LSB reduces after waves from the KH instability have advected downstream of the LSB, before the LSB recovers [11]. In the context of the current work (transonic compressor), this would mean that the blockage through the passage changes, which would cause a shock movement. For a detailed literature review on the stability of laminar separation bubbles, the reader is referred to Michelis et al. [11].

6.2 The Use of RANS/URANS

In the case of RANS/URANS, an uncertainty study is performed in both Fine/Turbo and the in-house Rolls-Royce Hydra CFD codes. The results suggest that RANS/URANS does not capture the physics required to resolve the oscillation.

An exhaustive attempt was made to determine whether the oscillation can be resolved using RANS/URANS [13], since it is of relevance for industry.

To avoid confusion in the interpretation of RANS and URANS results, numerical sources of oscillation are discussed. These include multi-grid sweeps and sensitivity to CFL, as well as the mesh itself which may cause oscillations in URANS even if RANS shows sufficient mesh independence. In the case of URANS, numerical sources of oscillation are damped out [13] by the inertial term. In the case of RANS, it may be suggested that numerical sources of oscillation show the region where the physics are sensitive (when a passage shock is present), however perhaps for the wrong reasons. Another problem in the case of URANS is the inability to capture interactions between stochastic and deterministic unsteadiness which are possibly essential for resolving such an oscillation mechanism [1].

6.3 Validation Test Cases

The mesh resolutions for all cases in the current work can be found in Table 6.1. The lower value of Δy^+ corresponds to the first cell outside of the boundary layer. In all cases, wall resolved LES is used and $y^+ = 1$.

Table 6.1 Mesh resolution for LES cases

Case	Span/chord	Reynolds	Mio. cells	Δx^+	Δy^+	Δz^+	y^+
TFAST Tr. casc. Re350k 20p	20%	350,000	28	25	4–45	25	1
TFAST Tr. casc. Re350k 30p	30%	350,000	42	25	4–45	25	1
TFAST Tr. casc. Re1.4M	20%	1,400,000	152	40–50	11–55	55	1
Tr. chann. Re2.5M Priebe	16% (SS)	2,500,000	330 DOF	40–60	N/A	22–32	1.3–1.7
Tr. chann. Re2.5M coarse	16% (SS)	2,500,000	52	65	45–66	65	1
Tr. chann. Re2.5M refined	16% (SS)	2,500,000	186	55	11–51	46	1
Lufthansa Technik Re1.4M	10%	1,400,000	57	55	11–55	55	1

6.3.1 LES of Transitional SBLI in Transonic Channel

A transonic channel numerical test case with compressor-like curvature and flow conditions, is used for validation of the LES settings. This is because no experimental test case for the shock oscillation mechanism existed. In a subsequent work by the author, the oscillation mechanism is validated experimentally [14]. While performing LES on the TFAST cascade, a larger upstream influence and significantly different flow structure in comparison to RANS was noticed for the transitional passage shock case. Therefore, the transonic channel with transitional SBLI is a suitable test case.

The test case geometry and boundary conditions can be found in Priebe et al. [16]. In this numerical test case, no turbulence is present at the inlet. The boundary layer interacting with the shock is laminar. The total pressure, temperature, exit static pressure, and Reynolds number are: P0 = 202,705.9 Pa, T0 = 351.3 K, PS_{exit} = 137,550.5 Pa, Re = 2,500,000.

The domain can be seen in Fig. 6.1. The chord length on the suction side is 155.96 mm. Numerical shadowgraph images for the RANS initial solution as well as LES (fine and coarse) can be found in Figs. 6.2, 6.3, 6.4, 6.5, 6.6 and 6.7. Indeed, the upstream influence of the SBLI in the case of LES is much greater than the RANS.

A comparison of time averaged pressure distribution for the fine mesh case in Hydra to that of the high-order GENESIS code used in Priebe et al. [16] can be found in Fig. 6.8. Note that at the inlet region, the curvature is not defined in Priebe et al. [16], hence a notable difference of the pressure distribution in this region. Although the time averaged pressure distribution and snapshots show adequate agreement between

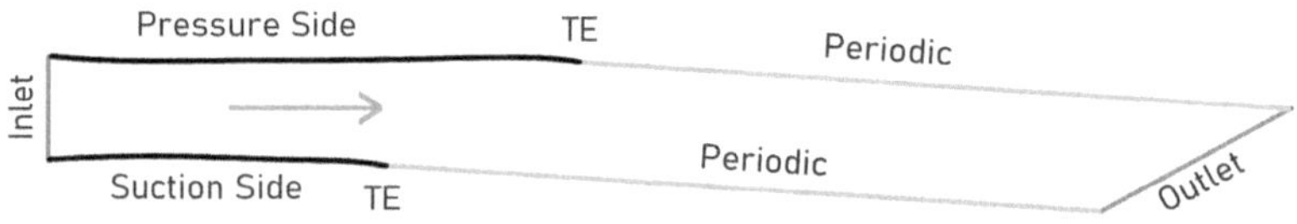

Fig. 6.1 Transonic channel domain nomenclature

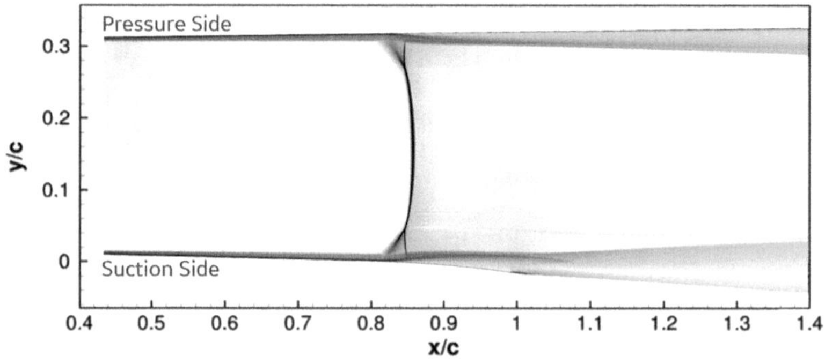

Fig. 6.2 Transonic channel: HYDRA RANS contours of $|\nabla\rho|/\rho$ on spanwise plane on high resolution mesh (186M). Note small upstream influence

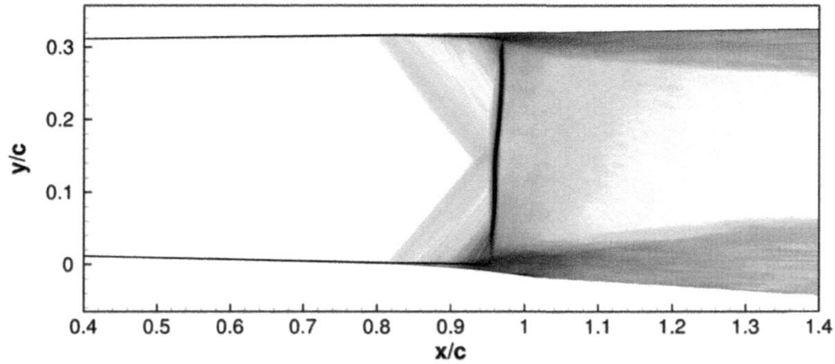

Fig. 6.3 Transonic channel: HYDRA time averaged contours of $|\nabla\rho|/\rho$ on spanwise plane for low resolution mesh (52M)

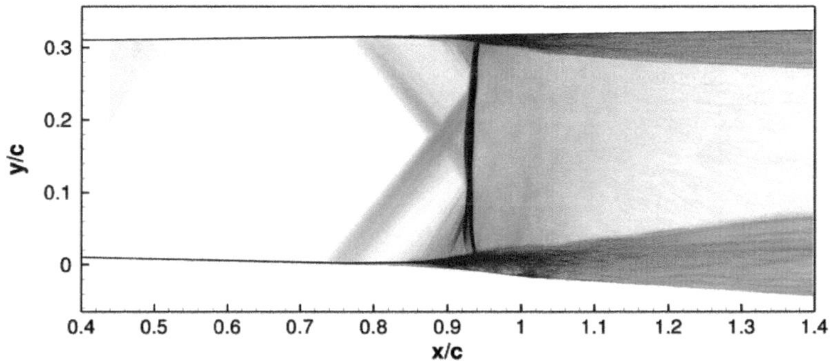

Fig. 6.4 Transonic channel: HYDRA time averaged contours of $|\nabla\rho|/\rho$ on spanwise plane for high resolution mesh (186M)

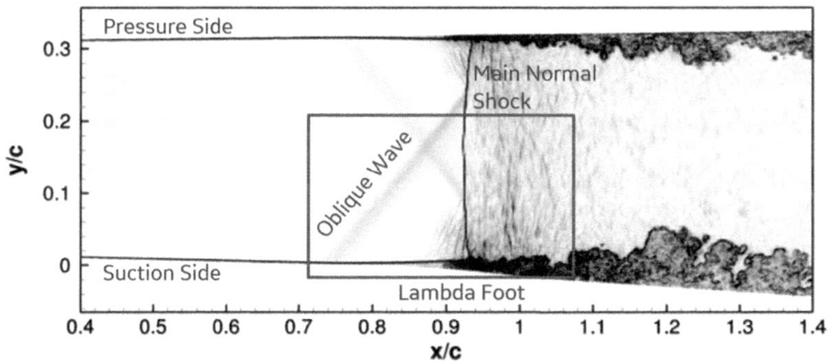

Fig. 6.5 Reproduced from Priebe et al. [16]. Transonic channel: instantaneous contours of $|\nabla\rho|/\rho$ on spanwise plane for 330 mio DOF GENESIS LES simulation

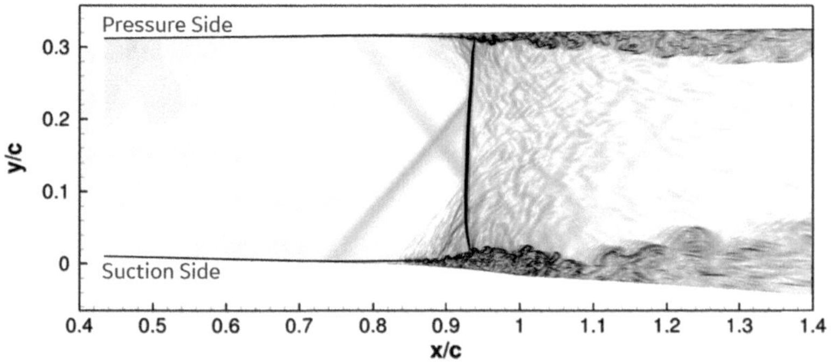

Fig. 6.6 Transonic channel: HYDRA instantaneous contours of $|\nabla\rho|/\rho$ on spanwise plane for high resolution mesh (186M)

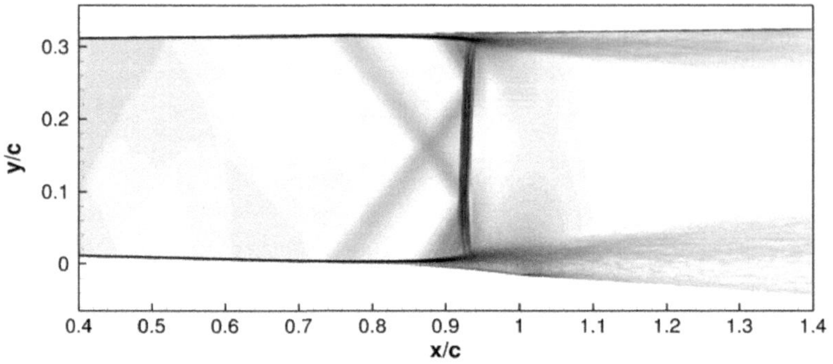

Fig. 6.7 Reproduced from Priebe et al. [16]. Transonic channel: time averaged contours of $|\nabla\rho|/\rho$ on spanwise plane for 330 mio DOF GENESIS LES simulation

Fig. 6.8 Transonic channel: suction side time averaged unsteady pressure distribution comparison

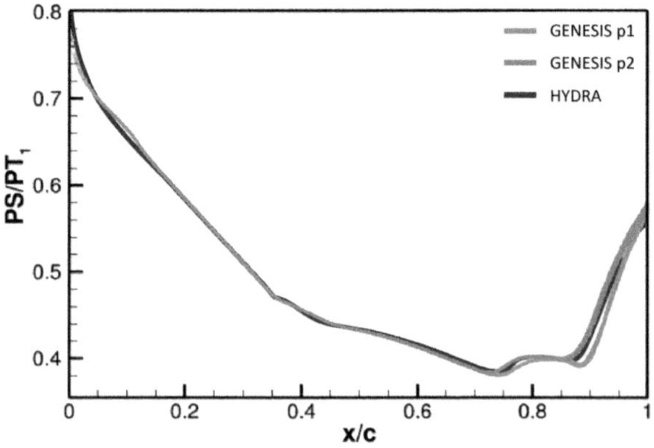

the HYDRA and GENESIS setups, the important time-dependent behaviour of the shear layer collapsing and recovering could not be captured on the suction side in the Hydra setup, even though the pressure side shear layer was showing the expected LSB collapse and growth behaviour. The simulation was stopped as it did not show a significant shear layer collapse 1e-3 seconds after moving into the right position. This had lead to the conclusion that the grid is not fine enough, or that a sub-grid model was required. Due to the relatively low grid resolution in this case, some artificial viscosity may be needed to capture the correct shearing physics. Therefore, the current work mainly focuses on reduced Reynolds cases for which the Reynolds number at hand is as resolved as that of Priebe et al. [16], and two values of artificial viscosity are tested (C_S) of 0.1 as well as 0 (corresponding to an implicit LES).

Observing the time averaged HYDRA solution for the coarse grid (Fig. 6.3), it can be seen that the shock is positioned too far downstream (95% of chrod as opposed to 92%). For the refined simulation, the solution definitely improves (Fig. 6.4), since the shock moves closer to that of the high-order GENESIS solution (Fig. 6.7). It can be seen that the upstream influence is, on average slightly smaller for the HYDRA simulation than the GENESIS solution. This also hints toward the possible need for some additional artificial dissipation in order to get a more representative solution. Subsequent works by the author have shown that the pressure-time footprint of the Hydra solution shows the same LSB collapse frequency as the GENESIS setup [12]. The required grid sizes and subgrid-scale models which show accurate growth and collapse of the LSB, which dominates the shock oscillation mechanism, have also been investigated [13]. Furthermore, the shock oscillation mechanism has been investigated experimentally and validated against LES results [14].

6.3.2 *RANS of Lufthansa Technik Cascade*

Since a RANS initial solution is used for LES, and since it was noted that RANS and LES results differ significantly for the passage shock condition, a brief RANS validation is performed. A cascade from Lufthansa Technik which was investigated at the German Aerospace Centre (DLR) is used. The Reynolds number is 1,400,000, and the chord length is 70 mm [7]. An attempt to match the isentropic Mach number distribution for two operating points (Mach 1.05 and 1.12) is made. This is shown in Fig. 6.9. At both operating points, a bow shock is present, and a laminar SBLI takes place. The k-ω SST turbulence model with γ-Re-θ transition model is used. A comparison of experimental and numerical Schlieren for the Mach 1.12 case can be seen in Fig. 6.10. It is concluded that an adequate agreement could be obtained between numerical RANS and experimental results.

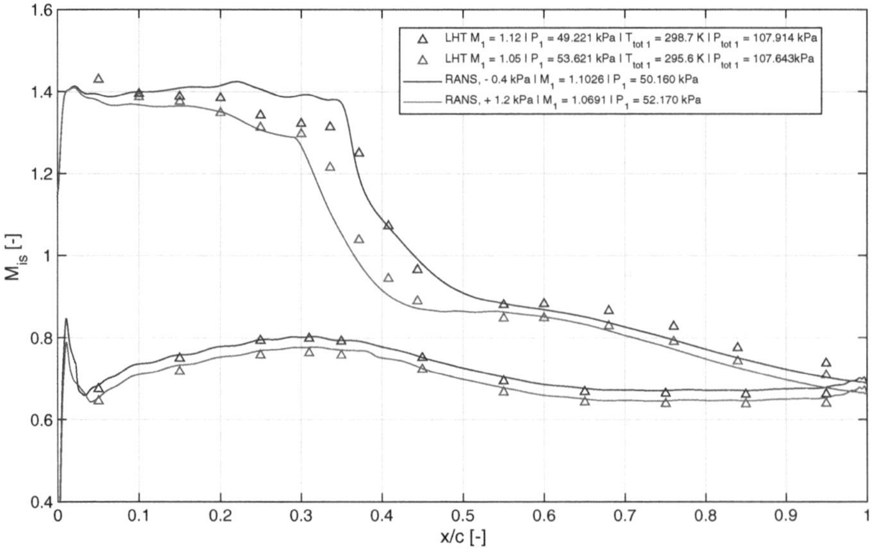

Fig. 6.9 Isentropic Mach distribution for Lufthansa Technik cascade RANS validation at two operating points

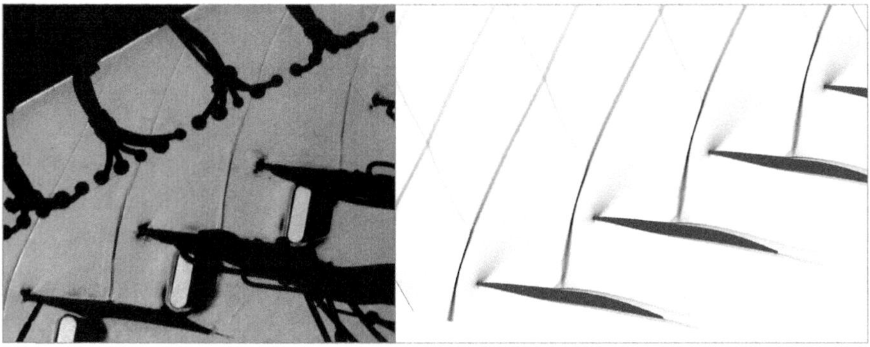

Fig. 6.10 Lufthansa Technik cascade, comparison of experimental Schlieren (left) with RANS contours of $|\nabla \rho|/\rho$ on spanwise plane (right)

6.3.3 Lufthansa Technik Cascade: LES of Upstream Influence

The wall static pressure footprint and shock location with time on the Lufthansa Technik Cascade can be seen in Fig. 6.11. Since predicting the correct extent of the upstream influence is important for the mechanism of interest, a test is perfomed with LES on the LHT cascade: From Fig. 6.12 it can be seen that RANS fails to predict a large upstream influence, whereas LES predicts a significant upstream influence.

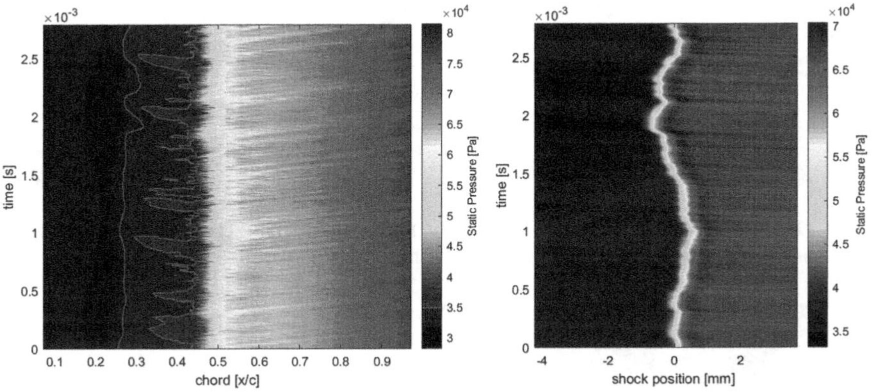

Fig. 6.11 Left: unsteady pressure distribution on suction side of Lufthansa Technik cascade, showing low-frequency unsteadiness in the upstream influence region. Right: shock position with time, half a pitch from blade

The upstream influence can be observed from the Schlieren recording, at passages where the position relative to the knife edge is such that it is visible. A comparison is made in Fig. 6.13. In the figure, note that the experimental upstream influence is visible for the passage at the top, although the shock at the top passage is positioned too far upstream. It is concluded that the LES is able to predict the extent of upstream influence with reasonable accuracy, whereas RANS fails in this regard. Indeed, since the laminar separation is very small in this case due to the low Mach number and relatively small (5% chord) shock oscillation observed in experiments—the origin of which is still unclear, the RANS performs well in predicting the time averaged isentropic Mach distribution regardless. The LES does not show a shock oscillation of any significance in this case, as the movement is only about 1 mm for a chord length of 70 mm (right side of Fig. 6.11). This is consistent with numerous LES studies of transonic compressor flows where a bow shock is present [2, 4, 5, 8, 9, 15, 17]. Note that shock unsteadiness may be seen in the immediate region of interaction with an unsteady tip-clearance vortex [5]. Note that a low frequency unsteadiness of the lambda shock foot (laminar separation shock) can be shown in the current work, as is also observed in Priebe et al. [17]. For the current simulation, this is shown on the left of Fig. 6.11.

Note that in Priebe et al. [17], even though a low-frequency unsteadiness at the lambda shock foot was shown, the shock waves themselves are relatively steady at the operating points which were tested (bow shock conditions). This is analogous to findings in the current work, and is due to the relatively small separation in these cases. It was seen that at lower Reynolds numbers where the separation is more extreme, shock oscillations of larger amplitude occur [13].

Fig. 6.12 Lufthansa Technik cascade: RANS (left) versus LES (right) contours of $|\nabla \rho|/\rho$ at Ma = 1.12 annotated with colour to highlight lambda shock (red) and upstream influence (green)

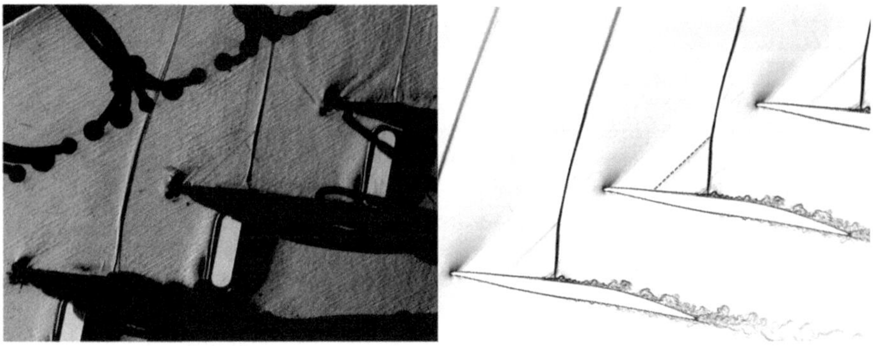

Fig. 6.13 Lufthansa Technik cascade: experiment versus LES, contours of $|\nabla \rho|/\rho$ at Ma = 1.05

6.4 Investigation of Shock Oscillation at Representative Condition in Transonic Compressor

The full span compressor from which the TFAST cascade was derived, was simulated with RANS at the condition at which the strain gauge vibrational measurements showed the largest excitation. The resulting flow structure in RANS shows a passage shock with pre-shock Mach number of 1.6 (Fig. 6.14).

Therefore, LES is performed on a Q2D slice with pre-shock Mach 1.6 and passage shock condition. The domain has inviscid endwalls and a span of 20% of chord. The relatively large span was chosen due to the large expected separation and shock oscillation, in order to prevent constraining turbulent structures. A 30% case is also tested to verify this. The AVDR distribution imposed on the endwalls is based on previous work performed on the same cascade during the TFAST project [6]. The chord length is 100 mm at the original Reynolds number of 1.4 Mio. In order to fulfill the mesh requirements of Priebe et al. [16], the Reynolds number is reduced to 350,000. The current objective is to investigate the oscillation mechanism. For the purpose of orientation and convenience, an overview of such a flow field can be found in Fig. 6.15.

6.4.1 TFAST Cascade at Re = 350,000

The passage shock condition with pre-shock Mach 1.6 is selected as a first attempt, following the conclusion that this is likely the most unstable condition in the real fan. The comparison of shock oscillation in terms of wall static pressure evolution (Fig. 6.16) and formation of instabilities (Fig. 6.17) of a 20% vs. 30% chord in span domain setup, allows us to continue with 20% chord in span for the rest of the simulations due to the similarity between the cases, suggesting an independence of the extent of span on the shock oscillation. The cascade is simulated at Re = 350,000, with explicit LES ($C_S = 0.1$) as well as with with implicit LES. Indeed, an oscillation is present for the $C_S = 0.1$ case (Fig. 6.18). In Fig. 6.18, the time steps are not equi-spaced, rather the phases within the cycle are shown. The cycle starts out with a disturbance-free laminar shear layer upstream of the shock impingement. The upstream influence of the shock

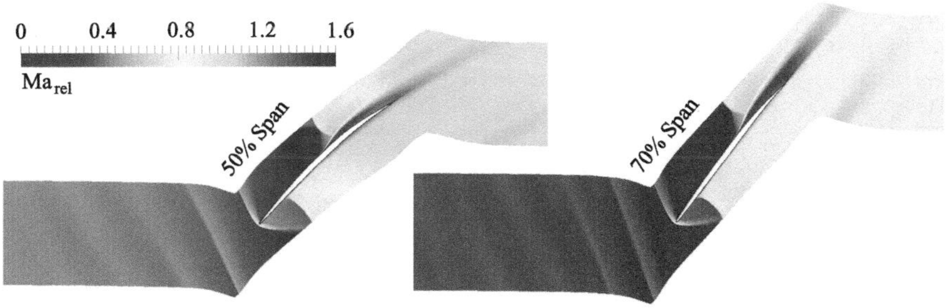

Fig. 6.14 Flow in transonic compressor: full span RANS simulation, cuts at 50% and 70% span showing contours of Mach number

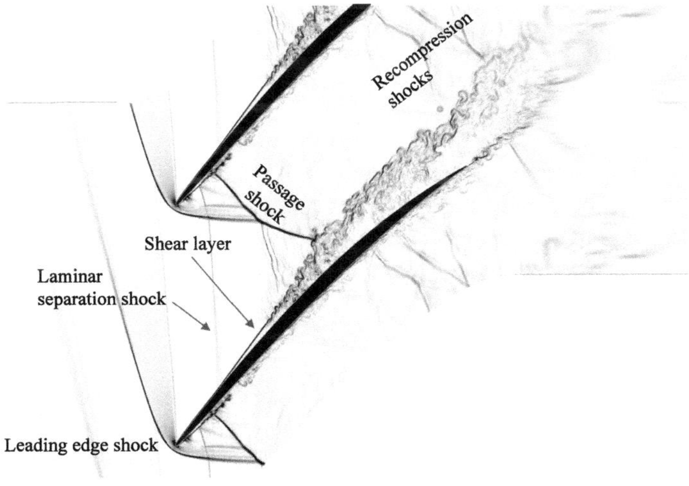

Fig. 6.15 Nomenclature for shock structure in transonic compressor. Re 1.4 Mio. LES simulation

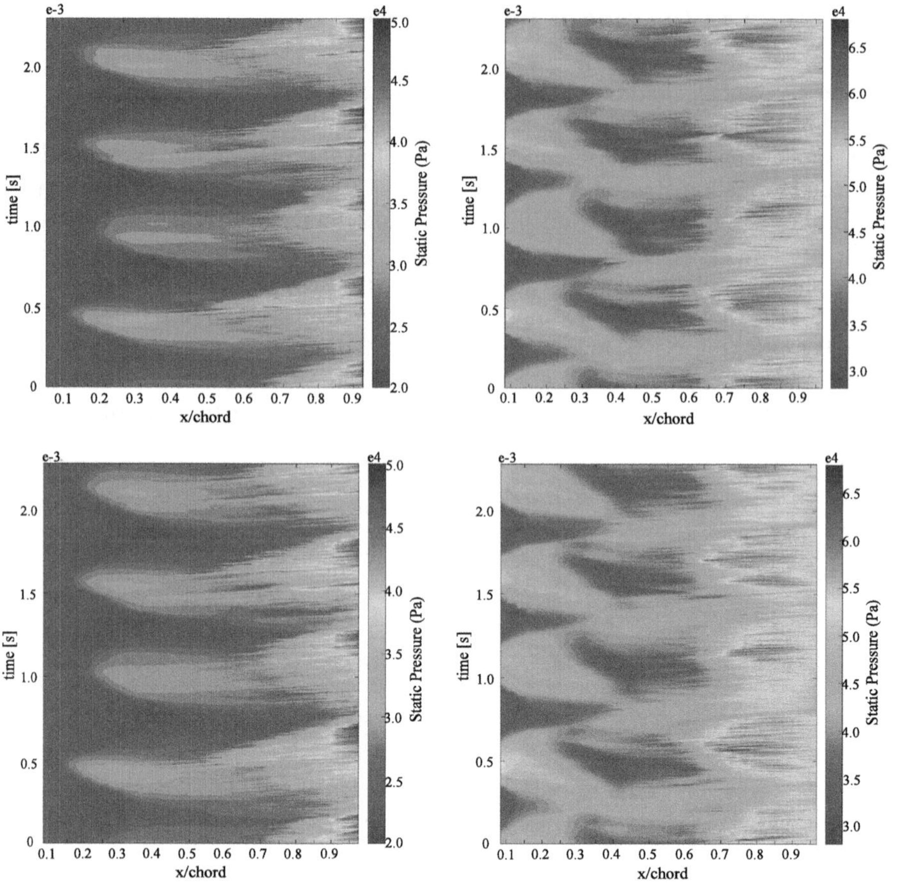

Fig. 6.16 TFAST cascade at Re = 350,000 comparison: time evolution of pressure distribution at passage shock condition (largest shock oscillation) for 20% chord in span (top) and 30% (bottom). Left: suction side. Right: pressure side

grows and eventually, spanwise waves which likely result from the amplification of a KH instability, form and advect downstream, suppressing the separation. As the separation is suppressed, the shock moves downstream again and following this, the LSB starts to recover and hence the cycle can repeat. Before the KH instability forms, there is an increase in the difference of shearing velocities at a certain density of the fluid on the shear layer interface. This can be seen from Fig. 6.18, since the colouring is by velocity magnitude and the iso-surface is of constant density. Hence it is suspected that the formation of the KH instability is dictated by a threshold needing to be satisfied in this regard. In subsequent work, it is seen that the upstream growth velocity of the laminar separation bubble is in the order of the acoustic wave propagation velocity in all investigated cases and at different Reynolds numbers [13].

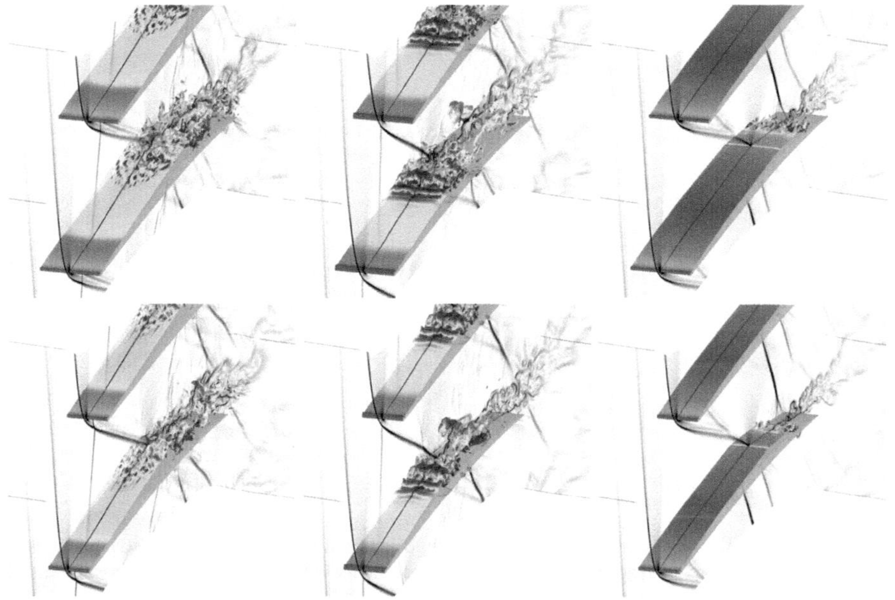

Fig. 6.17 TFAST cascade at Re = 350,000: laminar separation shear layer/LSB growth phase (left). Onset of significant advective instabilities leading to separation suppression (middle). Resultant collapsed shear layer (right). top: 30% chord in span. bottom: 20%

Figure 6.19, provides a side-view of the LSB dynamics. Bubble growth (note the lambda shock foot position difference between t = 0 and t = 1), formation of KH instability, its downstream advection and collapse of the bubble for an emphasised case (high artificial viscosity—($C_S = 0.2$) can be observed. The observed behaviour of the LSB is similar to that observed experimentally by Michelis et al. [11], with regard to the spanwise waves resulting from the KH instability advecting downstream of the LSB and following this, the bubble collapses.

In addition to the 20% chord in span case of Fig. 6.16, the pressure evolution in time can be compared for more cases in Figs. 6.20, 6.21, 6.22 and 6.23. The shock oscillates at a frequency of approximately 1750 Hz, when the oscillation is present at Re = 350,000. This is considerably higher than the expected experimental frequency in the order of 200–500 Hz, however the Reynolds number is four times lower than the experimental case. For comparison, an initial test at Re 1.4 Mio. shows a shock oscillation at 470 Hz implicit LES (Fig. 6.24), has been carried out, although the resolution may be marginal. More validation work is needed and an experimental investigation is planned.

When the sub-grid model is turned off (Fig. 6.21), the shock stops oscillating, and no KH instability arises. This is possibly due to the change in blockage, which changes the operating point. The Re = 350,000 case is particularly sensitive to this, due to the size of the separation and laminar shear layer relative to the geometry, and turning off the SGS model effectively reduces the viscosity. The back pressure needed

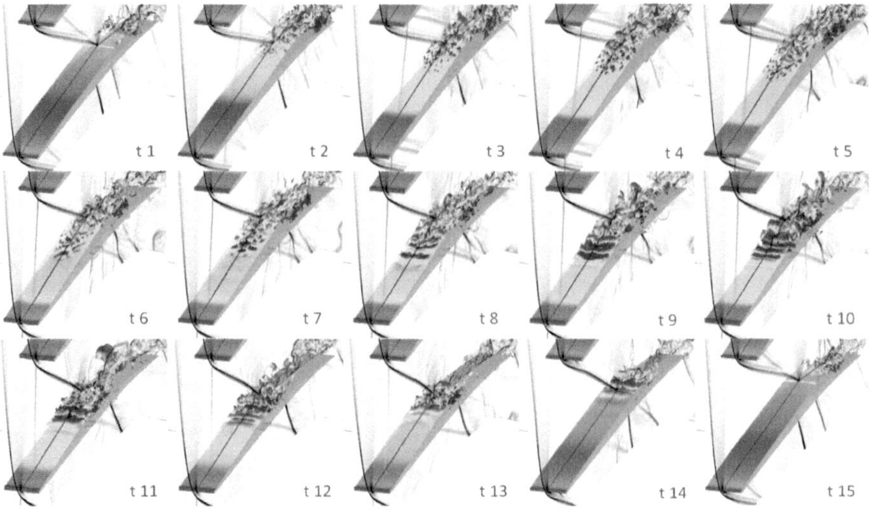

Fig. 6.18 TFAST cascade at Re = 350,000: image grid showing oscillation mechanism using numerical shadowgraph and an ISO surface of constant density with colouring by velocity magnitude

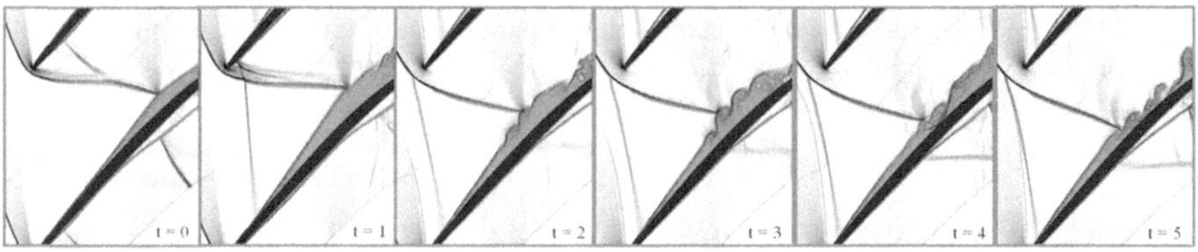

Fig. 6.19 TFAST cascade at Re = 350,000: laminar separation bubble dynamics for (C_S = 0.2) (emphasised case)

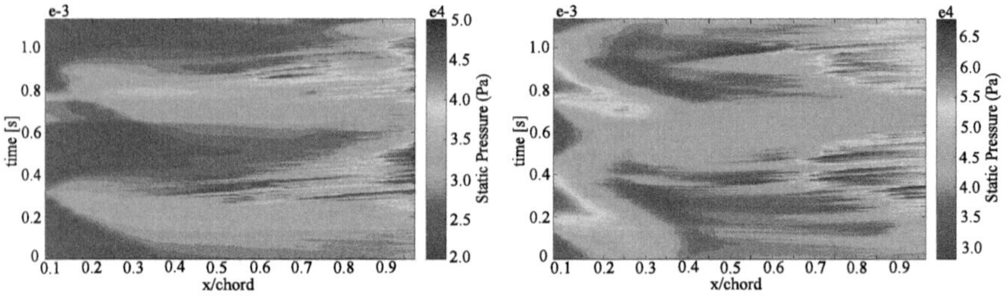

Fig. 6.20 TFAST cascade at Re = 350,000: unsteady pressure distribution for (C_S = 0.2) at P_b = 53 kPa

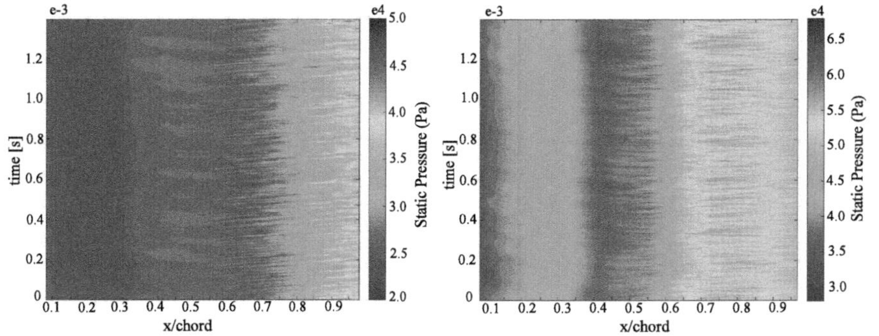

Fig. 6.21 TFAST cascade at Re = 350,000: unsteady pressure distribution for (C_S = 0) at P_b = 53 kPa

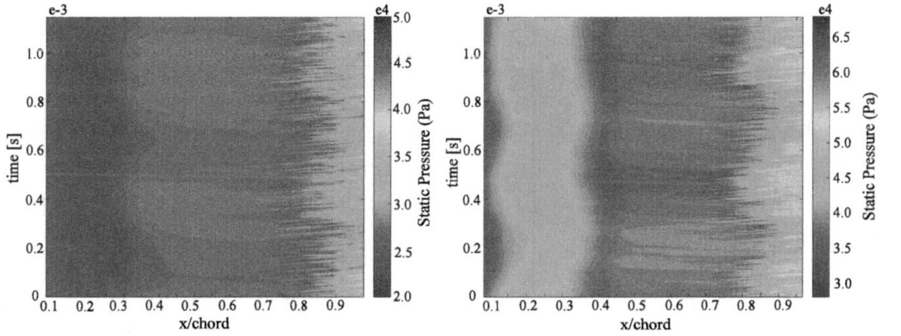

Fig. 6.22 TFAST cascade at Re = 350,000: unsteady pressure distribution for (C_S = 0.1) at P_b = 50 kPa

to start the oscillation is affected and it can be shown that a significant oscillation is not present in both cases (C_S = 0 & 0.1) due to a lack of KH instability for a swallowed shock (Figs. 6.21 and 6.22). The results suggest the reason for the KH instability not arising is possibly that the LSB is not allowed to grow far enough upstream to meet the conditions for the KH instability to be satisfied. This needs to be investigated further.

Indeed it can be shown that when the back pressure is increased from 53 to 55 kPa for the case when C_S = 0, a KH instability returns to the shear layer and the shock starts to oscillate at the familiar frequency of 1750 Hz (Fig. 6.23), but with a different amplitude than the C_S = 0.1 case. Similarly, the C_S = 0.1 case produces KH instabilities at P_b = 53 kPa, but not at P_b = 50 kPa. The C_S = 0.2 case (Figs. 6.19 and 6.20) also shows a shock oscillation at 1750 Hz. It can therefore be concluded that regardless of the Smagorinsky constant of the sub-grid modelling, the frequency of oscillation is constant, although the amplitude is not.

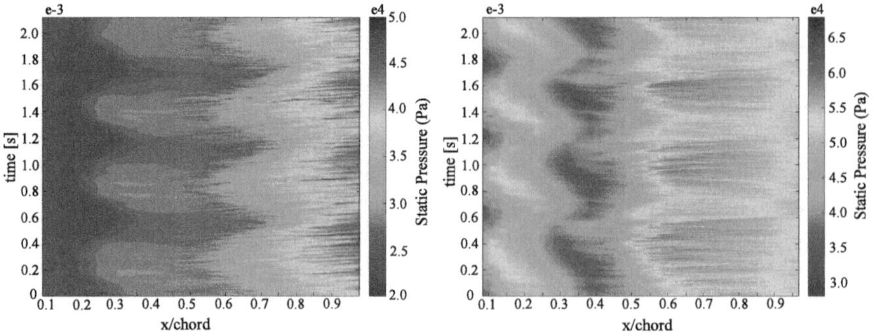

Fig. 6.23 TFAST cascade at Re = 350,000: unsteady pressure distribution for (C_S = 0) at P_b = 55 kPa

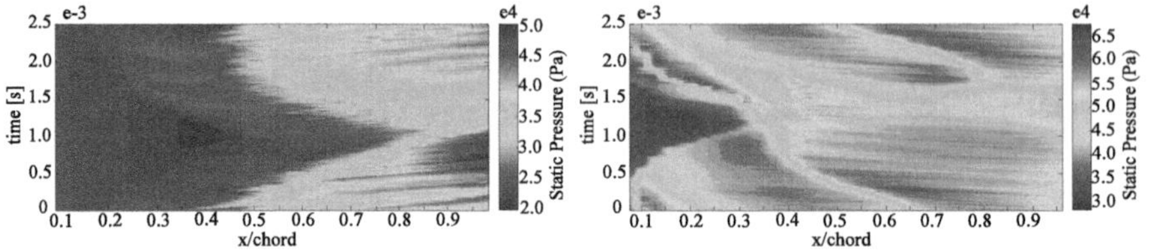

Fig. 6.24 TFAST cascade at Re = 1,400,000: unsteady pressure distribution for (C_S = 0) at P_b = 55.6 kPa

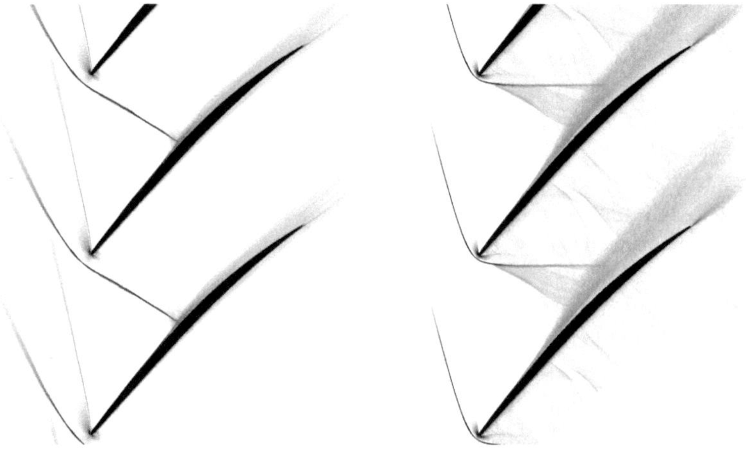

Fig. 6.25 TFAST cascade at Re = 1,400,000: time averaged contours of $|\nabla\rho|/\rho$ at pre-shock Ma = 1.3 (bow shock) (left) and pre-shock Ma = 1.6 (passage shock) (right)

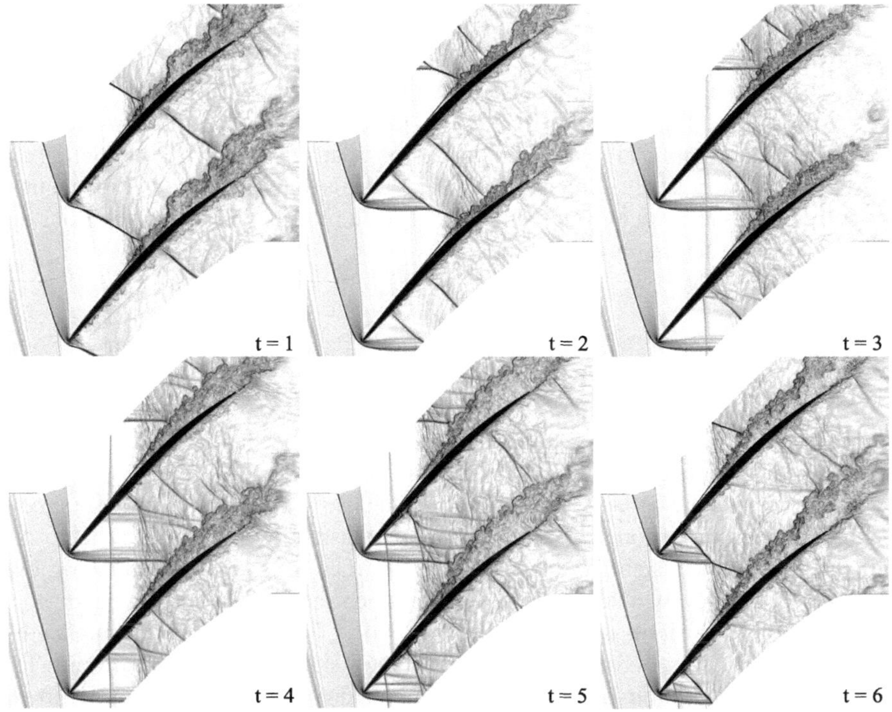

Fig. 6.26 TFAST cascade at Re = 1,400,000: instantaneous contours of $|\nabla \rho|/\rho$ at pre-shock Ma = 1.6 showing shock oscillation

6.4.2 TFAST Cascade at Re 1,400,000

As described, it is the difference in the separation bubble size from collapsing to recovering, as well as the sensitivity of the passage shock to blockage, which is causing the shock to oscillate, given that the instabilities form upstream of the shock. In the bow shock case, the separation bubble is significantly smaller than the passage shock case due to the reduced Mach number/shock strength, and no passage shock is present. From the time averaged shadowgraph comparison in Fig. 6.25, it can be seen that the leading-edge shock wave is stable in both bow and passage shock cases.

It was mentioned before that the Re 1,400,000 passage shock case shows a shock oscillation at 470 Hz (Fig. 6.24). A numerical shadowgraph sequence of this shock oscillation can be seen in Fig. 6.26. Note the significant difference in the separation size between the snapshots at t = 2 and t = 5 during the course of the oscillation.

Indeed, the leading-edge shock wave itself does not oscillate in the LES, neither for the Lufthansa Technik cascade, nor for the TFAST cascade (although there is unsteadiness of the upstream shock foot) as in accordance with literature mentioned in Sect. 6.3.3.

Even though this simulation is run without sub-grid model with a coarser relative grid than at Re = 350,000, the shear layer manages to collapse, which gives rise to the oscillation. As per the logic of Sect. 6.3.1, the sub-grid model should not significantly affect the oscillation frequency. During subsequent work, was found that the Smagorinsky sub-grid scale model performs relatively poorly for predicting the correct extent of upstream influence [12]. The reader is referred to another subsequent work by the author, where different sub-grid scale models and grid resolutions are compared to a DNS resolution solution [13].

6.5 Conclusion

The current work sheds light on the possible oscillation mechanics which are present in the real fan at altitude when a laminar SBLI is present.

It is found that the collapse and recovery of the laminar separation bubble (shock-induced separation) through a Kelvin-Helmholtz instability is the root of the oscillation, and that this instability most likely forms when the shock induced separation is sufficiently large, which happens to occur when the suction side passage shock impingement is positioned more to the fore side of the blade. Furthermore, this oscillation is not present for the swallowed and bow shock operating conditions. The passage shock condition is very sensitive to blockage changes with regards to the shock position. The Kelvin-Helmholtz instability forming on the LSB shear layer causes a suppression of the separation bubble, as soon as it had grown far enough upstream, which affects the blockage and causes significant shock movement. The higher the sub-grid modelling, the longer the shear layer and hence the more shock movement is caused when the larger LSB collapses and grows. The level of sub-grid modelling does not affect the oscillation frequency, but affects the amplitude.

Acknowledgements The author is participating in the TEAMAero project, which has received funding from the European Union's Horizon 2020 research and innovation programme under grant agreement No 860909. The Lufthansa Technik cascade tests were performed under the German Aviation Research Program (LuFo V-3, grant number 20X1706) and is funded by the Federal Ministry for Economic Affairs and Climate Action of Germany. The experiments were carried out by Dr. Alexander Hergt and his associates.

References

1. Diwan S, Ramesh O (2009) On the origin of the inflectional instability of a laminar separation bubble. J Fluid Mech 629:263–298. https://doi.org/10.1017/S002211200900634X
2. Gomar A, Gourdain N, Dufour G (2011) High-fidelity simulation of the turbulent flowin: a transonic axial compressor. In: 9th European conference on turbomachinery: fluid dynamics and thermodynamics, ETC 2011—conference proceedings of vol 1

3. Gropengiesser H (1969) Beitrag zur Stabilität freier Grenzschichten in kompressiblen Medien. Deutsche Luftund Raumfahrt, Forsch. Ber :69–25
4. Hah C (2009) Large eddy simulation of transonic flow field in NASA rotor 37. In: 47th AIAA aerospace sciences meeting including the new horizons forum and aerospace exposition. https://doi.org/10.2514/6.2009-1061
5. Hah C, Bergner J, Schiffer H-P (2008) Tip clearance vortex oscillation, vortex shedding and rotating instabilities in an axial transonic compressor rotor. https://doi.org/10.1115/GT2008-50105
6. Hergt A, Grothe P, Flaszynski P, Szwaba R, Piotrowicz M, Kaczynski P, Tartinville B, Hirsch C (2020) Transition location effect on shock wave boundary layer interaction, WP3 internal flow compressors 144
7. Hergt A, Klinner J, Willert C, Grund S, Steinert W (2022) Insights into the unsteady shock boundary layer interaction. In: Proceedings of the ASME turbo expo 2022: turbomachinery technical conference and exposition, vol 10A: turbomachinery—axial flow fan and compressor aerodynamics. Rotterdam, Netherlands, 13–17 June 2022 (V10AT29A033, ASME). https://doi.org/10.1115/GT2022-82720
8. Joo J, Medic G, Philips DA, Bose S (2014) Large-eddy simulation of a compressor rotor
9. Li Z, Ju Y, Zhang C (2022) Quasi-wall-resolved large eddy simulation of transitional flow in a transonic compressor rotor. Aerosp Sci Technol 126:107620. ISSN 1270-9638, https://doi.org/10.1016/j.ast.2022.107620
10. Maslowe SA, Kelly RE (1971) Invisoid instability of an unbounded heterogeneous shear layer. J Fluid Mech 48:405
11. Michelis T, Yarusevych S, Kotsonis M (2017) Response of a laminar separation bubble to impulsive forcing. J Fluid Mech 820:633–666. https://doi.org/10.1017/jfm.2017.217
12. Nel PL, Janke C, Vasilopoulos I, Swoboda M, Schreyer A-M, Hady A, Flaszyński P (2024) Effect of transition on self-sustained shock oscillations in highly loaded transonic rotors. AIAA J 62(6):2063–2075. https://doi.org/10.2514/1.J063378
13. Nel PL, Schreyer A-M, Schrijer FFJ, van Oudheusden BW, Janke C, Swoboda M. Research configuration to study shock oscillation mechanism in highly loaded transonic fans. AIAA J, manuscript submitted for publication
14. Nel PL, Schreyer A-M, Schrijer FFJ, van Oudheusden BW, Janke C, Vasilopoulos I, Swoboda M. Shock oscillation mechanism of highly separated transitional shock-wave/boundary-layer interactions. AIAA J, manuscript submitted for publication
15. Papadogiannis D, Garnaud X (2017) Unstructured large eddy simulations of the transonic compressor rotor 37. https://doi.org/10.2514/6.2017-3612
16. Priebe, S, Wilkin, D, Breeze-Stringfellow A, Jothiprasad G, Cheung LC (2020) Large eddy simulation of laminar and turbulent shock/boundary layer interactions in a transonic passage. In: Proceedings of the ASME turbo expo 2020: turbomachinery technical conference and exposition, vol 2A: turbomachinery. Virtual, Online, 21–25 Sept 2020 (V02AT32A005, ASME). https://doi.org/10.1115/GT2020-14244
17. Priebe S, Wilkin D, Breeze-Stringfellow A, Mousavi A, Bhaskaran R, d'Aquila L (2022)Large eddy simulations of a transonic airfoil cascade. In: Proceedings of the ASME turbo expo 2022: turbomachinery technical conference and exposition, vol 10A: turbomachinery—axial flow fan and compressor aerodynamics. Rotterdam, Netherlands, 13–17 June 2022 (V10AT29A010. ASME). https://doi.org/10.1115/GT2022-80683

Chapter 7
Reduced Order Computational Methods for the Development of Propulsive Technologies for Supersonic Aviation to Achieve Climate Neutrality

Karel Van den Borre, Ali C. Ispir, Bora O. Cakir, and Bayindir H. Saracoglu

Abstract In recent years, supersonic aviation has regained attention by the research community and industrial actors alike. Certainly, the environmental constraints and hypersonic cruise opportunities add new underexplored dimensions. The MORE&LESS consortium builds upon previous European efforts to develop high-speed civil air transportation by assessing a broad spectrum of high-speed flight, ranging from Mach 2 up to Mach 8. There exists no single propulsion concept that is able to power an aircraft over this wide range. This chapter investigates propulsion concepts at both ends of this velocity spectrum. At the low end, this consists of a turbojet with afterburning which powers a Concorde-like airframe. On the high end, a combined cycle engine merges six air-turbo rockets and a dual-mode ramjet/scramjet (DMR) engine to power a waverider concept. The tools used to characterise the engine performance require a high computational efficiency in order to allow for quick design iterations and automated optimisation procedures. Furthermore, to ensure high simulation accuracy without jeopardising computational efficiency, a reduced-order modelling framework built on high-fidelity simulations is used.

Keywords Reduced order methods · Supersonic aviation · Propulsion

K. Van den Borre · A. C. Ispir · B. O. Cakir · B. H. Saracoglu (✉)
von Karman Institute for Fluid Dynamics, Waterloosesteenweg 72, Sint-Genesius-Rhode, Belgium
e-mail: bayindir.saracoglu@vki.ac.be

K. Van den Borre
e-mail: karel.van.den.borre@vki.ac.be

A. C. Ispir
e-mail: ali.can.ispir@vki.ac.be

B. O. Cakir
e-mail: bora.orcun.cakir@vki.ac.be

T. Tuovinen et al. (eds.), *Advanced Computational Methods and Design for Greener Aviation*, Computational Methods in Applied Sciences 59,
https://doi.org/10.1007/978-3-031-61109-4_7

7.1 Introduction

In 2005, the European Commission launched the *Long-Term Advanced Propulsion Concepts and Technologies* (LAPCAT) project to explore advanced propulsion technologies and innovative concepts for sustained hypersonic flight in civil applications [1]. LAPCAT-II, which began in 2008, focused on further developing the most promising concepts from LAPCAT, specifically a Mach 5 and a Mach 8 aircraft design [2]. Starting from 2018, the STRATOFLY consortium assessed the feasibility of high-altitude stratospheric flight in terms of economic viability, environmental sustainability, and social acceptance. This research involved the integration of propulsion, aerodynamics, airframe, and systems into a detailed Mach 8 aircraft configuration, as well as mapping the stratospheric climate impact [3]. The *MDO and Regulations for Low-boom and Environmentally Sustainable Supersonic aviation* (MORE&LESS) project, initiated in 2021, continues to review the environmental impact of supersonic and hypersonic aviation, applying a multidisciplinary framework to verify compliance with environmental requirements. This effort contributes to the EU goal of reaching climate neutrality by 2050 through the introduction of renewable fuels such hydrogen, biofuels and e-fuels. The project covers a broad range within the supersonic regime with multiple promising aircraft configurations, including those studied in the previously mentioned projects [4]. In this chapter, two particularly promising designs of the MORE&LESS project will be explored. The first one is a Concorde-like concept, able to cruise at Mach 2. The second is a waverider concept capable of reaching Mach 5 and eventually Mach 8.

7.2 Turbo-Based Propulsion Concept for Mach 2 Cruise

For the Mach 2 concept, inspiration is drawn from the only two historical supersonic passenger aircraft to ever operate commercially. The first is the TU-144, which saw limited operation from 1975 to 1978 [5]. The second one is the well-known Concorde, operating from 1976 to 2003 [6]. Both concepts were powered by a military-derived turbojet engine which incorporated reheat for take-off, climb and acceleration. Firstly, to design and size the propulsion plant for the new Mach 2 concept, the Olympus593 engine is recreated as a validation case [7]. Here the goal is to match its reported performance as closely as possible. This case can be adapted in a later stage to match the new airframe. This process starts with reproducing the Olympus 593 performance in steady on-design conditions using the Gasturb software [8]. The steady on-design model fixes the engine design parameters, which are moved to the transient EcosimPro tool to simulate an entire flight to assess the full engine performance, efficiency, operability and controllability.

Table 7.1 Summary of the Olympus 593 engine characteristics [6, 9–14]

	Units	Takeoff	Cruise	Peak performance
Mach number	[–]	0.302	2.0	1.2
Altitude	[m]	0	16000	12000
Wet thrust	[kN]	179.46	–	78.70
Dry thrust	[kN]	149.55	44.62	60.54
TSFC	[g/kN/s]	39.37	33.71	39.94
Max TIT (cont.)	[K]	1350	1350	1350
Max TIT (peak)	[K]	1450	1450	1450
Max pressure ratio	[–]	15.5	15.5	15.5

7.2.1 Steady Design Based on the Olympus 593 Heritage

The steady on-design model of the Olympus 593 engine is created using the Gasturb software. This software calculates cycle performance in an analytical way, using a 0D thermodynamic description of the components, contributing to its excellent computational efficiency. More information regarding the 0D modelling approach in Gasturb can be found in the work of Van den Borre, Ispir and Saracoglu [7]. Table 7.1 summarises key performance characteristics of the engine as found in the open literature, such as Turbine Inlet Temperature (TIT), uninstalled thrust (F_u) and Thrust Specific Fuel Consumption (TSFC). This includes the conditions at the end of the runway, at cruise conditions and at peak performance. The cruise point is the reference that will be matched as closely as possible. In this first step, engine bleeds, power off-take and cooling flows are neglected for simplicity.

The steady design created in Gasturb delivers 44.58 kN of thrust, which is within 0.1 % margin of the literature value reported in Table 7.1 [13]. It has a TSFC of 32.4 g/kN/s, which is 4 % lower than the literature value [13]. However, as we neglected the losses generated by the cooling and bypass flows, this is within an acceptable range.

7.2.2 Transient Engine Simulation

After establishing a suitable engine design in Gasturb, a corresponding transient model is created in EcosimPro. This is an object-oriented framework for the simulation of continuous systems composed of individual components. Each of these components contains a mathematical formulation to describe their physical behaviour in the form of a set of analytical/differential equations. After closing the system by providing boundary conditions, the software assembles the differential equations describing the components into a single large system, which is solved using

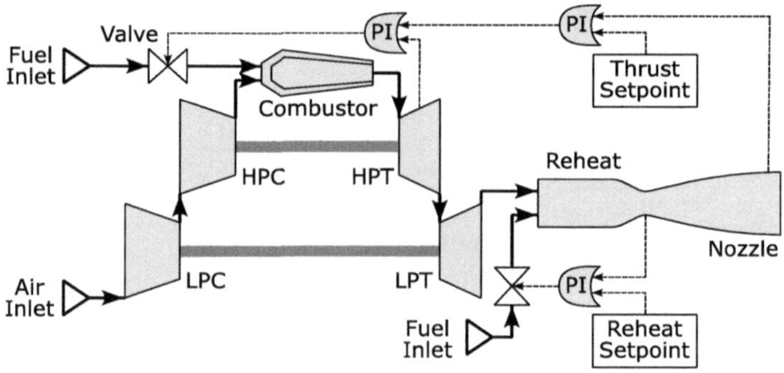

Fig. 7.1 Transient Olympus593 engine schematic showing the physical engine model, the nested thrust control loop and the reheat control loop

DASSL (Differential/Algebraic System Solver) [15]. The solver approximates the time derivatives using a backwards differentiation of order k. The resulting algebraic system is solved at each time step with an implicit Newton-Rhapson method. The order k and the time-step are automatically adjusted during runtime as a function of the evolution of the solution. This allows to efficiently solve a system containing both fast and slow transients. The European Space Propulsion System Simulation (ESPSS) library is used to model the behaviour of the fluid components (combustion chamber, afterburner, nozzle, valves, etc.) and adds tabulated data for the fluid properties, derived from the NIST REFPROP database [16].

The model consists of two parts, the physical engine model and the control scheme. The schematic of the engine can be seen in Fig. 7.1. For the physical model, all design variables were derived with the Gasturb reference. The thrust controller consists of two nested PI controllers. The inner loop controls the TIT by varying the mass flow rate of fuel. The TIT set-point of this inner loop is determined by the outer loop, which controls the thrust by varying the requested TIT. The afterburner is controlled using a separate loop, controlling the exhaust temperature by varying the fuel flow to the afterburner. This rudimentary control strategy would not be suited for implementation in real aircraft, but suffices for this transient model to track a pre-defined thrust profile.

The resulting transient model is used to simulate the engine performance along an entire reference flight, which is described in terms of Mach number, altitude and thrust requirement (Fig. 7.2). This figure also shows a test without afterburner where a thrust deficit during the climb phase can be seen. This means the aircraft can not complete the prescribed mission. By switching on the afterburner during climb, the thrust requirements could be met for the entire mission.

The transient simulation allows to examine the operation of the turbomachinery. Figure 7.3 depicts the turbomachinery maps for the compressors and turbines, together with the locus of their operating point. This shows that during the climb phase all components except the HPT are pushed beyond their capabilities, with the operating point moving off the map. To solve this issue, the turbomachinery maps

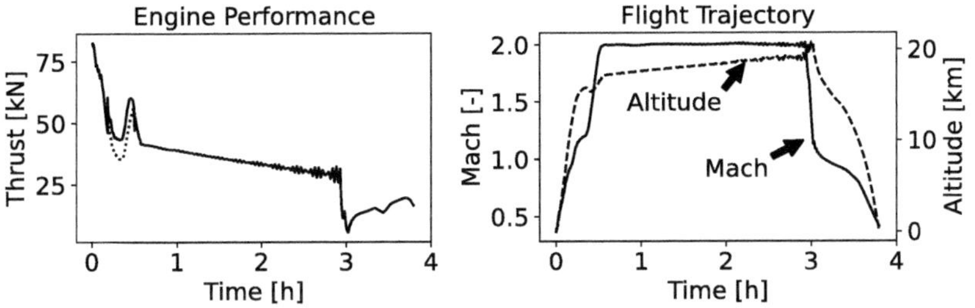

Fig. 7.2 The right figure shows the reference Mach 2 trajectory. The left figure shows the thrust requirement (——), thrust output with reheat (- - -) and the thrust output without reheat (······). The trust requirement and thrust output with reheat overlap, adapted from prior studies [7]

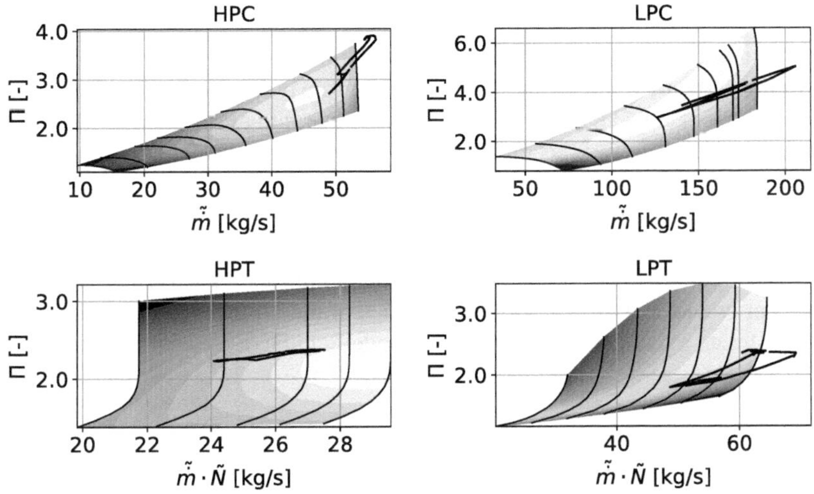

Fig. 7.3 Locus of the operation point for all turbomachinery components for the full flight trajectory, adapted from prior studies [7]

could be rescaled to better fit the operational domain. More practically, this would entail physically enlarging the cross-sectional area of the turbomachinery, which enables them to process a larger mass flow rate. This results in a horizontal stretch of the map. More information regarding the turbomachinery map scaling can be found in the Gasturb manual [8].

7.3 Mach 5 Propulsion Concept

Increasing the flight Mach number (M_∞) to high-supersonic levels results in excessively high static temperatures of the incoming flow. The flow temperature can easily reach the melting temperatures of the advanced materials used in the manufacturing

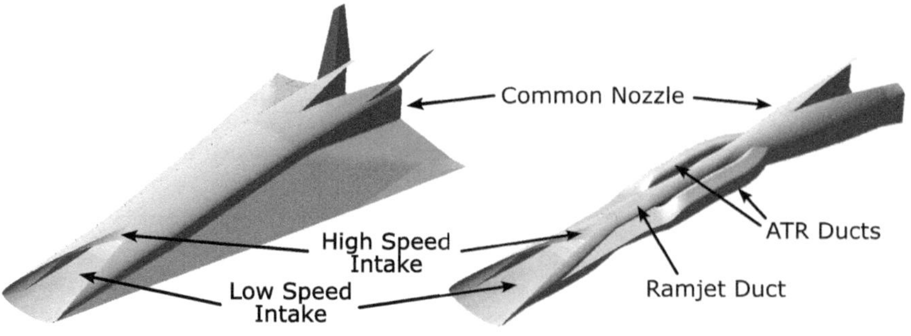

Fig. 7.4 External layout and internal flow path of the MR5 waverider concept

of modern turbomachinery components. The MORE&LESS Mach 5 concept, MR5, solves this issue by implementing a combined cycle propulsion plant consisting of 6 Air-Turbo-Rocket (ATR) engines and a dual-mode ramjet/scramjet (DMR) engine. During take-off, climb and acceleration the aircraft uses the ATR engines. After Mach 4.5 the ramjet starts to operate as well. During Mach 5 cruise the ATR engines are shut down. A visualisation of the integration of the propulsion plant in the airframe is depicted in Fig. 7.4. In the following sections, more information about the ATR engines and how their operating parameters are optimised for a given flight trajectory will be provided first. The working principles of the DMR engine and the required steps to create a reduced-order surrogate model will be described afterwards.

7.3.1 Off-Design Optimisation of ATR Engines

An ATR engine is a turbo-based engine that helps to increase the maximum flight Mach number by removing the thermal limit of the turbine in conventional turbojet cycles. This is achieved by turning the turbine with a high-enthalpy fuel flow instead of hot combustion gasses. A schematic of the ATR engine is depicted in Fig. 7.5. Air enters the ATR engine at the low-speed intake. The airflow is compressed by two counter-rotating fans, after which it moves directly to the combustion chamber. The liquid hydrogen fuel is stored in the tanks at 21 K and 3 bar [17]. It is pumped into the system by the fuel pumps and heated up when it passes through the regenerative units at the nozzle and combustion chambers. The high-enthalpy hydrogen gases expand and drive the turbine where the released mechanical power is used to run the counter-rotating fans and fuel pump. The expanded hydrogen then enters the ATR combustion chamber where it is mixed with the compressed air and burned. The hot combustion gasses enter the nozzle, where they expand and accelerate in order to provide thrust. A steady numerical model of this case was created in EcosimPro, and was investigated in detail on prior studies [18].

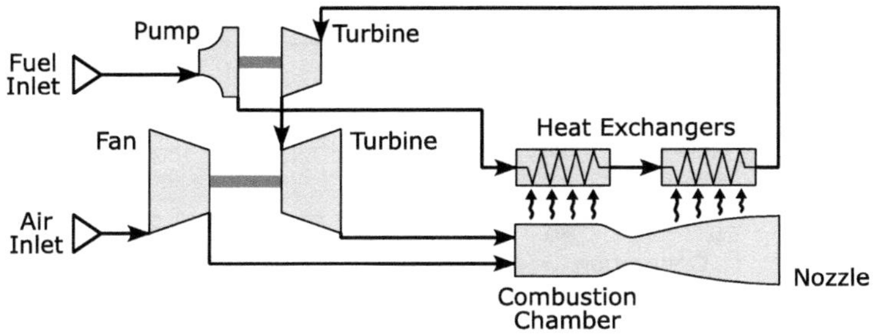

Fig. 7.5 Schematic layout of the hydrogen-powered ATR engines

The engine throttling is characterised by three variables. The first one is the combustion chamber pressure which is determined by the compression ratio of the fans. The second variable is the nozzle throat area. The last variable, the mixture ratio, is defined as the air mass flow rate divided by the fuel mass flow rate. In order to determine the optimal settings of these three throttle variables for the operating conditions along the entire flight trajectory, a single objective optimisation is performed which aims at achieving the lowest fuel consumption while meeting the minimum thrust requirement. This optimisation is performed using the in-house developed Computer-Aided Design and Optimisation tool (CADO) [19]. This non-gradient-based optimisation implements a differential evolution algorithm where a population of individuals adapts to better fit the environment, in analogy to Darwinian natural selection. These algorithms are based on a numerical implementation of the natural processes such as mutation, cross-over and selection.

To illustrate the working principle, let's take a look at a single operating condition, more specifically, flying at Mach 0.5 and 4000 m of altitude. At this point of the mission, a thrust level of 2300 kN is required to accelerate the aircraft. After running the optimisation, a clear convergence can be observed when looking at Fig. 7.6, which depicts the populations in terms of the fuel consumption ($\dot{m}_{fuel}$) and thrust output. The optimisation tends towards designs which are close to the minimum thrust constraint, which is to be expected as producing excess amounts of thrust will incur a fuel consumption penalty. Initial generations are not able to provide enough thrust, consequently, the best fuel consumption per generation is increasing at the start of the process. When performing this optimisation at multiple points, an engine throttling schedule is created, which determines how the three throttle parameters need to be set along the entire flight trajectory. This throttle schedule can be seen in Fig. 7.7, which also depicts the actually delivered thrust along the trajectory. A more in-depth analysis of this work is presented in the prior studies of Goncalves et al. [20].

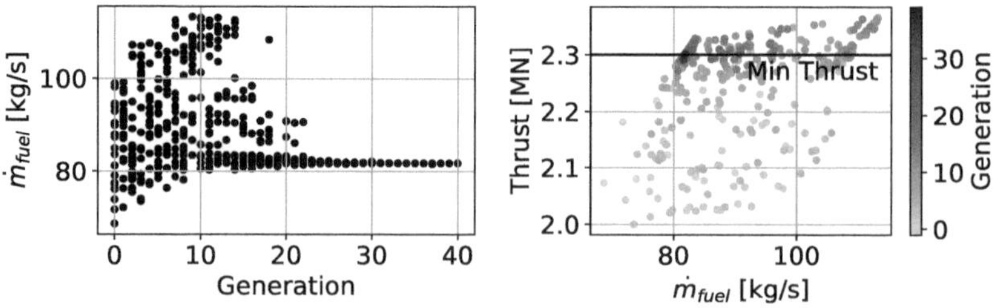

Fig. 7.6 Convergence monitors of the throttle schedule optimisation for a single operating point. The right figure shows the thrust output and fuel consumption ($\dot{m}_{fuel}$) for all individuals. The left figure shows how the individuals converge as the generation progress

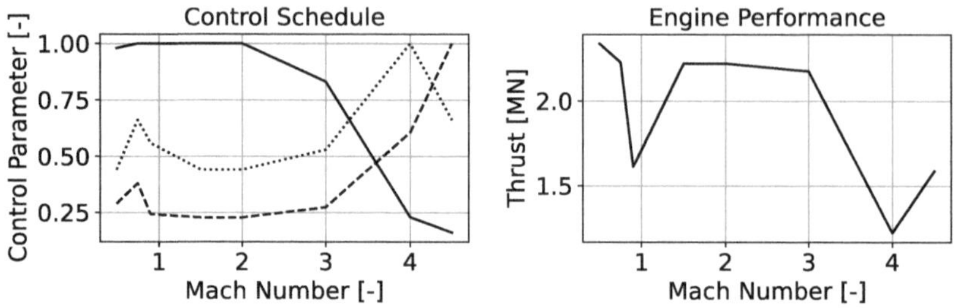

Fig. 7.7 Optimised control schedule and performance of the ATR engine. The control parameters, throat area (——), chamber pressure (- - -) and mixture ratio (······), are normalised with respect to their maximum value

7.3.2 Neural-Network-Based Scramjet Surrogate Model

Ramjet and scramjet propulsion are the preferred way to power supersonic and hypersonic vehicles cruising faster than Mach 3. This elegant solution involves a lean architecture without any rotating parts. The absence of rotating turbomachinery increases the maximum flight Mach number by removing the thermal limit these fragile components impose. In ramjet engines, the air is slowed down to subsonic velocities before it enters the combustion chamber. In scramjet engines, the airflow remains supersonic throughout the whole engine, including the combustion process. Although the geometry of ramjet engines is simple compared to turbo-based engines, the flow physics through the engine duct are quite complex, with the flow modulating between the supersonic and subsonic regimes. The design and performance analysis of such engine configurations is vital to ensure the propulsive system is capable of meeting the mission requirements in terms of thrust and fuel consumption. In the following sections, a reduced order analysis method is introduced to investigate the propulsive performance characterisation of different design choices on intake configurations, starting from complete propulsive flow path simulations of a ramjet. The 1D model consists of 4 modules, the inlet, the isolator, the combustor and the exhaust.

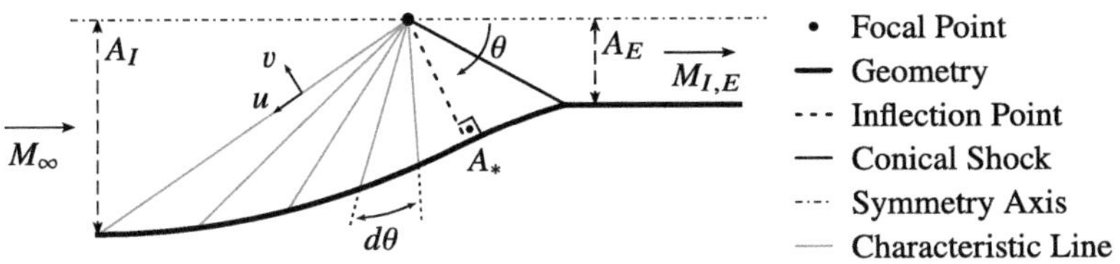

Fig. 7.8 Schematic of the Busemann intake [21]

The intake module is based on axisymmetric flow contours provided by internal compression flows of Busemann diffusers. The structure of this flow consists of an isentropic compression terminated by a conical shock, which is depicted in Fig. 7.8. For brevity, the details of the design of these inlets are omitted, but more information can be found in the publications of Cakir et al. [21, 22]. The modelling of the isolator depends on the outflow conditions of the intake. If the flow is subsonic after the terminating conical shock, the flow conditions at the exit of the inlet are imposed directly to the subsonic combustion chamber. If the flow is still supersonic at the exit of the intake, the isolator is modelled with a single normal shock [20]. For the combustion chamber and nozzle, the 1D steady inviscid flow equations, coupled with a finite-rate detailed chemistry approach and JANAF tables, are solved along the entire constant area combustor duct and the diverging nozzle [23]. To model the detailed chemistry of the hydrogen-air combustion, the Jachimowski mechanism, incorporating 33 reactions among 13 species [24], is used. The incoming air and fuel flows are assumed to be perfectly mixed when entering the combustor. The flow is perfectly expanded to the ambient conditions through the contoured nozzle, accounting for the Summerfield criteria that states that the minimum nozzle exit pressure needs to be higher than 30% of the ambient pressure [25].

The resulting numerical model is used to examine the multidimensional design space of a DMR engine. The design space consists of 5 design variables which are flight Mach number, flight altitude (h), intake wedge angle, intake exit Mach number ($M_{C,I}$) and fuel-air equivalence ratio (ER). The output performance parameters are computed and used to train an Artificial Neural Network (ANN) regression allowing for a comprehensive exploration of the design space. To gain more information regarding the followed methodology and a more in-depth analysis of the results, the reader is referred to the work of Ispir et al. [26, 27].

The design space of the scramjet engine under investigation is summarised in Table 7.2 The output parameters consist of total pressure recovery in the intake (CPR), captured air mass flow rate (MCR), startability index, uninstalled thrust (F_u), specific impulse (I_{sp}) and unburned hydrogen mass fraction at the outlet (Y_{H_2}). In total, 3350 observations are gathered to train the ANN. For each objective function, a separate ANN was trained. A schematic representation of such an ANN can be seen in Fig. 7.9.

The surrogate ANN network is used to compute the DMR performance maps, which can be used to draw some conclusions related to the the engine characteristics. Figure 7.10 depicts that higher flight Mach numbers lead to a higher uninstalled thrust output of the engine. One of the main reasons is that at higher flight velocities, a higher

Table 7.2 The design variable ranges of the scramjet engine under investigation

Variables	Range
Cruise mach number (ϕ_1)	[6–8]
Cruise altitude [km] (ϕ_2)	[25–35]
Equivalence ratio (ϕ_3)	[0.14–0.5]
Wedge angle (ϕ_4)	[0–8]
Pre-combustion mach number (ϕ_5)	[1.3–4.5]

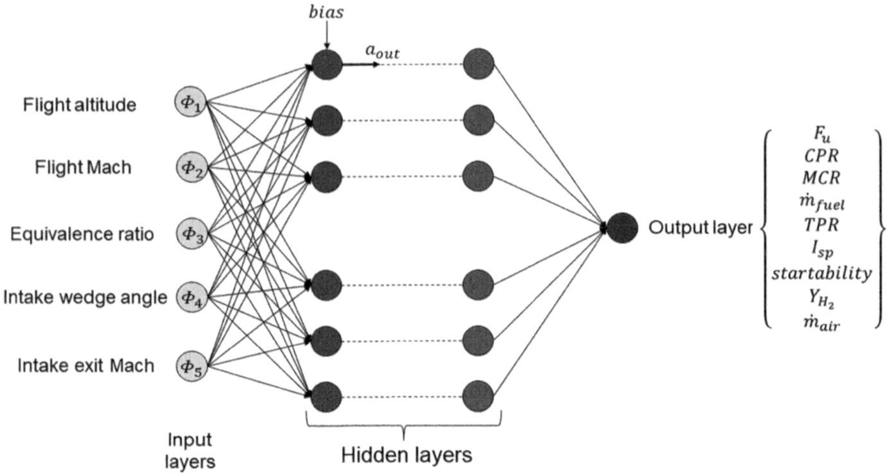

Fig. 7.9 Schematic representation of the ANN DMR surrogate model. Image adapted from [26]

mass flow rate of air is captured at the intake. For the same reason, the thrust output of the engine is higher at lower flight altitudes, where the air is denser. In general, it is observed that an increase of equivalence ratio (ER) leads to a larger thrust output, which is according to expectation in the lean combustion region, where adding fuel leads to a higher heat release.

Figure 7.10 shows that for all flight Mach numbers a negative thrust zone exists when the $M_{C,I}$ exceeds a certain limit. At these high $M_{C,I}$ levels, the pressure and temperature rise in the intake are limited because the flow's Mach number is not significantly reduced. This results in poor pre-combustion conditions that are unable to auto-ignite the fuel-air mixture. As a result, there is no thrust production.

For flight Mach numbers of 6, there exists an additional region of negative thrust production when $M_{C,I}$ is too low. This is caused by thermal choking of the flow, which severely limits the mass flow rate through the engine, to the detriment of the thrust output. At low equivalence ratios this only occurs at low values of $M_{C,I}$, but as the equivalence ratio increases, so does the heat release, causing the region of $M_{C,I}$ where thermal choking occurs to expand.

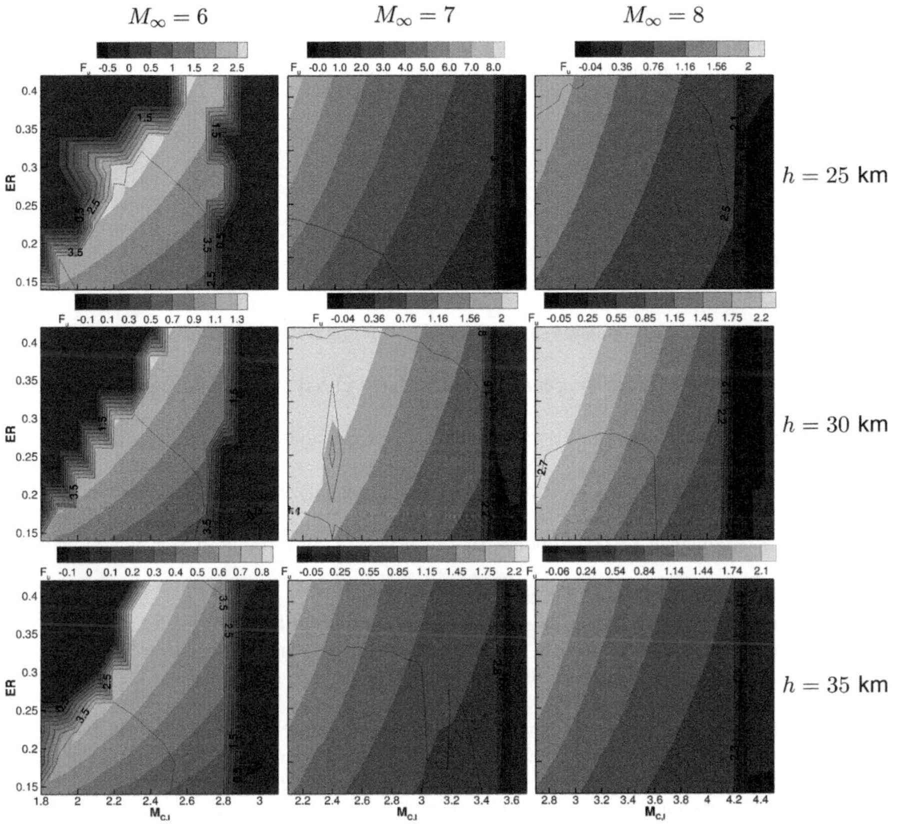

Fig. 7.10 Uninstalled thrust (F_u [MN]) maps of the propulsion system under investigation designed with the zero wedge angle scenario. The flight Mach number (M_∞) ranges from 6 to 8 and the flight altitude (h) from 25 to 35 km. The black lines represent the specific impulse (I_{sp} [ks]), adapted from [26]

7.4 Conclusion

This chapter aims to provide an insight into the reduced-order numerical tools used to study three propulsion cycles, ranging from Mach 2 to Mach 8. Firstly, the turbojet was simulated, where conventional numerical methods were used to study both the steady and transient engine behaviour. The second concept was the air-turbo-rocket, where conventional numerical methods were coupled with a gradient-free differential evolution optimiser to determine a suitable engine throttle schedule along a predetermined flight trajectory. The third cycle was the ramjet/scramjet engine, where an analytical reduced order model was used to train an artificial neural network which allowed for a comprehensive design space examination.

Acknowledgements The MORE&LESS project leading to this publication has received funding from the European Union's Horizon 2020 Framework Programme for Research and Innovation under grant agreement number 101006856. Additionally, gratitude is expressed to all the project contributors not explicitly mentioned here, including colleagues who were involved in experiments, simulations, or other aspects.

References

1. Long-term advanced propulsion concepts and technologies. https://cordis.europa.eu/project/id/12282
2. Long-term advanced propulsion concepts and technologies II. https://cordis.europa.eu/project/id/211485
3. Stratospheric flying opportunities for high-speed propulsion concepts. https://cordis.europa.eu/project/id/769246
4. MDO and REgulations for low-boom and environmentally sustainable supersonic aviation. https://cordis.europa.eu/project/id/101006856
5. Gorden Y, Rigmant V (2005) Tupolev Tu-144: Russia's concorde. Red Star, vol 24. Midland Publishing
6. de Sausmarez S (2010) www.heritageconcorde.com
7. Van den Borre K, Ispir A, Saracoglu B (2022) System level transient model of a propulsion plant for a Mach 2 Vehicle. In: 2nd international conference on high-speed vehicle science and technology. council of European aerospace societies. Bruges, Belgium
8. Gasturb GmbH: Gasturb 14 User Manual (2016)
9. Calder P (1971) Olympus 593 powerplant in concorde. In: National aeronautical and space engineering and manufacturing meeting, p 710775. https://doi.org/10.4271/710775
10. Trubshaw E (1991) Concorde flight testing—powerplant and performance flying. In: Aerospace technology conference and exposition, p 912192. https://doi.org/10.4271/912192
11. Heaton G, Franchi J, McKinlay R (1976) Operational experience on concorde. International Council of the Aeronautical Sciences, pp 563–571
12. Eames JD (1991) Concorde operations. In: Aerospace technology conference and exposition, p 912161. https://doi.org/10.4271/912161
13. Halliwell I (2022) Let's re-engine the concorde!—request for proposal
14. Calder PH, Gupta PC (1976) Future SST engines with particular reference to Olympus 593 evolution and concorde experience. The Aeronautical J 80(786):235–252. https://doi.org/10.1017/S0001924000029080
15. Petzhold L (1982) A description of DASSL: a differential/algebraic system solver. In: Proceedings of the 10th IMACS world congress
16. Lemmon E, McLinden M, Huber M (2002) NIST Standard reference Database23: reference fuid thermodynamic and transport properties—REFPROP, version 7.0. National Institute of Standards and Technology, Standard Reference Data Program
17. Fusaro R, Vercella V, Ferretto D, Viola N, Steelant J (2020) Economic and environmental sustainability of liquid hydrogen fuel for hypersonic transportation systems. CEAS Space J 12(3):441–462. https://doi.org/10.1007/s12567-020-00311-x
18. Fernández Villacé V (2013) Simulation, design and analysis of air-breathing combined-cycle engines for high speed propulsion. Ph.D. thesis, Universidad Politécnica de Madrid. https://doi.org/10.20868/UPM.thesis.23340
19. Verstraete T (2010) CADO: a computer aided design and optimization tool for turbomachinery applications. In: 2nd international conference on engineering optimization. Lisbon, Portugal
20. Goncalves PM, Ispir AC, Saracoglu BH (2019) Development and optimization of a hypersonic civil aircraft propulsion plant with regenerator system. In: AIAA propulsion and energy 2019

forum. American Institute of Aeronautics and Astronautics, Indianapolis, IN. https://doi.org/10.2514/6.2019-4421
21. Cakir BO, Ispir AC, Saracoglu BH (2022) Low fidelity design and analysis of propulsion systems for high-supersonic cruiser concepts. In: 2nd international conference on high-speed vehicle science and technology. Council of European Aerospace Societies, Bruges, Belgium
22. Cakir BO, Ispir AC, Saracoglu BH (2022) Reduced order design and investigation of intakes for high speed propulsion systems. Acta Astronautica 199:259–276. https://doi.org/10.1016/j.actaastro.2022.07.037
23. Ispir AC, Goncalves PM, Saracoglu BH (2020) Thermodynamic efficiency analysis and investigation of exergetic effectiveness of STRATOFLY aircraft propulsion plant. In: AIAA Scitech 2020 forum. American Institute of Aeronautics and Astronautics, Orlando, FL. https://doi.org/10.2514/6.2020-1108
24. Jachimowski CJ (1988) An analytica study of the hydrogen-air reaction mechanism with application to scramjet combustion. Technical Paper 2791, National Aeronautics and Space Administration
25. Fernández-Villacé V, Paniagua G, Steelant J (2014) Installed performance evaluation of an air turbo-rocket expander engine. Aerosp Sci Technol 35:63–79. https://doi.org/10.1016/j.ast.2014.03.005
26. Ispir AC, Cakir BO, Saracoglu BH (2022) Design space exploration for a scramjet engine by using data mining and low-fidelity design techniques. In: 2nd international conference on high-speed vehicle science and technology. Council of European Aerospace Societies, Bruges, Belgium
27. Ispir AC, Zdybał K, Saracoglu BH, Magin T, Parente A, Coussement A (2023) Reduced-order modeling of supersonic fuel–air mixing in a multi-strut injection scramjet engine using machine learning techniques. Acta Astronautica 202:564–584. https://doi.org/10.1016/j.actaastro.2022.11.013

Part III
Aeronautics High-Lift Application and Wing Design

Chapter 8
Lattice Boltzmann Simulation of a Deploying Krueger Flap Device

Jorge Ponsin and Carlos Lozano

Abstract We describe the performance of a lattice Boltzmann method for predicting the flow features of the turbulent unsteady flow generated during the motion of a Krueger flap designed for laminar wings. The simulations are performed using the Simulia XFlow solver. A wall-modeled large-eddy simulation approach (WMLES) is used to predict the turbulent flow effects. The numerical setup is described and the simulation results are compared with the high-quality experimental data obtained in the ONERA-L1 wind tunnel (W/T) during the H2020 UHURA project.

Keywords Lattice Boltzmann · High lift system · Krueger flap · WMLES

8.1 Introduction

Laminar flow wing technology is a promising avenue that can lead to a significant reduction in drag and, consequently, in fuel consumption. It requires specific high-lift (HL) concepts that are subject to the main constraint that in cruise conditions, and with the device in retracted position, the flow on the upper surface must remain unperturbed. One possible solution is to replace the classical slat element with a folded Krueger device [1]. The unsteady flow present during the fast deployment and retraction of a Krueger device is quite complex, with massive turbulent flow separation in some phases of the motion, and strong dynamical effects are expected. Therefore, the unsteady CFD flow prediction is a big challenge that requires improved standard tools based on URANS solvers complemented by high-fidelity CFD tools

J. Ponsin (✉) · C. Lozano
Computational Aerodynamics Group, National Institute of Aerospace Technology (INTA), Carretera de Ajalvir, Km 4, 28850 Torrejón de Ardoz, Spain
e-mail: ponsinj@inta.es
URL: http://uhura-project.eu

C. Lozano
e-mail: lozanorc@inta.es
URL: http://uhura-project.eu

T. Tuovinen et al. (eds.), *Advanced Computational Methods and Design for Greener Aviation*, Computational Methods in Applied Sciences 59,
https://doi.org/10.1007/978-3-031-61109-4_8

based on scale-resolving simulation methods. Furthermore, detailed experimental data is required to validate the CFD predictions and to understand the flow physics produced by the motion of the Krueger device.

The EU funded UHURA project [2] aims chiefly at performing experimental research on the unsteady complex flow behaviour associated to the deployment/retraction of a Krueger device mounted on a laminar wing. Additionally, it seeks to improve and validate unsteady CFD simulation methodologies [3]. The focus is set on URANS solvers and the special grid techniques required to deal with large element motions (chimera grids, immersed boundaries, etc.). Additionally, other unsteady flow methodologies are investigated, such as advanced hybrid RAN-LES turbulence models as well as alternative CFD techniques such as the Lattice Boltzmann Method (LBM).

Several previous publications have shown the potential applicability of LBM to moving HL configurations. In the framework of the sliding mesh approach, [4] investigates lift hysteresis effects in a trap wing with a conventional HL system subject to a time varying angle of attack, while [5] is a proof of concept for the application of an immersed boundary LBM scheme to the deployment and retraction of a conventional HL system. Aside from these works, the literature is scarce in these kind of applications and, as far as we know, the present paper presents the first application and validation of the LBM methodology to the prediction of the unsteady flow around a deploying/retracting Krueger flap.

8.2 Simulation Methodology

8.2.1 Lattice-Boltzmann Methods

The LBM is a promising alternative method for the simulation of unsteady flows due to several reasons: first, it is intrinsically unsteady and has low numerical dissipation/dispersion errors, which makes it suitable for joint use with high-fidelity turbulence models based on scale resolving methodologies. Second, its underlying adaptive Cartesian octree mesh approach, together with the way wall boundary conditions are imposed, is particularly well suited to dealing with complex and/or moving geometries. Finally, the approach is extremely efficient from the computational viewpoint, even though the time step advancement is limited by the explicit stepping nature of the time integration method. The LBM is a mesoscopic CFD approach that is based on the numerical resolution of a discretized version of the Boltzmann transport equation for the probability distribution function of particles at a given spatial location and velocity. Macroscopic flow properties such as density, velocity and pressure arise as the successive (velocity) moments of the distribution functions. It can be shown theoretically that the isothermal weakly compressible NS equations are recovered from a Chapman-Enskog (multiscale) expansion of the moments of the LBM equations.

8.2.2 Computational Methodology

Within the UHURA Project, INTA used the commercial LBM code Simulia-*XFlow* [6]. *XFlow* is based on a three dimensional 27-velocity lattice model (D3Q27), which seems to be the most appropriate set of velocities for turbulent computations due to the higher level of symmetry of the velocity lattice. The collision operator is based on a *central moment* model that has a high degree of accuracy and stability, and benefits from a slightly extended Mach number application range compared to standard collision operators. For the computation of turbulent flows *XFlow* uses a LES approach with the wall-adapting local eddy viscosity (WALE) subgrid scale (SGS) model. This is supplemented with a wall shear-stress based wall-modelled LES (WMLES) approach with a non-equilibrium algebraic wall function in order to reduce the meshing requirements in the near-wall regions.

The meshing approach is based on an adaptive Octree Cartesian technique, where the number and arrangement of Octree levels depends on the far-field and minimum resolutions specified in the geometry as well as the prescribed buffer layers among different levels. On-the-fly adaptive mesh refinement is used in dynamic motions. The mesh is adapted in the near wall region to guarantee sufficient resolution for the wall turbulence model. In addition, a dynamic adaptive algorithm based on the local level of vorticity has been used to resolve the wakes and large separation areas. These two adaptive options are only available for parallel shared-memory architectures in the version of the code used for the computations, which limits the size of the meshes that can be simulated.

Dynamic motion is handled with an immersed moving-boundary method based on a modified partially saturated cell method (PSM) that amounts, roughly speaking, to a modification of the collision operators with an additional term proportional to the solid volume fraction of the cell. This algorithm has (roughly) a 40% CPU time overhead with respect to static computations (fixed geometry).

8.3 LBM-WMLES Numerical Setup

During the first part of the UHURA project, research activities were dedicated to studying the sensitivity of the computations to grid resolution and numerical setup, with the ultimate goal of performing accurate static/dynamic LBM-WMLES simulations of the Krueger flap with a reasonable CPU time cost. Here, we give a brief overview of the work performed in that stage. A thorough account of these activities can be found in [7, 8]. The investigations were mainly focused on the following points:

(1) Establish the minimum boundary layer resolution for LBM-WMLES.
(2) Find the appropriate mesh resolution in high curvature regions (i.e., near the leading edge of the main element).
(3) Find the appropriate mesh resolution to adequately resolve large scales in wakes.

Fig. 8.1 Iso-surface of Q parameter coloured by vorticity for the Krueger device in perpendicular position

(4) Fix the appropriate spanwise domain extension L_z, as an extruded 2.5D approach will be used.
(5) Find the appropriate time steps, simulation times for statistics (static cases), averaging time, etc.

Two main lessons were extracted from these preliminary analyses. The first one is that the deploying Krueger test case is very challenging for LBM and, thus, it requires a careful numerical set up and mesh design, paying special attention to mesh resolution in certain areas.

The second lesson is that XFlow's shear-stress-based WMLES approach has difficulties to properly model the flow physics close to the wall. It was observed that transition to turbulence is not always achieved on boundary layers (especially at the upper main wing), leading to under-resolved turbulent boundary layers (Fig. 8.1). In fact, turbulence is present on the upper side of the main wing only in some phases of the deployment/retraction cycle, when it is triggered by strong incoming flow perturbations. In any other case, lack of turbulent content results in the under-prediction of turbulent boundary layer (TBL) effects along the upper main wing, and also affects the shear layers downstream of the main element, resulting, overall, in an over-prediction of lift compared to RANS simulations.

This type of problem is not unheard of in general WMLES models (not necessarily LBM), and different strategies have been devised to trigger turbulent fluctuations in modelled boundary layers. In our case, an exploratory study on tripping transition using cylindrical roughness elements [9] was carried out (Fig. 8.2). The elements produce a rapid switch to turbulence of the boundary layer, resulting in improved Cp and lift predictions at the cost of increased computational cost and an undesirable influence on the flow upstream of their location, so they were not used in the validation stage of the project.

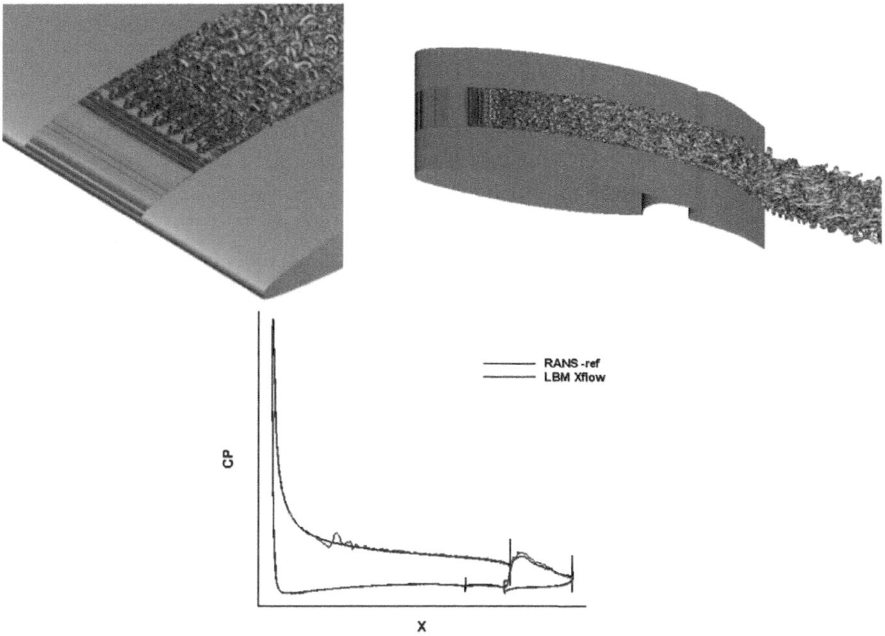

Fig. 8.2 Iso-surface of Q parameter coloured by vorticity and Cp distribution for the Krueger device in retracted position with roughness elements

8.4 Assessment and Validation of the LBM Approach

The third step of the UHURA project was dedicated to the validation of the CFD tools against experimental data obtained during W/T campaigns. Due to computational and time constraints, we focused in reproducing the conditions of the ONERA L1 wind tunnel campaign [10]. Overall, one dynamic and 10 static computations were carried out. The dynamic computation simulates a complete deployment/retraction cycle at free-stream velocity U = 30 m/s with deployment/retraction and holding times 1 s and 0.25 s, respectively, for a total of 2.25 s. The motion of the Krueger device was obtained from experimental measurements of the isolated drive engine. Static computations, on the other hand, were performed for five fixed positions of the Krueger device at two free-stream velocities U = 30 m/s and U = 45 m/s. Static computations are used to assess the influence of the Reynolds number on the results and, additionally, to assess the influence of dynamic effects by comparing the static cases at U = 30 m/s to the dynamic simulation. A more thorough account of these activities can be found in [11].

8.4.1 Computational Set-up for the ONERA-L1 Experiment

The geometry used in the simulations corresponds to the DLR-F15 configuration modified with a full span laminar leading-edge Krueger device immersed in a simplified rendition of the ONERA L1 tunnel. The CAD model of the wing and the ONERA-L1 wind tunnel section were provided by DLR. High resolution STL surface meshes were generated from these geometries using an in-house tool. With the available computational resources, it is not possible to simulate the complete wind tunnel test section with LBM-WMLES. Therefore, as shown in Fig. 8.3, the computational domain has been designed to reproduce the effect of the ONERA-L1 wind tunnel section walls in a simplified manner. The upper and lower walls have been placed at their real location relative to the model, and have been prescribed as slip walls. The longitudinal size of the W/T has been enlarged up to 4 times the original W/T section length in order to decrease the impact of the inlet and outlet boundary conditions on the flow at the location of the configuration. The span-wise size, L_z, is set to 0.24 m, which is about twice the length of the Krueger panel and 1/10th of the width of the ONERA L1 wind tunnel. This should be large enough to contain the largest (3D) turbulent structures produced in the wake area behind the Krueger device. Notice that, with this settings, our simulations do not distinguish between the part-span or full-span models tested in the ONERA-L1 W/T.

Mesh resolutions are carefully designed in each area so as to capture the relevant structures. On the main wing, the minimum resolution $\Delta_{\min}$ is set to $\Delta_{\min}/C_{ref} \sim 2 \times 10^{-3}$ where C_{ref} is the length of the wing in retracted position, resulting in roughly 10 cells within the boundary layer. The resolution is doubled at the leading edge of the wing and at the rear flap area, while it is set to $\Delta_{\min}/C_{ref} \sim 6.25 \times 10^{-4}$ at the Krueger device. Dynamic mesh adaptation with a target resolution $\Delta_{\min}/C_{ref} \sim 2.25 \times 10^{-3}$ is prescribed at the wake behind the Krueger device. Overall, these choices result in mesh sizes ranging from 12 M cells for the retracted position up to 16 M cells for the Krueger at 75°.

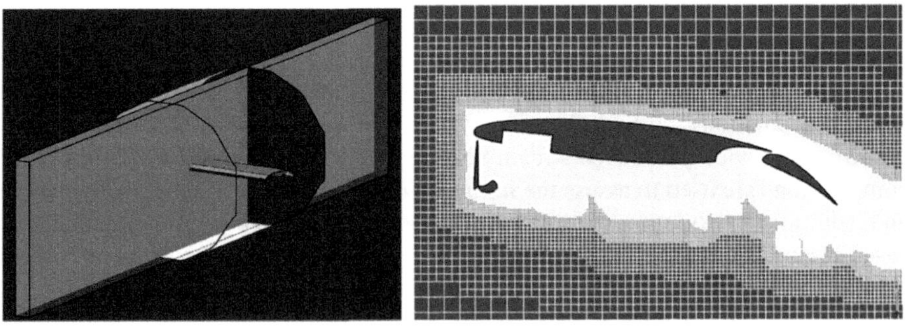

Fig. 8.3 Left: schematics of the computational domain. The ONERA-L1 wind tunnel section (maroon octagonal shape) is shown for reference. Right: snapshot of the instantaneous computational mesh

Concerning the boundary conditions, an inlet velocity has been prescribed at the inlet of the domain, while characteristic outflow non-reflecting boundary conditions have been applied at the outlet. Periodicity has been used for the side walls, and slip boundary conditions have been used for the upper and lower walls of the domain. No synthetic turbulence is imposed at any station and no roughness elements have been used to trip turbulence.

The time step of the simulations is given in terms of the Mach number and spatial resolution by an acoustic scaling argument. Choosing Mach numbers $M = 0.1$ for $U_\infty = 30$ m/s and $M = 0.15$ for $U_\infty = 45$ m/s yields time steps at the finest grid level $\Delta t_{\min} U_\infty / C_{ref} = 4.5 \times 10^{-5}$ and $\Delta t_{\min} U_\infty / C_{ref} = 6.75 \times 10^{-5}$, respectively.

Static simulations were run for about 15–25 convective time units (CTU), depending on the test case, after which averaged statistics (mean values of global quantities of interest and pressure distributions) were collected for about 5–10 additional CTU's.

8.4.2 Results

Figure 8.4 shows a breakdown by components of the lift history for a dynamical simulation spanning a full deployment-retraction cycle (1 s deployment time, 0.25 s holding time, 1 s retraction time). Conditions are aoa = 6°, $U = 30$ m/s and Re = 1.23 Mill. We note that there is a significant loss of total lift that comes basically from the rear flap and the main wing.

This loss of lift occurs at different positions of the Krueger panel in deployment and retraction phases. This is a significant dynamical effect (hysteresis) caused by the stall of the rear flap at different deflections of the Krueger device during motion (Figs. 8.5 and 8.6) and their associated circulation effect on the main wing lift. The hysteresis is not observed on the Krueger device itself (Fig. 8.6, right), in agreement with ONERA [12] and DNW-NWB [13] experimental results.

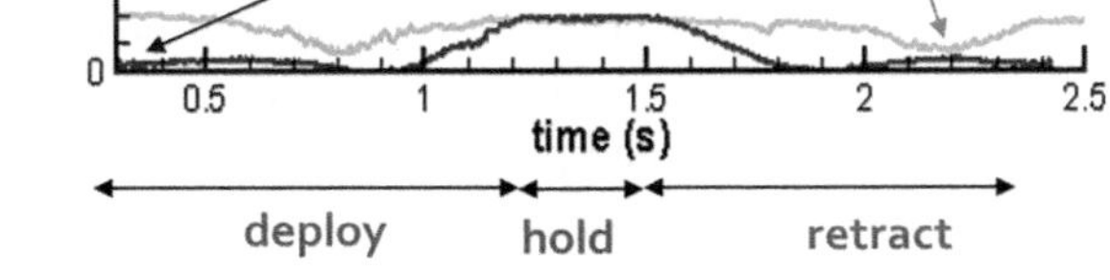

Fig. 8.4 Lift history for a full deployment/retraction cycle. Deployment starts at t = 0.25 s

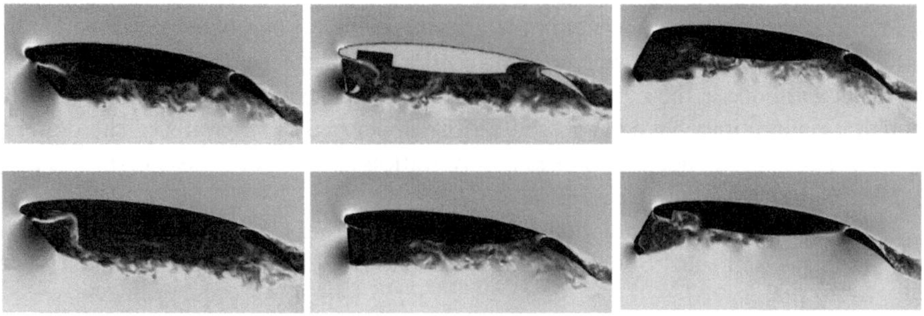

Fig. 8.5 Snapshots of the instantaneous velocity field during a complete cycle. Upper row: deployment phase, left to right (37.5°, 75°, 112.5°). Lower row: retraction phase, right to left (112.5°, 75°, 37.5°). Lift minima occur at 75° during deployment and 37.5° during retraction

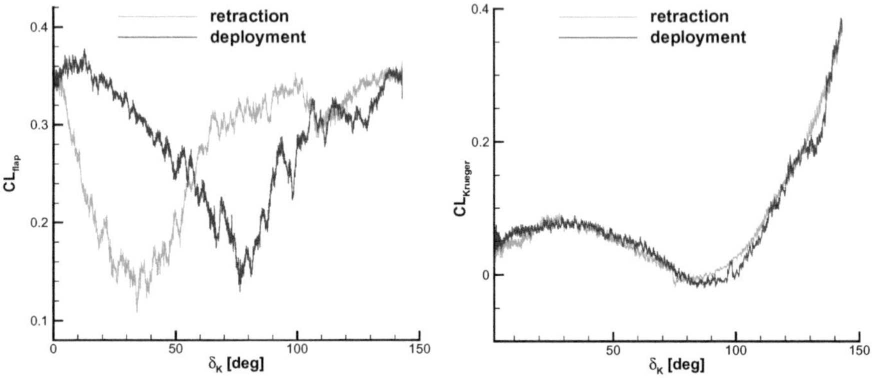

Fig. 8.6 Evolution of lift and drag with respect to the Krueger panel deflection angle for the rear flap (left) and the Krueger panel (right)

Static simulations at $U = 30$ m/s are quite similar to dynamic simulations during the deployment phase (Fig. 8.7), except at $\delta_{Krueger} = 37.5°$, where the static simulations predict a reduced lift on the flap and main elements. The opposite occurs for the retraction phase. In this case, a large discrepancy occurs at $\delta_{Krueger} = 75°$, when the Krueger is in nearly-perpendicular position relative to the wing. The static simulations predict the maximum loss of lift, while for the retracting dynamic simulation this is delayed until the Krueger panel reaches an angle close to 37.5°.

Figure 8.8 shows the comparison of the numerical and experimental pressure distributions at 5 positions of the Krueger device in the low-velocity case $U = 30$ m/s for the dynamic deployment phase. The figures show instantaneous data from dynamic simulations and phase-averaged data from MEMS sensors located on the wing leading edge and Krueger panel. We see that there is a good agreement between numerical and experimental results.

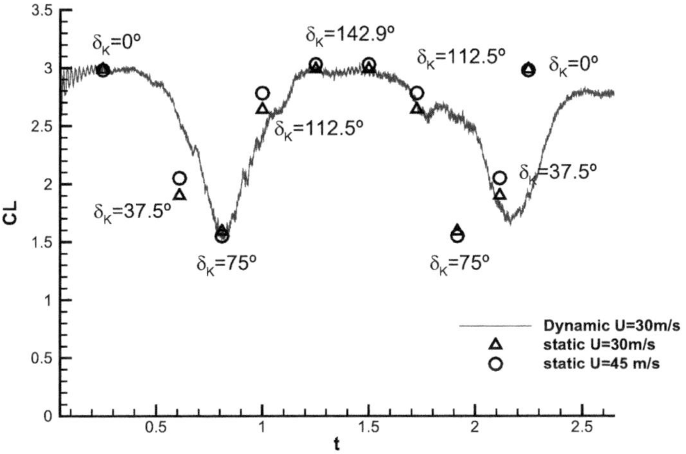

Fig. 8.7 Comparison of predicted lift for dynamic ($U = 30$ m/s, red) and static time-averaged computations (Re = 1.23 Mill vs. Re = 1.85 Mill)

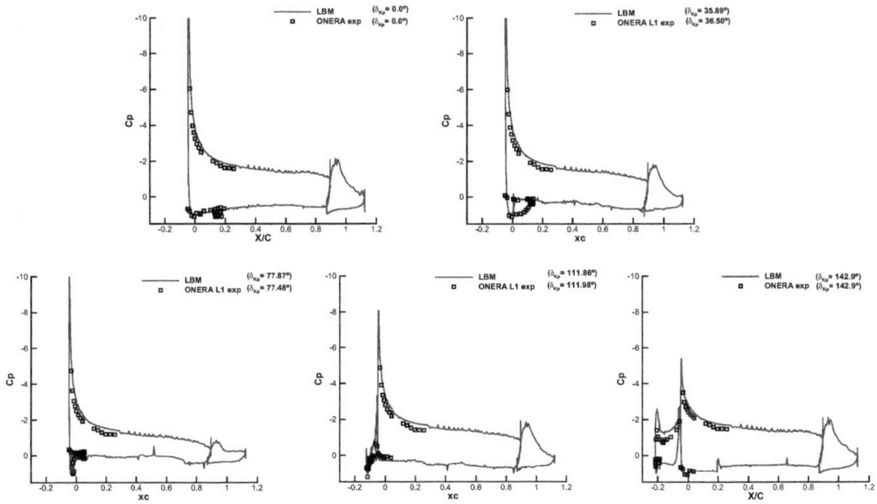

Fig. 8.8 Numerical versus experimental pressure distributions (ONERA L1) during deployment

Figure 8.9 focuses on dynamical effects by comparing experimental and numerical data for the total lift of the wing. The experimental value for lift is obtained by integrating the main wing leading edge Cps (MEMS), and is used as a proxy for the total wing lift history during deployment/retraction, while simulation data corresponds to the total main wing lift. Notice that experimental results confirm the hysteresis effect between deployment and retraction. Both results agree in the prediction of

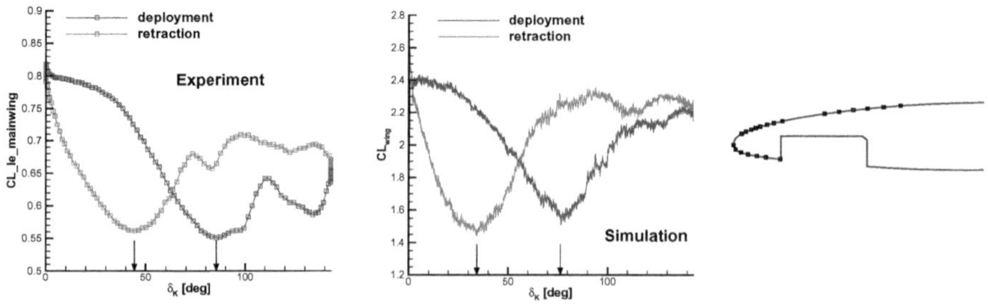

Fig. 8.9 Computed lift versus experimental lift for the front section of the main wing

the values of the Krueger panel deflections at which minimum lift occurs at deployment/retraction phases, with a maximum of $+10°$ difference between simulation and experiment, and the behavior observed during the retraction phase is qualitatively well reproduced by the simulations.

Finally, Figs. 8.10 and 8.11 compare time-averaged numerical results with PIV data [14] from the ONERA L1 wind tunnel at several static positions of the Krueger device for the high-velocity case ($U = 45$ m/s). Qualitatively, there is a fair agreement between numerical results and experimental PIV data, particularly in the identification of zones of reverse flow and areas with high turbulent kinetic energy.

The total CPU wall time for the complete dynamic simulation (lasting 2.5 s) was about 70,000 CPUh. The limitation of the dynamic LBM algorithm used in the computations to parallel shared-memory architectures (already mentioned in 2.2) results in high turn-around computational times. This makes the method impractical at present for the design phase in industrial environments. Nevertheless, considerable speed-up has been achieved in the new versions of the code using GPU acceleration. This makes the approach competitive with traditional CFD for this kind of simulations.

8.5 Conclusions

Lattice Boltzmann methods show a great potential for the robust and efficient simulation of deploying high-lift geometries. There is an inherent limitation in the accuracy of the flow prediction near the walls due to the use of wall functions for turbulent flow (which are mandatory both for RANS and/or WMLES on computational efficiency grounds), and this can impair the prediction in cases involving smooth turbulent flow separation (e.g. rear flap separation). In spite of this limitation, the results obtained within the UHURA project for fast deployments of the full-span configuration compare well with the experiments performed in the ONERA L1 W/T, especially at the Krueger device area, where the separation is mostly induced by the geometry. The strong dynamical effects present in the experiments for the 2D

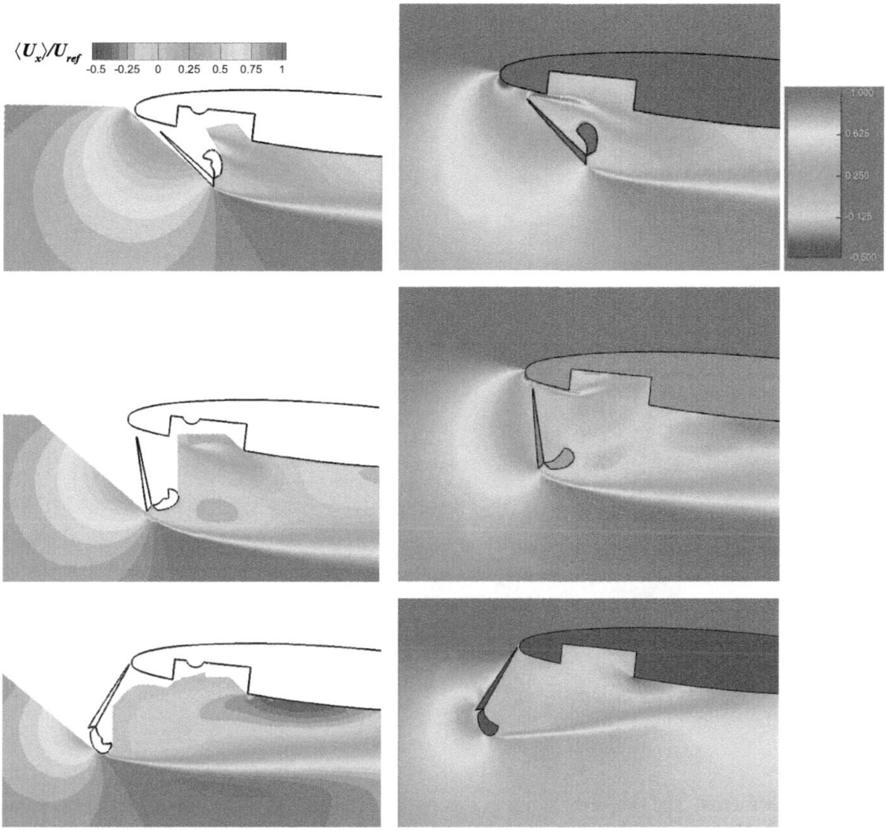

Fig. 8.10 Averaged velocity field for three Krueger deflections (37.5° upper row, 75° middle row, 112.5° lower row). Left column: PIV ONERA L1 results. Right column: LBM results

full span configuration are qualitatively predicted by the simulations and the averaged flow field obtained from the LBM simulations compares well with the PIV experimental results obtained in the ONERA W/T. The LBM approach has proved to be computationally efficient for scale-resolving turbulence simulations, but the computational resources required are still considerably high. This drawback can be overcome with GPU acceleration, making this approach an attractive alternative to traditional CFD approaches.

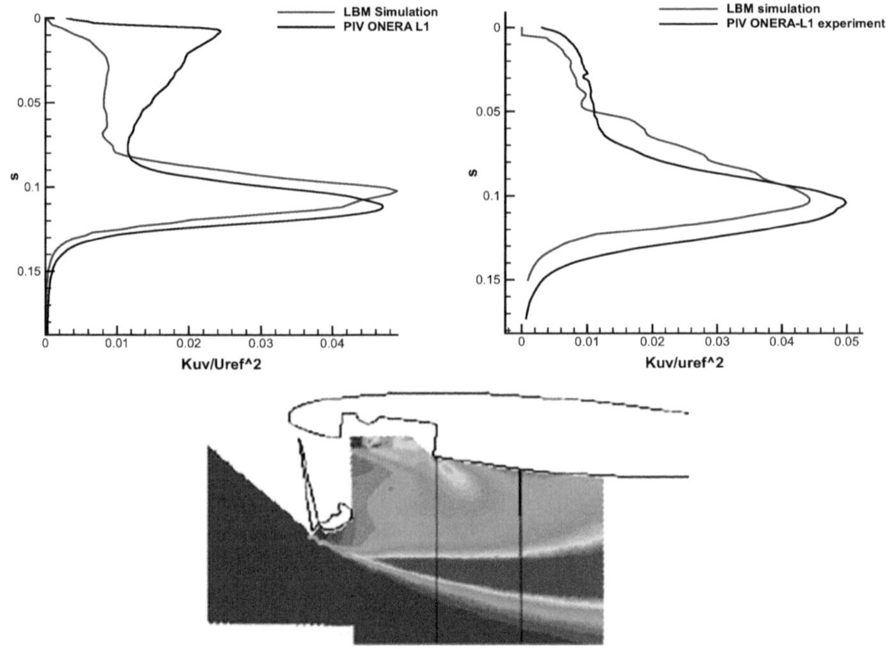

Fig. 8.11 Averaged turbulent kinetic energy k_{uv}/U_∞^2 profiles (upper figures) along the two lines indicated in the bottom figure for the "barn door" Krueger flap position

Acknowledgements The project leading to this publication has received funding from the European Union's Horizon 2020 Research and Innovation Programme under grant agreement N° 769088.

References

1. Iannelli P, Wild J et al (2012) Analysis and applications of suitable CFD-based optimization strategies for high-lift system design. In: ECCOMAS 2012 proceedings, paper 2833
2. Hasabnis A, Wild J, Strüber H, Moëns F, van Rooijen B, Maseland H (2023) The UHURA project at a glance—motivation and objectives. IOP Conf Ser: J Phys. 2523, 012005
3. Moëns F, Wallin S, Maseland H (2022) UHURA project CFD methods for unsteady high lift simulation. In: 12th EASN international conference (EASN 2022)
4. Fabiano E, Fares E, Martinez F, Nolting S (2012) Unsteady flow simulation of high-lift stall hysteresis using Lattice Boltzmann approach. In: AIAA 2012-2922, 2012 AIAA applied aerodynamics conference
5. Holman D, Brionnaud R, Martinez F, Mier-Torrecilla M (2012) Advanced aerodynamic analysis of the NASA high-lift trap wing with moving flap configuration. In: AIAA 2012-2923, 2012 AIAA applied aerodynamics conference
6. Trapani G, Brionnaud R, Holman D (2018) XFlow contribution to the third high-lift prediction workshop. In: AIAA 2018-2847, 2018 AIAA applied aerodynamics conference

7. Ponsin J, Lozano C (2020) Technical report on assessment of LBM and best practices for the prediction on unsteady aerodynamics during deployment/retraction phases of high lift systems. UHURA deliverable D22-2
8. Ponsin J, Lozano C (2021) Progress towards simulation of Krueger device motion with Lattice Boltzmann methods. IOP Conf Ser: Mater Sci Eng 1024:012050
9. Deck S, Weiss P, Renard N (2018) J Comput Phys 363:231–255
10. Wild J, Strüber H, Moens F, Rooijen B, Maselan H (2022) A validation program for dynamic high-lift system aerodynamics. In: 8th European congress in computational methods in applied sciences and engineering, ECCOMAS congress 2022
11. Ponsin J, Lozano C (2022) Summary of lattice Boltzmann LES technique results for aero-loads prediction. UHURA deliverable D41-5
12. Tanguy G (2022) Data of ONERA L1 wind tunnel test. UHURA deliverable D33-3
13. Löser T. Wind tunnel test report DNW-NWB. UHURA deliverable D33-5
14. Tanguy G, Monnier J, Verbeke C, Wild J (2022) Experimental aerodynamic investigation of a Krueger flap device using particle image velocimetry. In: 9th European conference for aeronautic and aerospace sciences, EUCASS 2022

Chapter 9
Lessons Learnt from Chimera Method Application to a Deploying Krueger Device

Apurva Hasabnis, Hans Maseland, Frédéric Moens, Aleš Prachař, and Jochen Wild

Abstract The motion of the Krueger flap—as the most promising leading edge high-lift system device for laminar wing technology—is characterized by a relatively large movement (about 140° deflection) at a relatively high deflection speed (up to 200°/s) compared to classical leading-edge devices. The configuration varies throughout the motion from a nearly sealed retracted position to a gapped flow in fully deflected position, comparable to a slat device. During its deployment or retraction phase, highly unsteady flow is observed, leading to both performance degradation and loads fluctuations on the different elements (Krueger panels for instance). For proper design and analysis, a validated numerical simulation tool considering both the change in external shape and the unsteady flow feature is necessary. The evaluation of such numerical tools and their validation based on dedicated experiments is the core of the UHURA EU funded project. In terms of numerical simulation, the Chimera method is an established method for simulation of moving bodies by employing overlapping grids. In the framework of the UHURA project, several partners applied their CFD capabilities, mainly URANS, based on the Chimera technique in order to validate the method for this specific application in comparison to wind tunnel tests.

A. Hasabnis (✉) · J. Wild
German Aerospace Center (DLR), Lilienthalplatz 7, 38108 Braunschweig, Germany
e-mail: apurva.hasabnis@dlr.de

J. Wild
e-mail: jochen.wild@dlr.de

H. Maseland
Royal Netherlands Aerospace Centre (NLR), Amsterdam, The Netherlands
e-mail: hans.maseland@nlr.nl

F. Moens
French Aerospace Lab (ONERA), Meudon, France
e-mail: frederic.moens@onera.fr

A. Prachař
Czech Aerospace Research Centre (VZLU), Prague, Czech Republic
e-mail: prachar@vzlu.cz

T. Tuovinen et al. (eds.), *Advanced Computational Methods and Design for Greener Aviation*, Computational Methods in Applied Sciences 59,
https://doi.org/10.1007/978-3-031-61109-4_9

This paper outlines the different Chimera approaches ranging from structured/2D to hybrid/3D in steady and unsteady simulations for the different types of setups investigated, namely straight and swept wing with full-span and part-span Krueger flap. It summarizes common challenges and best practices for the application of the Chimera approach for such a device.

Keywords Aerodynamics · Unsteady · High-lift systems · CFD · Chimera · Krueger flaps

9.1 Introduction

Werner Krueger was the first to define the 'nose split flap' in Göttingen [1], which would later be known solely by his name as 'The Krueger Flap'. This leading-edge high-lift device was intended to increase the stall angle, thereby increasing the maximum lift coefficient and reducing the flight speed at take-off, approach and landing. These flaps are unique as they are deployed from the pressure side of the wing, against the flow direction. The Krueger flaps were extensively used by Boeing on all transport aircraft from B707 to B747 at the beginning of the age of the jet liners [2, 3]. Although the original Krueger flap was simply a hinged panel, the B747 already introduced a small slot to improve the aerodynamic performance. Krueger flaps gained attraction again as the laminar wing technology was flourishing [3], as the technology offers the highest potential in aerodynamic drag reduction by keeping the wing boundary layer flow in a laminar state and thus reducing frictional forces. Apart from that, they also serve as shielding devices against contaminants such as dirt, insects and other pollutants. As the Krueger flap needed to be re-visited to be at-par with the current aerodynamic requirements, it served as motivation for the UHURA project. During the deployment from the pressure side of the wing against the flow, the Krueger flap is perpendicular to the flow and produces large area of separation. Thus, it is crucial to pass this position as fast as possible and also avoid the risk of jamming in case of the full-span configuration. These requirements introduce multiple challenges while studying the flow behaviour around the Krueger flap.

In order to accurately simulate the aerodynamics associated with the motion, this paper concentrates on the CFD analysis applying the Chimera technique, compares the results with the experimental findings from the wind tunnel tests and also tries to understand the limitations and challenges experienced. The Chimera—or overset mesh—method [4] is an established method for simulation of overlapping grids. With this method, volumes around parts of the geometry are meshed separately and the complete geometry is combined by overlapping those meshes. The separate domains are then coupled by appropriate interpolation algorithms to get an equilibrium state that the interfaces. Meshing parts independently has made this method popular for moving bodies like propellers and rotors (e.g. [5]) or control surfaces (e.g. [6]). Also, for high-lift system aerodynamics it is an appropriate method to mesh complex geometries. It partly allows to overcome restrictions in mesh topology due to the complicated topology of complicated bounds. Therefore, the chimera method has been used for some of the first full 3D high-lift transport aircraft simulations [7–9]. While early simulations suffered an expensive search of the proper interpolation schemes, significant progress has been made to achieve affordable computational requirements [10]. It is still an industrially used method competitive to domain conforming mesh methods, as established even in the last high-lift prediction workshop [11, 12]. It is thus a promising method to simulate deflecting high-lift systems as shown for a 2D wing section in [13].

The challenging factor for the targeted application of the Krueger flap deflection is its motion and the corresponding proper definition of the variable interface regions. For the complex geometry cases mentioned above, the Chimera method was applied to a steady geometry that allows for distinct well-aligned sizing prescriptions. For the rotary cases of rotors or propellers, the shape of the interfaces of the domains are well defined by the rotational (mostly cylindrical) fluid volumes. Now, for the case of the deflected high-lift device, the interpolation zones must be appropriate throughout the deployment path. This is even more pronounced for the Krueger flap type high-lift system as this performs an about 150° rotation around the wing leading edge.

Within the UHURA project, a key activity was to qualify the validity of Chimera method top simulate the dynamic behaviour of a fast moving Krüger flap in comparison to wind tunnel data. In the following paper we first explain the type of motion considered, followed by a description of the different Chimera approaches followed by the various partners involved. Finally, some validation is shown for simulations in comparison to wind tunnel data for static and dynamic cases.

9.2 Krueger Panel Motion

In order to address the flow physics of a Krueger flap during deployment and retraction, in a first step a Krüger flap was defined for the of the DLR-F15 airfoil with the laminar leading-edge variant [14]. The design incorporated the requirements on a feasible kinematics, which was further used to establish the exact motion path depending on the geometric relations inside the mechanical system.

The Krueger device movement can be best defined as a multi-body motion, wherein the main wing motion is described w.r.t. the wind-tunnel, that of the panel w.r.t. the wing and of the bullnose w.r.t. the Krueger panel. In order to simulate the very large deflections out of scope for any standard mesh deformation technique, most widely recognised overset mesh method—or the Chimera method was chosen to be validated for the designate purpose of high-lift system deflection. To implement the Chimera technique with overlapping meshes, each moving part is meshed individually and the grids are put together using various predefined approaches according to specifications of the underlying CFD codes to establish the desired overlap. Figure 9.1 shows an example of such a Chimera grid composition by one grid each around every

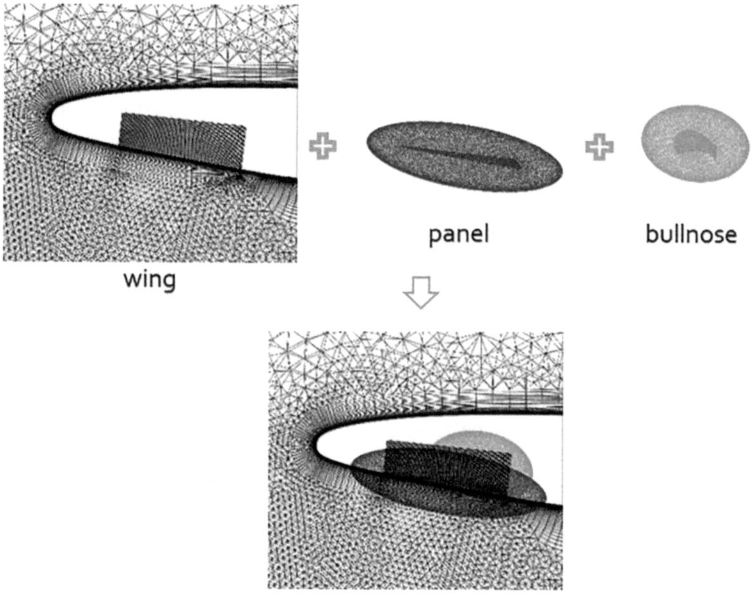

Fig. 9.1 Example of individual and overlapping grids set-up to establish Chimera boundaries

Krueger element and the wing. The mesh parameters are chosen depending upon the necessary overlap with the neighbouring moving part. Within the project we also addressed improvements in mesh technology by assessing the proprietary solver technology of the different partners.

It was observed during initial simulations that a minor gap was desired to establish Chimera boundaries. Thus, a gap of about 0.1% of reference chord was introduced (Fig. 9.2). This essentially helps provide an interpolation zone between the different moving components even in the fully retracted position and allows determining the interpolation points of the overlapping grids to implement the chimera technique. A fully closed surface would have isolated the cavity and removed the interpolation capability in time. Finally, the kinematics of the Krueger elements (panel and bullnose) has to be parameterized according to the manufactured systems (Fig. 9.3). After finalising this configuration, it was up to individual partners to take on the meshing and CFD simulations. As there are many established meshing strategies, the project partners chose their distinct approaches to tackle the task at hand, which will be discussed in this paper in a later chapter.

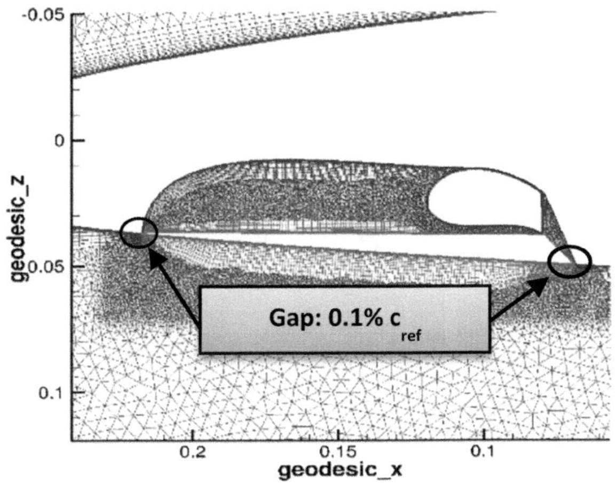

Fig. 9.2 Gaps introduced to the Krueger panel set-up to establish interpolation boundaries for the Chimera technique in fully retracted position

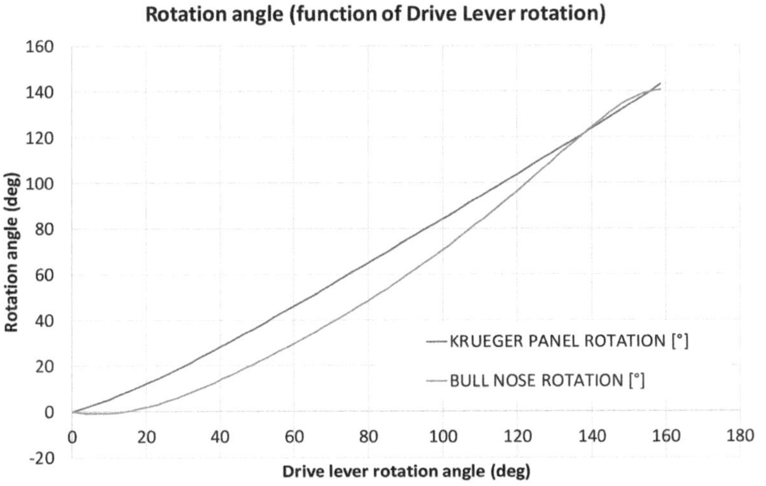

Fig. 9.3 Rotation angle of the Krueger panel (blue line) and of the bullnose (red line) as a function of the Drive lever rotation angle

9.3 Experimental Validation Data Base

A major block of activity in the second phase was the preparation and the execution of a series of wind tunnel tests to create a data-set on the unsteady aerodynamics of the Krueger flap deployment. We tested with two wind tunnel models of the DLR-F15 family in three separate facilities. Test matrices were set up to try various wind speeds, wing sweeps and panel configurations. The wind velocities were chosen at

Fig. 9.4 Set-ups of the DLR-F15 model family in the three wind tunnel facilities used within the UHURA project: (left) DLR-F15-LLE in ONERA-L1; (centre) DLR-F15-LLE in DNW-NWB; (right) DLR-F15*LS*-LLE in DNW-LLF open test section

30 and 45 m/s with respect to the wing leading edge normal direction. This was done to eliminate compressibility effects based on the wing sweep and to enable direct comparison of pressure levels independent of wing sweep.

The smaller model—the standard DLR-F15 model with 600 mm clean wing chord—was first setup as a 2D straight wing with a full span Krueger flap in the ONERA-L1 wind tunnel facility in Lille, France (Fig. 9.4 left). The various tests conducted include wing-pressure distribution, unsteady pressure measurements with MEMS and PIV measurements.

The DLR-F15 model was mounted in cantilever configuration in the DNW-NWB tunnel (Fig. 9.4 middle). Here, the emphasis was on the impact of wing sweep—we measured a straight and a 23° swept wing configuration—as well as the variation of deployment profiles in terms of time, speed and acceleration. The tests conducted include unsteady pressure measurements using Kulite dynamic pressure sensors and Bosch pressure sensors based on MEMS technology, steady pressure measurement using PSI-scanner, study of the wake rake, Krueger deformation (SPR) and Krueger deployment angle. Emphasis of this test was put on the variation of the motion profile, even with using different levels of acceleration ranging from 165 to 1000 m/s^2.

Finally, the Krueger flap was implemented on the large-scale model DLR-F15LS, which Is an exact 2:1 upscale of the DLR-F15 model with 1.2 m clean wing chord and a span of 7 m (Fig. 9.4 right). This model was measured as a 30° swept wing in the DNW-LLF, a low speed, high-Reynold's number facility at Marknese, The Netherlands. Emphasis here was to obtain data at even higher Reynolds number and higher wing sweep. In addition, the larger model size allowed for the implementation of more pressure sensors thus giving additional information on the spanwise variation. The wing pressure distribution was measured along with the unsteady

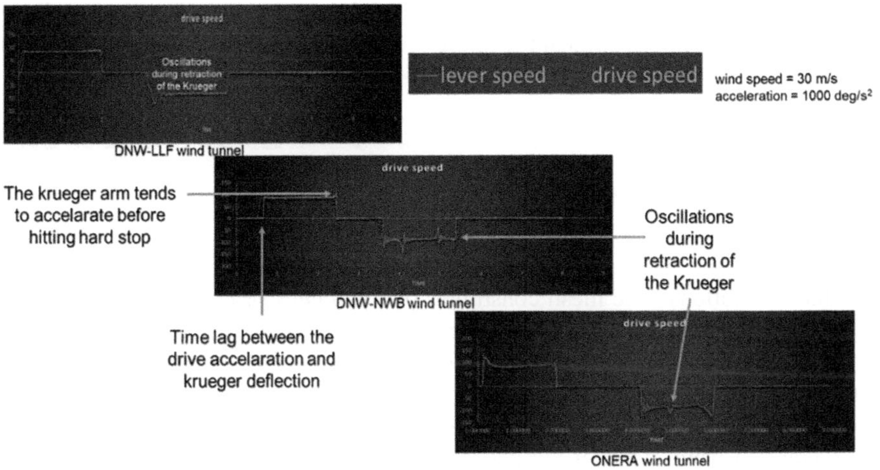

Fig. 9.5 Specific challenges for unsteady validation from comparison of commanded (blue) and measured (orange) Krueger flap motion

pressure on Krueger with MEMS and PSI. Krueger deformation, deployment angles and PIV flow field were studied. For the motion, the deployment times were doubled according to the model scale to keep the convective time constant. Further details of the wind tunnel models and the test set-ups can be found in [15].

During the analysis of the wind tunnel experiments, some critical motion features regarding the prescription of the motion were found at all the three wind tunnels. To highlight this, Fig. 9.5 shows the drive and lever speeds during motion. Please note that the rotational speed of the driving motor—represented with the colour blue—was directly measured at the driving engine, while the angular speed of the lever was derived as the time derivative of the measured angle at the kinematics station and is depicted in orange. A time lag was also observed between the drive motion (start of engine) and the Krueger deflection (start of Krueger panel motion). This is related to releasing a certain pre-tension inside the kinematics for keeping the Krueger in retracted position slightly tightened against a stop position. Further, it was observed that the Krueger arm tends to accelerate towards the end of extraction, which is caused by a load direction change just when the Krueger gest to the deflected position. Since a mechanical stop was introduced to ensure that the Krueger stopped at the exact desired position, the rotational speed of the lever suddenly stops when hitting the stop contact. On the retraction side, again a delay is seen between the start of the engine motion and the start of the Krueger deflection. This is due to the fact that again certain amount of torque is required to set the lever into motion from stationary position and additionally to overcome the aerodynamic forces. Furthermore, during early retraction, the lever speed tends to oscillate due to the load change during the acceleration phase.

In summary it was concluded that the use of the exact motion path is essential for the validation since otherwise the local positions at a certain time stamp would differ between CFD simulation and experimental data. Nevertheless, these effects were noted to be less significant in the DNW-LLF tunnel, due to the presence of drive motors on both ends of the wing.

9.4 Meshing Strategies

The meshing strategies have been studied by many to establish methods that result in reliable solutions. The mesh constitutes of small cells or elements that together represent the entire flow field. Each cell represents a discrete space in this flow field which is then simulated by applying various mathematical equations to correctly establish the flow physics. The quality of the mesh in turn is responsible for the reliability of obtained solution from the numerical simulations.

There are several types of meshes that can be chosen from, depending upon the application or preference. Few of them are listed here, more information can be obtained in [16]. An unstructured mesh in 2D or on the surface typically consists of elements, such as triangles or polygonal shapes. In 3D the unstructured elements filling the volume are consistently tetrahedrons or—in hybrid unstructured meshes—prism and pyramids. The hybrid unstructured meshes have gained popularity as they provide flexibility around complex geometries and help avoid the overheads and size of large pure Tetrahedral meshes. If implemented accurately, these meshes can prove to be very efficient in numerical simulations.

The structured mesh on the other hand, uses a Cartesian system at its computational core and is characterised by eight nodal points in three dimensions with regular connectivity. Although extremely convenient to implement in case of orthogonal geometry, it requires significant adjustment when the geometry exhibits curvatures and deviates from standard shapes, which is usually the case. Thus, in order to depict the boundary conditions accurately, precise mesh treatment is required. To mesh non-orthogonal geometries, multiblock or block-structured meshes are used. In other cases, a transformation function is used in order to create a body-fitted mesh for such geometries. A block structured mesh helps overcome the resulting skewed cells of a body fitted mesh as shown in Fig. 9.6. Rectangular within each block, the block-structured grid fits together in an unstructured manner. Such grid, that maintains continuity across various interfaces, allows representation of differences in the rest of the field. Complete continuity across different block interfaces can be achieved by treating the various interfaces as branch-cuts [17].

In case of mixed meshes, a combination of structured and unstructured meshes is used in order to derive the advantages of both—precision and structure of the first and flexibility of the later. This approach proves to be very efficient in case of complex and non-standard geometries. Especially the Chimera approach offers to use different kind of mesh types for the different domains without the need to care for conforming interfaces. Depending upon the boundary conditions, a Hex-dominant strategy creates a Hex-dominant mesh by utilising a semi-automated poly-cube method. The resulting mesh constitutes of Hexahedra, Tetrahedra and triangular

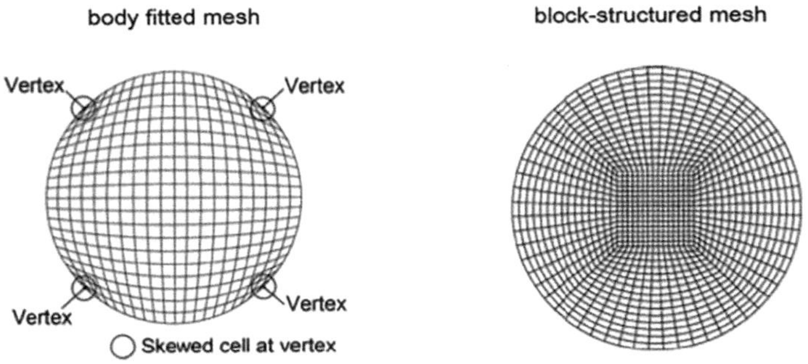

Fig. 9.6 Body fitted and block-structured mesh

Table 9.1 Meshing strategies used for sub-blocks of the chimera meshes

	DLR	ONERA	NLR	VZLU
Background	Hybrid unstructured	Cartesian	Block-structured	Hybrid unstructured
Wing/flap	Hybrid unstructured	Block-structured	Block-structured	Block-structured
Krueger flap	Hex-dominant	Block-structured	Block-structured	Block-structured

prisms. The combination of these multiple elements helps achieve better-quality hexahedra elements than all-hex meshes [18].

A Cartesian grid on the other hand, is non-body-fitted. It derives its advantage from the ability to characterise and compute intersections between the hexahedral cells and the surface mesh. Also eliminates the need for the cells to conform to the surface as in case of structured or unstructured meshes in case of complex geometries. In this approach, the hexahedral cells consisting of parallelepipeds and orthogonal grid system may extend through the solid boundaries into the computational domain [19].

Though both types of meshes present their own unique advantages, they are often used in combination in order to accurately cover the entire geometry and precisely represent the flow physics. For the study under consideration, various meshing strategies were adopted by individual partners in accordance with the established standards for the CFD analysis. This allowed for a thorough comparison of the obtained results at the end. The different meshing approaches considered by the partners are listed in Table 9.1. The remainder of this section details the approach of each partner.

9.4.1 DLR Meshing Strategy

The DLR meshing strategy constitutes of hybrid unstructured meshes with embedded structured blocks (Fig. 9.7). Standard prismatic meshes were used around single parts, whereas the global mesh is unstructured. Depending upon first CFD results obtained

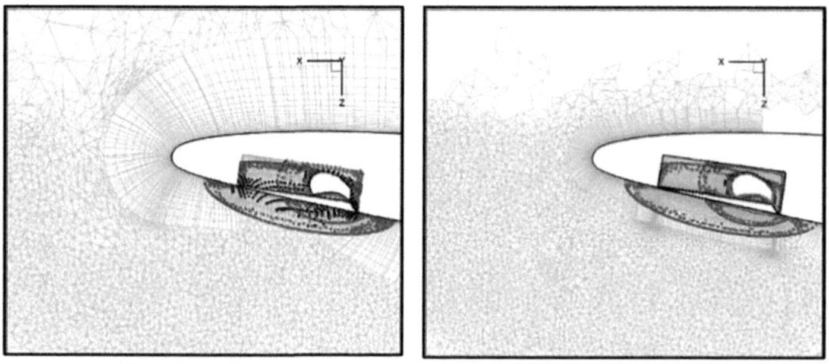

Fig. 9.7 DLR meshing strategy based on hybrid unstructured meshes: (left) standard mesh (v2.0) approach; (right) refined mesh (v3.3) with cavity meshed by structured elements

on state-of-the-art meshes (named v2.0), the meshes were further refined to better capture the flow physics in and close to retracted position, by meshing the wing cavity with a structured mesh block for better alignment of the cells (named v3.3).

The hole-cutting algorithm embedded in the DLR TAU code serves the purpose of establishing the chimera technique without necessitating extensive grid overlap. Once the interpolation boundaries are set-up by inserting grid boundaries of interacting grids, the Chimera implementation of the TAU code relies on two layers of interpolation points at the Chimera boundaries [20]. The comparison below (Fig. 9.7) highlights the differences between the earlier and refined grids. It was realised that the new mesh v3.3 agrees much better with the experimental results.

The impact of grid refinement can be seen through the pressure distribution graphs (Fig. 9.8). The dotted line represents the results from version 2.0 of the mesh and the thick line represents the final mesh. Snapshots at the bottom of each graph show the Krueger deflection pertaining to each graph. In case of the mesh v2.0, the pressure distribution shows non-physical behavior ($c_p >> 1$), e.g. in fully retracted position the flow gets trapped in the cavity resulting in undesirable pressure distribution. On the other hand, the v3.3 mesh captured the flow physics very reasonably apart the flow in the closed gap at fully deflected position. Further, comparison to experimental data as done in a later chapter verified the validity of the v3.3 mesh.

9.4.2 ONERA Meshing Strategy

ONERA adopts a more structured approach to meshing (Fig. 9.9). A standard structured mesh is chosen around single components. These individual structured meshes are in turn embedded in a Cartesian background grid [21].

Specific areas of the Krueger kinematics are considered critical and were further refined by introducing finer meshes (Fig. 9.10). These include the intentional small gap between the Krueger panel and the bull nose and collar grids when the Krueger

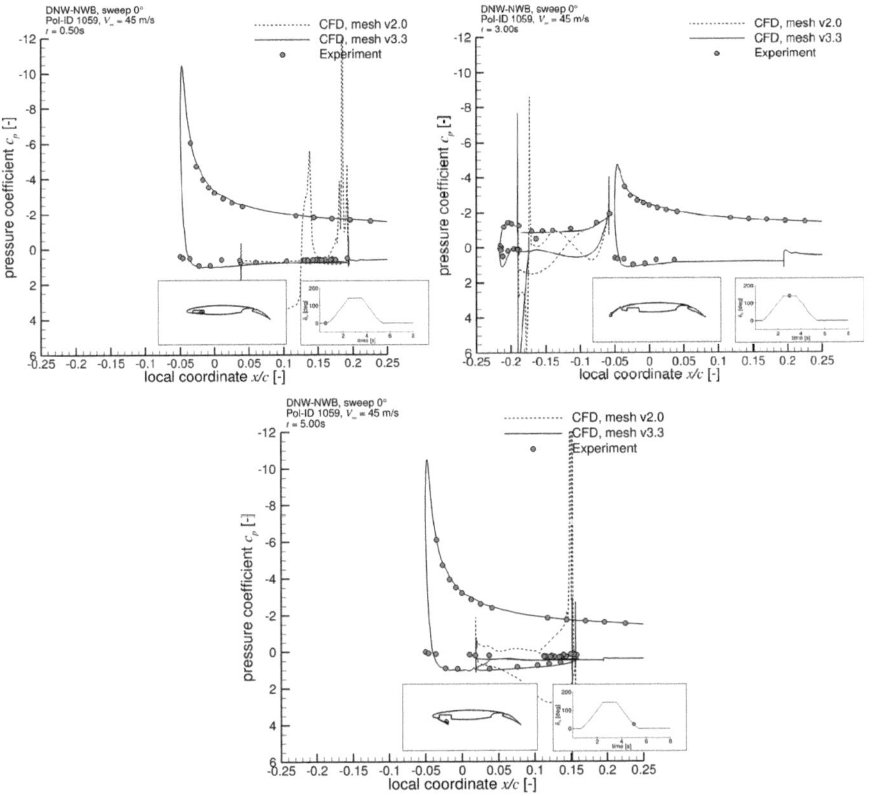

Fig. 9.8 Difference of pressure distributions throughout the Krueger movement path for different mesh strategies

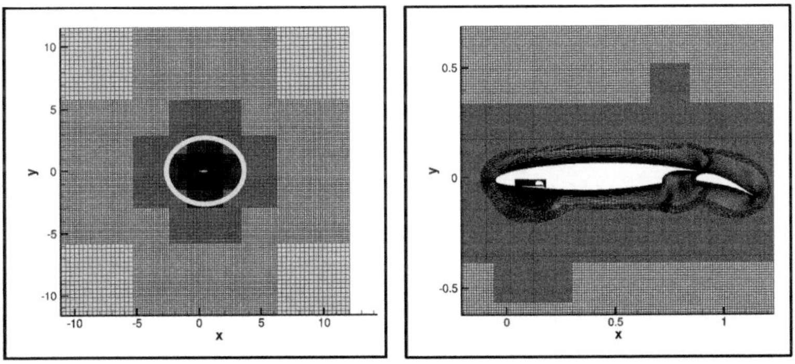

Fig. 9.9 ONERA meshing strategy based on structured meshes embedded in Cartesian background mesh

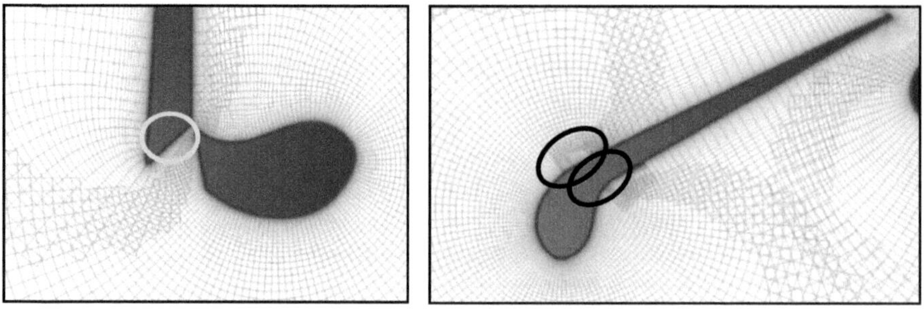

Fig. 9.10 Refined gap between the Bullnose and the Krueger panel

panel is fully deployed. Such refinement is necessary as it directly affects the CFD results. The refined grids were subjected to CFD analysis using the Chimera technique [22].

9.4.3 NLR Meshing Strategy

NLR also adopted structured meshes for individual components, which are in-turn embedded into a structured global mesh (Fig. 9.11). The figure on the left shows the dominant cells of the mesh, and the subordinate overlapping cells are highlighted on the right.

As a demonstration of the Chimera method, the DLR-F15-LLE multi-element airfoil with fully deflected Krueger device is considered. The dominant field cells are shown in Fig. 9.11. NLR meshing strategy based on solely structured multi-block meshes left, as determined by the Chimera method using the smallest wall distance as main criterion [23]. The other, subordinate cell types are depicted in Fig. 9.11

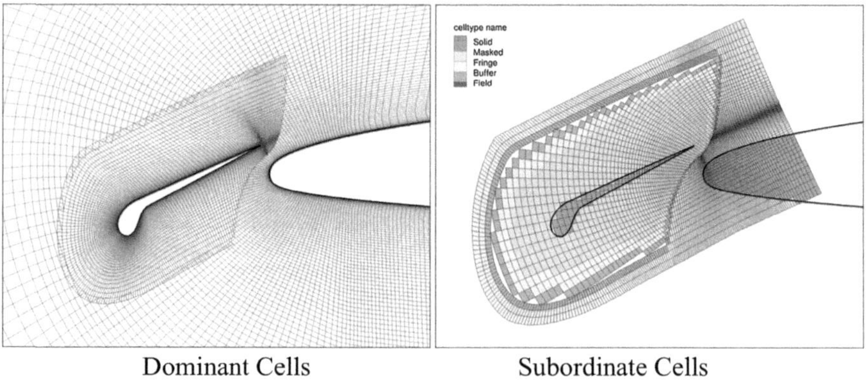

Fig. 9.11 NLR meshing strategy based on solely structured multi-block meshes

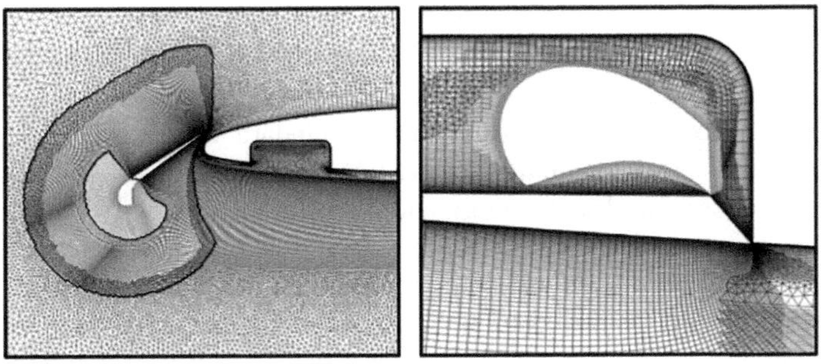

Fig. 9.12 VZLU meshing strategy based on hybrid unstructured meshes with embedded structured blocks around individual components

right, showing one layer of buffer cells and two layers of fringe cells that encircle the dominant field cells.

9.4.4 VZLU Meshing Strategy

VZLU implements cascade meshes, with a combination of structured and unstructured meshes (Fig. 9.12). The individual components, Krueger panel and bullnose, are meshed with structured quad grids. Whereas, the global grid consists of hybrid grid with refined structured grid to capture the wake of the Krueger, as represented in the image below. The hole-cutting algorithm for chimera method implementation is in this case based on wall distance, augmented for multiple parts [24]. The overlapped region needs to be detached from the solid walls, so a small gap is required between the parts. The most challenging regions are: Krueger flap in extended position (small gap between the Krueger panel and the Bull nose), folding of Bull nose (small gap between the Krueger panel and the Bull nose edge) and the fully retracted position (a gap between Krueger panel and the wing cavity). As the data communication occurs only between adjacent grid levels (main wing with Krueger panel, Krueger panel with the Bull nose), slightly higher overlaps may be required to achieve desired results.

9.5 CFD Simulations

The meshes generated through various approaches were then subjected to CFD simulations using individual CFD tools. In order to capture the very large deflection and movement of the Krueger out of scope for any other technique, the overset technique or the Chimera method was implemented. The method efficiently essentially helps establish the interaction between two bodies moving with respect to each other and the numerical simulations can thus cover the flow physics accurately. Apart from the standard RANS and uRANS approaches for the simulation, the scale resolving RANS/LES method was also tested for such a dynamic problem involving large scale separations [24]. To widen the spectrum, the less standard 'Immersed Boundary Layer (IBM)' and 'Lattice Boltzmann Methods (LBM)' [25] were investigated. Due to the underlying Cartesian mesh technology and the solution adaptivity it is expected to also provide computational time benefits. The results obtained were used to carry out a number of observations and comparisons with the experimental results from the three wind-tunnels by each partner. Some of the results are shared in this section.

The image on the left of Fig. 9.13 represents the DNW-NWB model in part-span configuration with a wing sweep of 23°. This particular simulation was performed at 45 m/s wind speed with a 1 s hold and 2 s Krueger deflection time. Whereas the image on the right of Fig. 9.13 showcases the straight wing in ONERA-L1 wind tunnel in full span configuration. This simulation was performed at a wind speed of 30 m/s with 1 s hold and 1 s deflection time. In the CL graph in the top-right corner, the variation of lift-coefficient can be observed along with the Krueger panel deflection. The bottom-right graph shows the pressure distribution at various stages for individual components.

As mentioned, the simulations were performed at various extension and retraction speeds, e.g. 1–4 s for the ONERA experiment. The following graphs in Fig. 9.14 show pressure distribution for 1 s cycle throughout the movement path during dynamic motion. The solid line represents the Cp distribution during retraction, and the dashed one during extension. It is interesting to note that the two movements follow different paths during motion. The passage of the Krueger flap induces loss of lift. There is a delay between the Krueger panel movement and the reaction of the pressure coefficient values on the main wing (in red). In the case of ONERA-L1 experiment

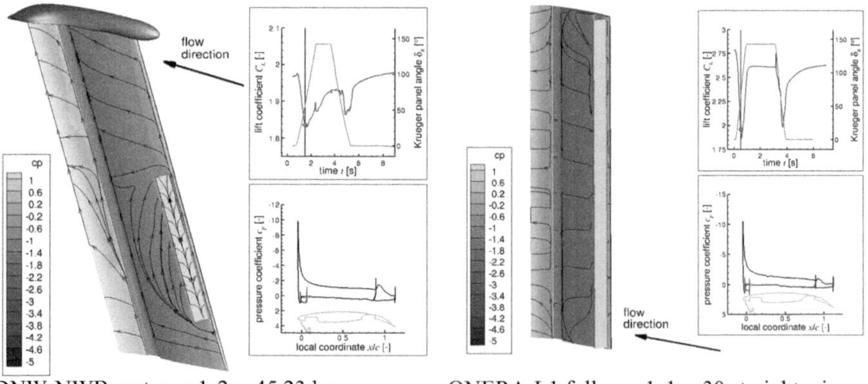

DNW-NWB partspan 1s2s u45 23deg sweep ONERA-L1 fullspan 1s1s u30 straight wing

Fig. 9.13 Examples of the results of CFD simulations: (left) DNW-NWB set-up with part span flap at 23° sweep; (right) ONERA-L1 set-up with full span flap

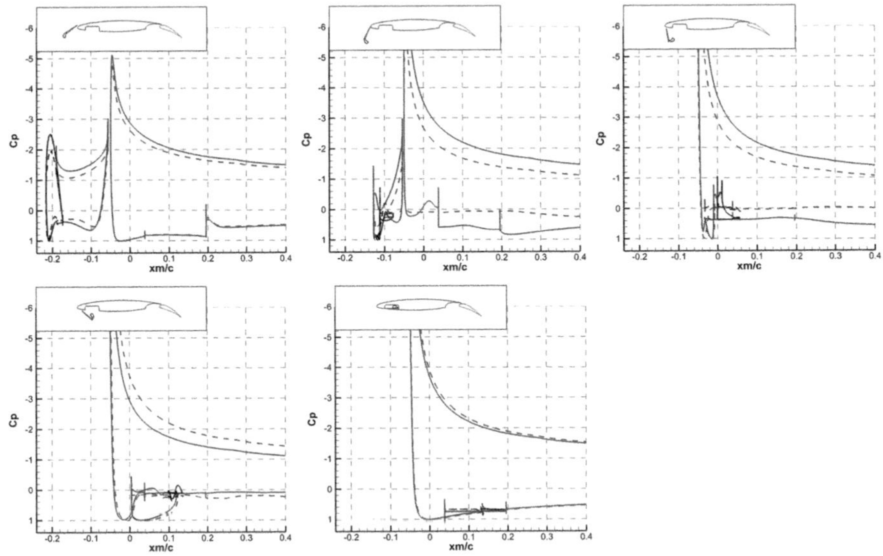

Fig. 9.14 Snapshots of the Cp distributions for various positions of the Krueger flap during 1 s dynamic movement cycle. Solid lines represent retraction phase, dashed lines represent extension of the Krueger flap

with full span Krueger flap motion introduces flow blockage near the pressure side of the leading edge, leading to significant loss of lift on the main wing.

9.6 Validation of CFD Results

Here we first consider the static deployment cases of the Krueger flap, i.e. retracted, fully deflected and partially deployed as depicted in Fig. 9.15 for the swept wing configuration in the DNW-NWB set-up.

The simulations in static cases consider the model in a free air environment where the experimental pressure distribution of the retracted configuration is adopted as a datum solution. The matching incidence angle reads $\alpha = 9°$ for the wind tunnel condition $\alpha = 8°$ at an onset velocity $V = 48.9$ m/s, which transforms to a wing normal wind velocity of 45 m/s to obtain comparable results from all the wind tunnels. The wing normal wind velocities were kept constant by adjusting the onset velocities in case of swept wing conditions wherever applicable.

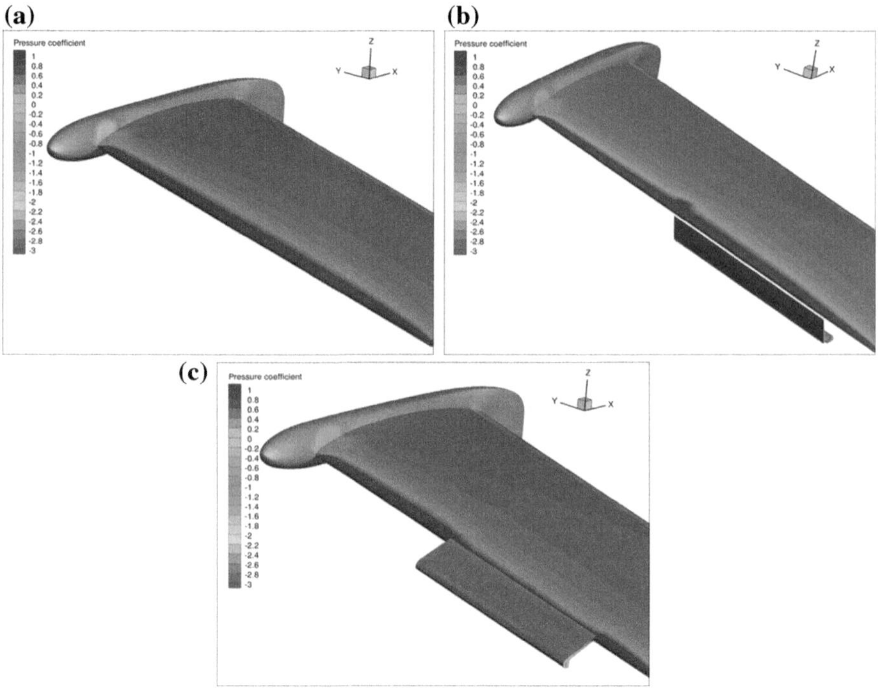

Fig. 9.15 Krueger configuration for DNW-NWB set-up: (**a**) retracted; (**b**) partly deployed; (**c**) fully extended

From flow physics point of view, the cross comparison of the results for the fully deflected and retracted configurations reveal the reduction of the suction level at the leading-edge of the main-wing due to the presence of the Krueger device in front of the wing (Fig. 9.16). Furthermore, the cross comparison of the results for the partially deflected and retracted configurations depict the reduction of stagnation pressure at the lower-surface of the main-wing due to a separation region downstream of the Krueger device. The good agreement of the numerical and experimental surface pressure distributions indicates a valid capturing of the flow physics and prediction of the static loads on the various components of the configuration.

Second, we consider a two-component Krueger device for a dynamic deployment case where the measured time-history of the deployment cycle is adopted, which is characterized by a variable rotational velocity. In case of a nominal deployment time of $T = 2$ s, the comparison of the numerical and experimental results is presented in the figure below (Fig. 9.17). This concerns the sectional pressure distribution, normal to the leading edge at the center span station of the part-span Krueger. Time instances for a deflection angle of $\Theta = 60°–140°$ have been selected. The Krueger leading-edge device moves in front of the wing and modifies the leading-edge pressure distribution.

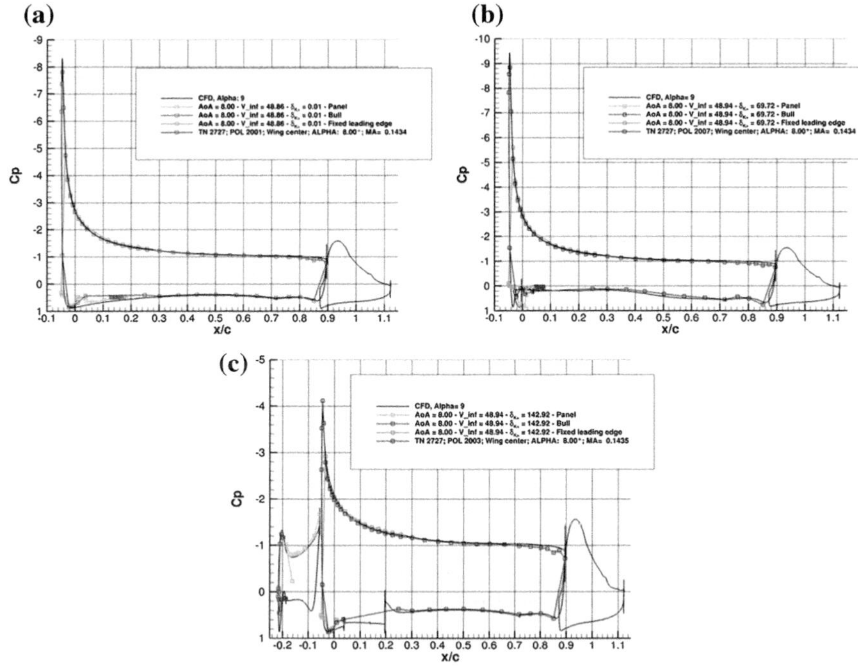

Fig. 9.16 Comparison of simulated and experimental pressure distribution at different static Krueger positions for the swept DNW-NWB set-up: (**a**) retracted; (**b**) partly deployed; (**c**) fully extended

The selected time instances show that the time-variation to the leading-edge pressure distribution is captured (and the time related loss of sectional lift). Furthermore, the sectional pressure distribution on the Krueger elements, i.e. the Krueger panel and the Bullnose, show a good agreement for the experimental and numerical results. This concerns the upper and lower pressure distribution and implies that the time variation of the loads on the Krueger leading-edge device is accurately predicted.

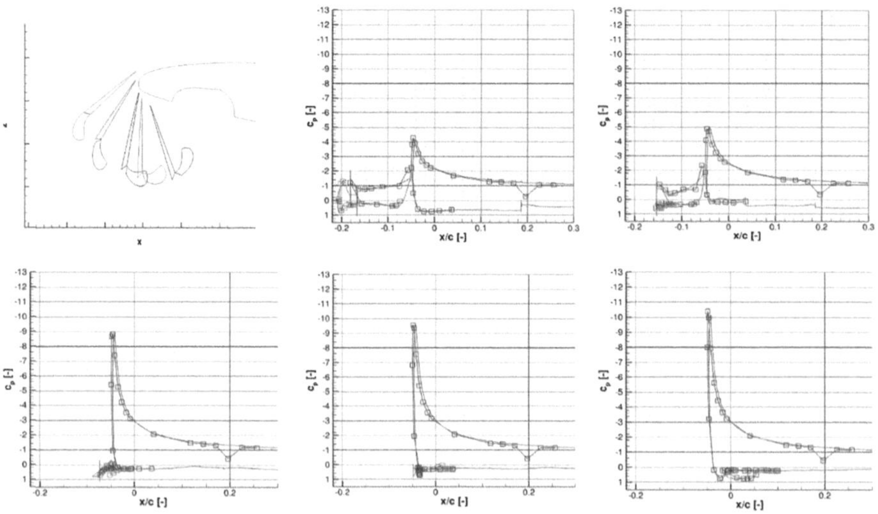

Fig. 9.17 Comparison of simulated and experimental pressure distributions at specific time instances during dynamic motion of the Krueger from data of DNW-LLF set-up: Krueger deflection angles top left to bottom right: 60°; 80°; 100°; 120°; 140°

9.7 Conclusion

The Aim of this paper is to present and discuss the challenges encountered as a result of implementing the Chimera technique with the target of capturing the unsteady flow behaviour around the Krueger deploying from the pressure side of the wing. Leading up to this paper, 9 project partners executed 5 CFD approaches including Chimera to compare results and establish validity. Overall 16 Static and 44 Dynamic cases were simulated varying in wind velocity, wing sweeps and deployment speeds. In total, 5 wind tunnel experiments were carried out with two model configurations at three different wind tunnel facilities. The campaign was carried out respecting the COVID-19 regulations, which meant no onsite model handling support and complete remote troubleshooting.

As mentioned above, the Chimera technique was chosen as a meshing strategy after considering multiple methods available out there. The Chimera method was found to be the most appropriate for the application in consideration. The CFD results obtained from this method have been compared with results obtained from the wind-tunnel campaigns in three different tunnels using two different models for validation. As a result of this study, it can be said that the Chimera method is capable to accurately predict the transient aerodynamic characteristics of a high-lift configuration as well as the changing device loads on the components of a deploying Krueger device. It has been observed that the results agree with the experimental data. Yet, significant efforts are required to match the experimental motion path, ensure availability of detailed experimental data and carry out thorough comparisons between the CFD and wind tunnel results.

Acknowledgements The project leading to this publication has received funding from the European Union's Horizon 2020 Research and Innovation Framework Programme under grant agreement No. 769088. Furthermore, special thanks are attributed to the entire contributing staff not explicitly named here, mainly all the colleagues contributing to the project—in simulations or experiment, all the colleagues at the testing facilities and all the colleagues running the highly innovative measurement technologies.

References

1. Krüger W (1943) Über eine neue Möglichkeit der Steigerung des Höchstauftriebes von Hochgeschwindigkeitsprofilen, AVA-Bericht 43/W/64
2. Rudolph PKC (1996) High-lift systems on commercial subsonic airliners, NASA-CR-4746
3. Boeing Commercial Airplane Group (1999) High Reynolds number hybrid laminar flow control (HLFC) flight experiment—II. Aerodynamic design, NASA CR-1999-209324
4. Steger JL, Dougherty FC, Benek JA (1983) A Chimera grid scheme. In: Ghia KN, Ghia U (eds) Advances in grid generation. ASME FED, vol 5
5. Meakin RL (1993) Moving body overset grid methods for complete aircraft tiltrotor simulations. AIAA paper 1993-3350
6. Blanc F, Roux F-X, Jouhaud J-C, Boussuge J-F (2009) Numerical methods for control surfaces aerodynamics with flexibility effects. In: International forum on aeroelasticity and structural dynamics
7. Rogers S, Cao H, Su T (1998) Grid generation for complex high-lift configurations, AIAA 1998-3011. In: 29th AIAA, fluid dynamics conference
8. Cao H, Su T, Rogers S (1998) Navier-Stokes analyses of a 747 high lift configuration, AIAA 1998-2623. In: 16th AIAA applied aerodynamics conference
9. Slotnick J, An M, Mysko S, Yeh D, Roth K, Rogers S, Nash S, Baker M (2000) Navier-Stokes analysis of a high wing transport high-lift configuration with externally blown flaps, AIAA 2000-4219. In: 18th applied aerodynamics conference
10. Rogers S, Roth K, Nash S, Baker M, Slotnick J, Whitlock M, Cao H (2000) Advances in overset CFD processes applied to subsonic high-lift aircraft, AIAA 2000-4216. In: 18th applied aerodynamics conference
11. Rumsey CL, Slotnick JP, Woeber CD (2023) Fourth high-lift prediction/third geometry and mesh generation workshops: overview and summary. J Aircr 60(4):1160–1177
12. Ollivier-Gooch CF, Coder JG (2023) Fourth AIAA high-lift prediction workshop: fixed-grid Reynolds-averaged Navier–Stokes summary. J Aircr. https://doi.org/10.2514/1.C037184
13. Renard N, Wild J (2012) Unsteady simulation of the flow during the deployment of high-lift systems. ECCOMAS 2012 (Vienna, Austria), Paper-ID 2833. In: Eberhardsteiner J et al (eds) Proceedings of the 6th European congress on computational methods in applied sciences and engineering. TU Wien
14. Iuliano E, Quagliarella D, Wild J (2022) Krueger high-lift system design optimization. In: The 8th European congress on computational methods in applied sciences and engineering—ECCOMAS congress 2022, Paper-ID 2241
15. Wild J et al (2022) A validation program for dynamic high-lift system aerodynamics. In: The 8th European congress on computational methods in applied sciences and engineering—ECCOMAS congress 2022, Paper-ID 2243. https://doi.org/10.23967/eccomas.2022.033

16. Lintermann A (2021) Computational meshing for CFD simulations, clinical and biomedical engineering in the human nose—a computational fluid dynamics approach. Springer Nature Singapore Pte Ltd., pp 85–115. https://doi.org/10.1007/978-981-15-6716-2_6
17. Thompson J, Soni B, Weatherill N (1998) Handbook of grid generation. CRC Press. ISBN 9780849326875
18. Sevilla R, Perotto S, Morgan K (2022) Mesh generation and adaptation: cutting-edge techniques. Springer International Publishing AG. ISBN: 9783030925390
19. Aftosmis MJ (1997) Solution adaptive Cartesian grid methods for aerodynamic flows with complex geometries: von Karman Institute for Fluid Dynamics. In: 28th computational fluid dynamics. Lecture series 1997-02
20. Spiering F (2016) Development of a fully automatic Chimera hole cutting procedure in the DLR TAU code. In: New results in numerical and experimental fluid mechanics X. Notes on numerical fluid mechanics and multidisciplinary design, vol 132. Springer International Publishing Switzerland, pp 585–595. https://doi.org/10.1007/978-3-319-27279-5. ISBN 978-3-319-27278-8
21. Péron S, Benoit C (2013) Automatic off-body overset adaptive Cartesian mesh method based on an octree approach. J Comput Phys 232
22. Benoit C, Jeanfaivre G, Canonne E (2005) Synthesis of ONERA Chimera method developed in the frame of CHANCE program. In: 31st European rotorcraft forum, pp 1–8
23. Boerstoel JW, Kassies A, Kok JC, Spekreijse SP (1996) ENFLOW, a full-functionality system of CFD codes for industrial Euler/Navier–Stokes flow computations, NLR TP 96286U. NLR, Amsterdam
24. Wallin S, Montecchia M, Eliasson P, Prachař A (2022) Scale-resolved simulations of the deployment and retraction of a Krueger high-lift device. In: The 8th European congress on computational methods in applied sciences and engineering—ECCOMAS congress 2022, Paper-ID 2245
25. Ponsin J, Lozano C (2022) Lattice Boltzmann simulation of a deploying Krueger device. In: The 8th European congress on computational methods in applied sciences and engineering—ECCOMAS congress 2022, Paper-ID 2213. https://doi.org/10.23967/eccomas.2022.163

Chapter 10
A Validation Program for Dynamic High-Lift System Aerodynamics

Jochen Wild, Henning Strüber, Frédéric Moens, Bart van Rooijen, and Hans Maseland

Abstract The feasibility of laminar flow control technology for future wing is bound to the development of a leading edge high-lift system that complies with the requirements on smooth surfaces to enable maintaining the laminar boundary layer flow, such as a Krueger flap. The European H2020 project UHURA, running from September 2018 to August 2022, has been focusing on the unsteady flow behavior around such high-lift system. It first time delivered a deeper understanding of critical flow features at this type of high-lift device during its deployment and retraction together with a validated numerical procedure for its simulation. UHURA performed detailed experimental measurements in several wind tunnels to obtain a unique data set for validation purposes of Computational Fluid Dynamics (CFD) software, including detailed flow measurements by Particle Image Velocimetry (PIV) and other optical measurement technologies. Advanced CFD methods promising significant improvements in the design lead time have been applied and validated against this database to obtain efficient and reliable prediction methods for design.

Keywords High-lift system · Aerodynamics · Krueger flap · Laminar technology wings

---

J. Wild (✉)
German Aerospace Center (DLR), Lilienthalplatz 7, 38108 Braunschweig, Germany
e-mail: jochen.wild@dlr.de

H. Strüber
Airbus Operations GmbH, Bremen, Germany
e-mail: henning.strueber@airbus.com

F. Moens
French Aerospace Lab (ONERA), Meudon, France
e-mail: frederic.moens@onera.fr

B. van Rooijen
German-Dutch Wind Tunnels (DNW), Marknesse, The Netherlands
e-mail: bart.van.rooijen@dnw.aero

H. Maseland
Royal Netherlands Aerospace Centre (NLR), Amsterdam, The Netherlands
e-mail: hans.maseland@nlr.nl

T. Tuovinen et al. (eds.), *Advanced Computational Methods and Design for Greener Aviation*, Computational Methods in Applied Sciences 59,
https://doi.org/10.1007/978-3-031-61109-4_10

10.1 Introduction

Laminar wing technology is expected to be the most significant aerodynamic contribution to aircraft drag reduction and fuel savings of next generation transport aircraft. Nevertheless, this assumption is based on some prerequisites: (i) it is assumed that the low speed performance in high-lift configuration is retained, thereby avoiding an increase in wing area to achieve the required approach speed; (ii) the high-lift system at the leading edge doesn't disturb the surface quality needed for laminar flow at least at the upper side, which prevents classical slat devices to be installed; and (iii) the wing surface stays clean from further disturbances introduced by insects, dust or dirt during ground and low altitude operations preventing laminar flow to develop.

For such aircrafts Krueger flaps are currently the most favored high-lift system concept at the leading edge of a transport aircraft wing. The Krueger flap originates back to the 1940s when Werner Krueger reported on a new high-lift device called the "nose split flap" [1]. The first Boeing jet airliners B707, B727, B737-100 and B747 all used Krueger flaps as their leading edge device [2].

Especially the vented Krueger flap with or without folding bull-nose is most likely the high-lift system enabling laminar wing technology. Aside the high-lift performance, it serves as a device for shielding the wing against contamination by insects or other pollutants during low altitude flight. Previous studies already showed the feasibility in terms of aerodynamics and wing integration [3].

A Krueger flap deploys from the lower side of the wing and moves around the leading-edge until it reaches a position in front and slightly above the wing leading edge. Figure 10.1 depicts the Krueger flap developed in the DeSiReH project [4] at several positions during this deployment. Especially the positions when the Krueger flap is perpendicular to the flow bears possible risks during operation. When the Krueger flap is passing the leading-edge it may shield the wing from the flow, thus causing a severe loss in lift. It is therefore required not to move all Krueger flaps along the span at the same time. Even older aircraft having sealed Krueger flaps operated in sequential mode — e.g. the Boeing B707 — or in groups like the Boeing B747.

In any case, the passage around the leading edge poses an additional criticality. For this reason, additional challenges arise that should be addressed during the design. The limited overall deployment time requires an increase of the deployment speed. The lower number of flaps is operated at the same time the higher is the required angular speed to deflect the full system within an acceptable time of about 20–30 s. At such high rotational speeds, unsteady effects on the flow start to get likely. In order to take these effects into account during design, it is necessary to validate the simulation methods in this flow regime. Since the movement of the Krueger flap significantly differs from other unsteady flow problems by the amount of movement and the characteristic speed, a new validation approach is needed to get confidence into the simulation methods.

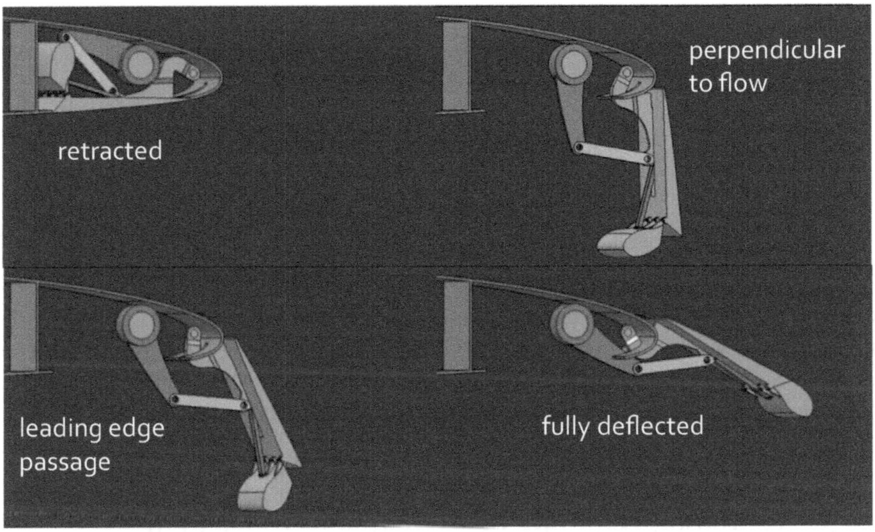

Fig. 10.1 Vented folding bull-nose Krueger flap at specific positions on the deployment path

10.2 Project Scope

The UHURA project (UHURA = **U**nsteady **H**igh-Lift—**U**nsteady **RA**NS Validation) is a Research Innovation Activity (RIA) funded within the Horizon 2020 Programme of the European Commission. It aims at validation of unsteady flow simulation methods for the designated type of unsteady flow problem. After an initial phase of designing the appropriate flow problem, the project in parallel matures methods for calculating the flow by different means of Computational Fluid Dynamics (CFD) and sets up a series of wind tunnel tests to obtain high-quality validation data.

The project has been structured in three phases (Fig. 10.2). The first phase concentrated on the specification of the investigated specimen. This included the aerodynamic design of a Krueger flap for the designated laminar wing airfoil [5], the definition of the kinematics for moving the Krueger flap from retracted to deflected position as well as establishing further requirements on the dynamics of the deflecting motion, especially the required deflection times. In the second phase, the numerical simulation methods were matured for the simulation of the deflecting Krueger device. This aimed to gain some best practice in terms of mesh generation and solver settings. In parallel, the wind tunnel experiments were setup and performed including modifying the wind tunnel models by incorporating a controlled moveable Krueger device into the leading edge of existing high-lift airfoil models, and maturing and synchronizing the time-resolved measurement techniques to be applied. The final phase was dedi-

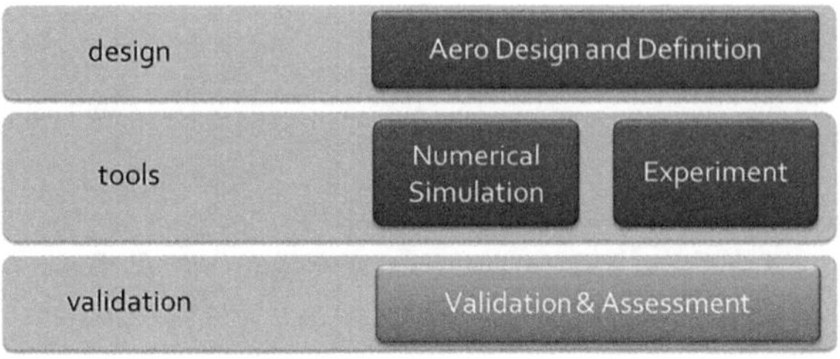

Fig. 10.2 Project structure of the UHURA project

cated to numerically simulate the performed experimental conditions including the measured deployment schedules of the Krueger flap motion and to compare with the detailed experimental database obtained in the wind tunnel experiments for validating the numerical tools.

10.3 Validation Database for Dynamic Krueger Flap Motion

A major objective of the UHURA project was the creation of a database with aerodynamic flow data for the unsteady flow properties of a fast-moving Krueger flap high-lift device. By combining the different models and tunnels, further influence factors were evaluated that cannot be easily done with a single entry. By using two models with the same shape at different scale, the Reynolds number effect can be differentiated from the effect of wind speed and Mach number. By comparing full-span and part span Krueger flaps, effects of the limited span during motion and its effect on the dynamic behavior is accessible. Further, the effect of wing sweep is evaluated by the different setups.

Wind tunnel tests were performed in three different wind tunnels with two models to cover a broad range of conditions including variation of Krueger span (full span—part span), wing sweep angle and Reynolds number. The used airfoil geometry is derived from the DLR-F15 high-lift geometry [6] equipped with a modified leading edge of a laminar wing airfoil denoted by the extension -LLE [7] (Fig. 10.3). This leading edge together with the Krueger flap designed in the project was adopted to two specific wind tunnel models. The baseline DLR-F15 wind tunnel model is a 600 mm chord high-lift model with a span of 2.4–2.8 m. This model was used in the ONERA L1 wind tunnel in a 2D wall-to-wall setup (Fig. 10.4 left) and the DNW-NWB wind tunnel as a cantilever wing with wing sweeps of 0° and 23° (Fig. 10.4 middle). The larger variant for swept installation DLR-F15*LS* is a 2:1 upscale with 1.2 m chord and 7 m span and was mounted at a sweep angle of 30° in the DNW-LLF wind tunnel (Fig. 10.4 right). Table 10.1 lists the test conditions of the different wind tunnel tests. With this wind tunnel strategy, the project was able to create a consistent database on the aerodynamics of a moving Krueger device including the effects of Reynolds number and wing sweep.

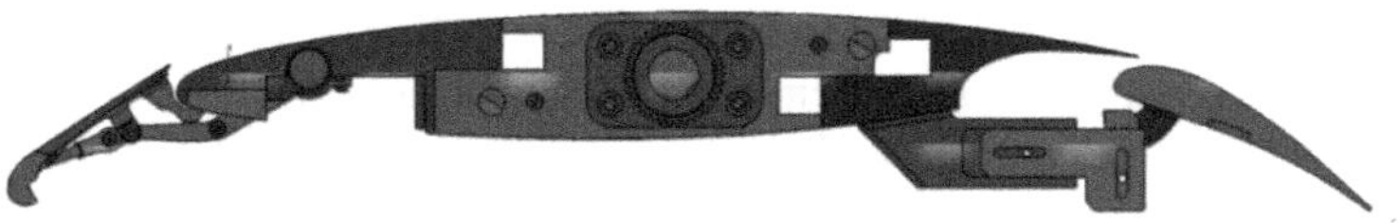

Fig. 10.3 Sectional view of the DLR-F15 wind tunnel model equipped with laminar leading edge (-LLE) and movable Krueger flap device on realistic kinematics

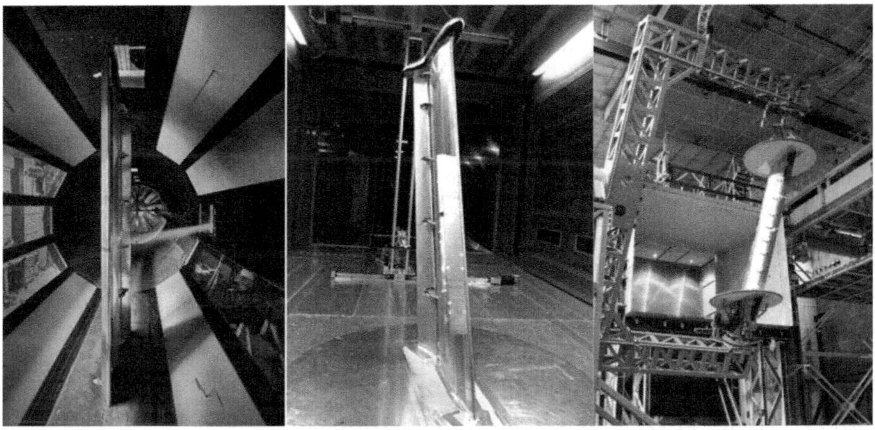

Fig. 10.4 Wind tunnel model installations used in the UHURA project: (left) DLR-F15-LLE mounted wall-to-wall in ONERA L1; (middle) DLR-F15-LLE mounted as swept cantilever wing in DNW-NWB; (right) DLR-F15LS-LLE mounted as swept wing in the open test section of DNW-LLF

Table 10.1 Test conditions of the wind tunnel tests

		ONERA-L1	DNW-NWB	DNW-LLF
General				
Wing sweep φ	(°)	0	{0; 23}	30
Wing normal wind speed $V_N = V_\infty \times \cos(\varphi)$	(m/s)	{30; 45}		
Static cases				
Angle of attack	(°)	[0 … $\alpha(C_{L,max}) + 5$]	[–5 … $\alpha(C_{L,max}) + 5$]	[−2 … 17]
Krueger flap angle	(°)	{0; 37.5; 75; 112.5; 142.9}		
Dynamic cases				
Angle of attack	(°)	6	8	15.7
Deflection time	(s)	{1; 2; 4}	{1; 2; 4}	{2; 4; 8}
Hold time	(s)	{1; 2}	{1; 2; 4}	1
Drive acceleration	(°/s^2)	1000	{165; 300; 500; 1000}	1000

In order to be comparable in all these data sets, it is important to establish a common basis and a common reference. This has been done by prescribing the pressure distribution of a 2D CFD simulation at an angle of attack of 6° as the designated operating point. During wind tunnel tests, first this condition has been established by tuning the angle of attack so that the corresponding target pressure distribution with retracted Krueger flap is obtained. The retracted Krueger flap pressure distribution has been used here to avoid uncertainties due to Krueger deformation und wind load, which are part of the data analysis.

Another aspect for being comparable between the different tunnel entries is the correlation of the dynamic behavior in terms of time scales. Starting with the nominal wind speeds of 30 and 45 m/s for the two-dimension flow, the wind speeds at the swept wing experiments were increased to obtain the same convective time over the wing, which is achieved by retaining the wing normal wind velocity. Doing so, no impact is expected due to Mach number variation and the time scales for the dynamic motion of the Krueger can be kept as is. For the DNW-LLF experiments with the 2:1 upscaled model, the motion time scales are two times those of ONERA-L1 and DNW-NWB experiments in order to achieve the same dimensionless time relations.

10.3.1 Tests at ONERA-L1 Wind Tunnel

In this test setup, the DLR-F15-LLE model wing was mounted vertically in the dodecagonal test section of the ONERA-L1 wind tunnel at ONERA in Lille, France (Fig. 10.5 left). A more detailed description of the test can be found in [8]. The wing incidence was controlled by a motor below the bottom wall, while on the top, the drive unit for the Krueger flap was installed. Just aside the Krueger drive unit, the two CMOS cameras for the 2D2C PIV setup were installed. The cameras recorded two adjacent windows on the pressure side of the wing capturing the motion path of the Krueger flap up to half the wing chord (Fig. 10.5 right).

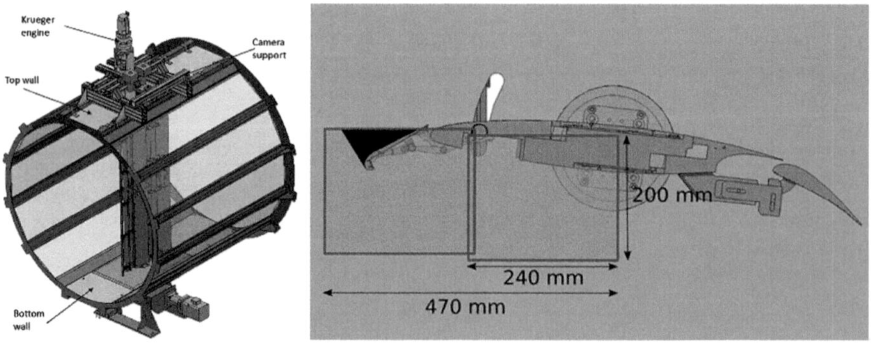

Fig. 10.5 (Left) General arrangement of wind tunnel setup of DLR-F15-LLE in ONERA-L1; (right) PIV image position relative to model cross section

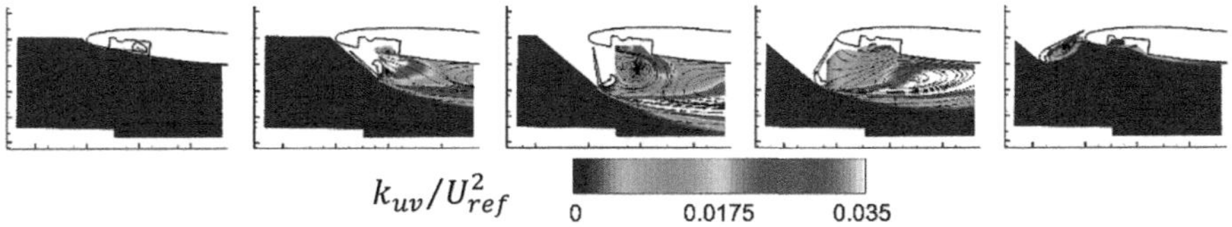

Fig. 10.6 PIV results at static Krueger flap positions left to right: 0°, 37.5°, 75°, 112.5°, 142.9°; turbulent kinetic in-plane energy at $V_\infty = 45$ m/s

Beside the PIV installation, the applied measurement techniques included static pressure measurements by 139 static pressure taps connected to multi-channel pressure scanners, dynamic pressure measurements by 26 Bosch BMP388 MEMS sensors and five Kulite LQ-080 dynamic pressure sensors for reference. The angular position of the drive shaft was transiently recorded from the drive system and an additional hall sensor.

For the two given wind speeds static data points have been acquired by an angle of attack sweep from zero incidence up to 5° beyond maximum lift coefficient. From this the reference angle of attack of 6° was established by comparison to the target pressure distribution for all further measurements with deployed Krueger flap. Second, at this incidence, statically deployed Krueger positions at 5 steps from fully retracted to fully deflected have been measured. For the turbulent statistics up to 2500 snapshots have been sampled by PIV. Figure 10.6 exemplarily shows the time averaged turbulent in-plane kinetic energy content for several fixed Krueger flap positions.

The major time of the test was spent with the recording of unsteady transient flow data by MEMS pressure sensors and PIV. PIV image acquisition was possible at a rate of 5 Hz. In order to get a sufficient angular resolution of the motion, the PIV image acquisition at high deployment speeds was repeated with a phase shift so that at least 20 snapshots per deployment cycle were obtained. This results in an angular resolution of about 7° Krueger deployment. Each phase angle of a motion cycle was recorded at least by 300 snapshots for a statistical evaluation of phase averaged data. Thus, for the shortest deflection time of 1 s, in total 1200 deployment and retraction cycles have been sampled. For the longest deflection time of 4 s the acquisition frequency was high enough to avoid the phase shift and repetition. Although this reduces the number of cycles, the overall measurement time is nevertheless the same.

A sufficiently high accuracy of data acquisition for this high number of runs was only possible with a sophisticated synchronization of all measuring systems, namely the Krueger drive controller, the PIV system, the MEMS system and the transient recording. This has been achieved by TTL handshake signals thus that the PIV acquisition and the motion initiation were synchronized at each deflection cycle. The ONERA-L1 test here also served as a check-out of the architecture of the overall measurement and test system setup and was used in the same or similar way in the other facilities DNW-NWB and DNW-LLF.

10.3.2 Tests at DNW-NWB Wind Tunnel

The test at DNW-NWB concentrated on obtaining pressure and transient data for the variation of the dynamic motion characteristics and to analyze the effect of wing sweep on a single deflected part-span Krueger element. This test was conducted in March/April 2021.

Due to the model installation, only the center part of the Krueger spanning 800 mm of span was actuated. No PIV was acquired in this test, but the deformation of the Krueger was recorded by use of a Stereo Pattern Recognition (SPR) system (see Fig. 10.7) that was targeted for the DNW-LLF entry, too. In addition, for the static conditions the wake of the airfoil model was measured by a traversable wake rake. The wing was installed at two different sweep angles, 0° and 23°, the latter being the mechanical limit of the model support.

It is emphasized again that in order to retain the dynamic characteristics, the wing leading edge normal velocity has been kept constant for tests with and without wing sweep. Thus, the wind speed for the swept wing was increased by a factor of $1/\cos(\varphi)$ compared to the straight wing. The reference angle of attack for all dynamic and static Krueger deflections were set to 8° based on a comparison of the pressure distribution measured in ONERA-L1 and DNW-NWB. This is somehow the angle of attack correction for the cantilever wing installation. Incidentally, this angle correction was the same of 0° and 23° sweep, which has been established separately.

The instantaneous data of the MEMS pressure sensors was phase averaged over 5–8 repetitions and then synchronized with the tunnel and drive data to obtain the dynamic pressure distributions with actual deflected positions as shown in Fig. 10.8. Clearly seen is the overshoot of the suction pressure in the second snapshot where the suction on the wing supersedes even the steady condition. This is attributed to the dynamic acceleration of the flow due to the Krueger pushing the flow around the leading edge.

Fig. 10.7 DLR-F15-LLE model with deflected Krueger flap mounted as cantilever wing with 23° wing sweep in the closed test section of DNW-NWB with marker arrangement for optical position and deformation measurement of the Krueger flap by SPR

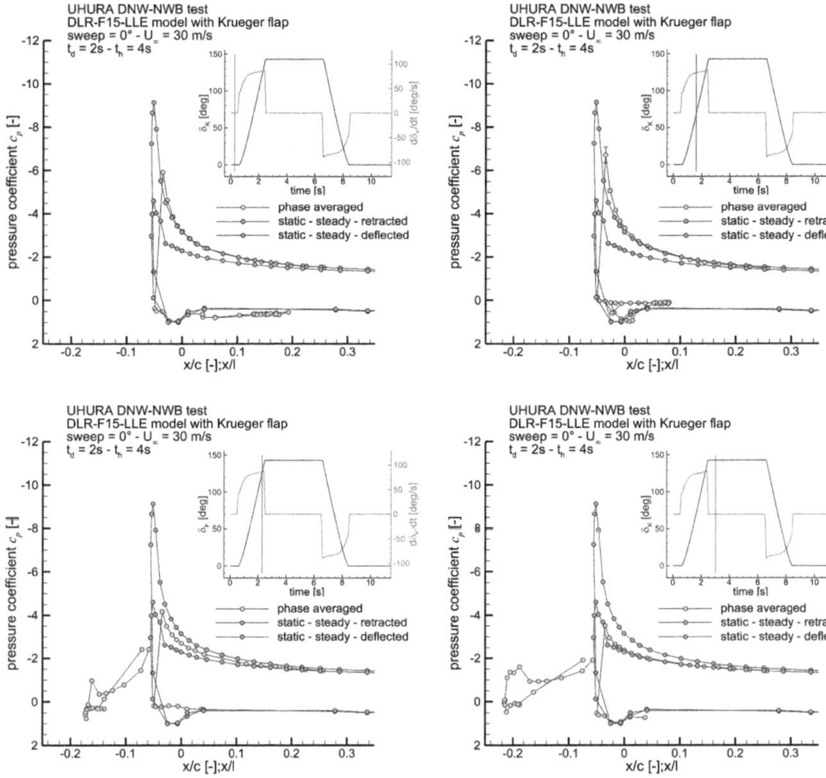

Fig. 10.8 Snapshots of the analyzed phase-averaged pressure distributions from MEMS data synchronized with tunnel data and position measurements

10.3.3 Tests at DNW-LLF Wind Tunnel

For the third test campaign, the larger DLR-F15*LS*-LLE model was mounted in the open $6 \times 8\ \text{m}^2$ test section in the Large Low-speed Facility DNW-LLF in Marknese, The Netherlands. The campaign was conducted in two entries. In April 2021 the data acquisition mainly concentrated on pressure, position and deformation measurements using SPR and discrete accelerometers on the Krueger flap, while the second entry in April 2022 concentrated on acquiring flow field information by PIV.

The larger model DLR-F15*LS*-LLE is a 2:1 upscale of the model used in ONERA L1 and DNW-NWB. Thus, at the same flow condition the double Reynolds number is obtained. The Krueger flap in this model spans 3 m and is like for the cantilever wing in DNW-NWB a part-span arrangement. Due to the larger size, the Krueger is now driven by two synchronized motors from both sides. Figure 10.9 shows on the left side the installed model with the SPR markers in the first entry and on the right side the PIV arrangement with the laser light sheet illuminating the wing in a center section.

Fig. 10.9 Installation of model and optical measurement techniques in the DNW-LLF facility: (left) SPR markers on the wing with deployed part-span Krueger flap; (right) laser system illuminating the PIV section in the center of the part-span Kreuger flap

Due to its size, the DLR-F15*LS*-LLE model allowed for a denser instrumentation for pressure measurements. While the MEMS arrangement has been retained from the small model, the number of static pressure tabs in the center section has twice the number of pressure ports increasing the overall accuracy of the pressure distribution. In addition, two more pressure sections are installed each 1 m to the left and right providing some information on the spanwise variation of the flow along the Krueger span. Further, in this model setup the static PSI pressure scanners were implemented into the wing leading edge providing only short pressure tubing. By this, the low frequency dynamics of the pressure were measured at a 100 Hz sampling rate simultaneously to the MEMS sensors in partly overlapping areas by a second system of the static PSI pressure scanners, thus providing additional reliability in the MEMS technique.

The test conditions in DNW-LLF were listed in Table 10.1 (right column). Like for the previous campaigns, the wind speed was selected to achieve a similar wing leading edge normal velocity. The deflection times are doubled in comparison to the previous campaigns accounting for the 2:1 upscale to achieve similar dimensionless time scales. The pressure distribution again has been adjusted to best match the target pressure distribution. Due to the wing sweep, the rotation axis along the wing span instead of perpendicular to the flow, and due to the open test section, in the DNW-LLF the angle of attack was adjusted to 15.7°.

As an example, Fig. 10.10 shows the derived geometric pressure tap positions colored by the actual pressure coefficient for the three pressure wing sections on the wing and the two on the Krueger marked by the size of the circle. Beside the variation of the pressure between retracted and deflected also the spanwise variation is seen to be minimal, which is a result of the relatively high aspect ratio of the 3 m Krueger panel span.

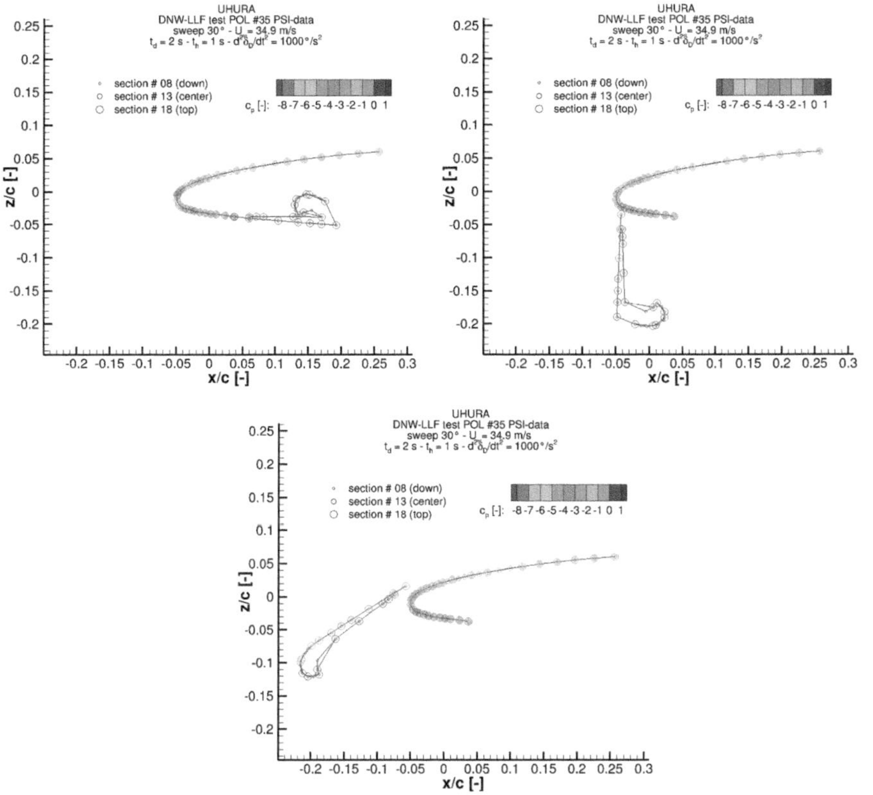

Fig. 10.10 Snapshots of the analyzed phase-averaged pressure distributions from PSI data synchronized with tunnel data and position measurements

In the second campaign four cameras were used to obtain 2D3C stereo PIV images at two acquisition windows similar to the ones in ONERA L1 at a sampling rate of 20 Hz using four laser sources. Due to the higher sampling frequency and the lowered deflection rates, no phase shifted repetition was required. In total 1000 samples have been obtained for each phase-locked position for the shortest deployment time of 2 s and 500 samples for the rest. 600 GB of raw PIV data have been recorded.

10.4 Simulation and Validation Strategy

The final project phase currently under progress is now dealing with the validation of the numerical methods using the described wind tunnel data. Several partners analyze their simulations with different methods of CFD. The methods range from application of state-of-the-art unsteady Reynolds-averaged Navier–Stokes solver (uRANS) based on the Chimera — or overset — approach, scale-resolving methods

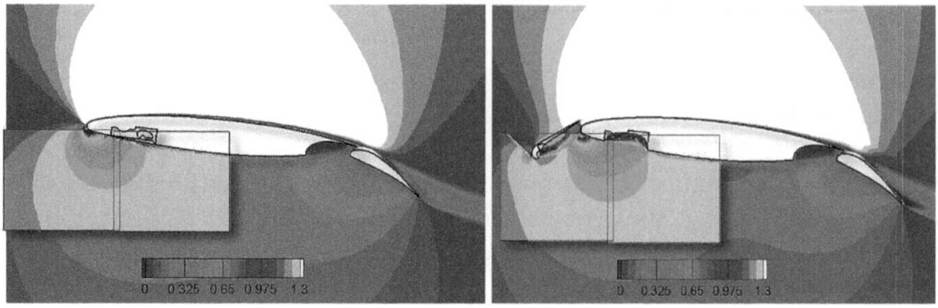

Fig. 10.11 Overlay of time averaged PIV image of static Krueger position with flow field obtained from CFD simulation: (left) retracted Krueger flap; (right) fully deflected Kreuger flap

of blended RANS and Large Eddy Simulation (LES) methods, Immersed Boundary Methods (IBM), and Lattice Boltzmann Methods (LBM). Some partners additionally couple with Computational Structure Mechanics (CSM) methods by Fluid–Structure Interaction (FSI) coupling to account for the model deformation.

It is not aim of this contribution to go into details here. This is done by other contributions to this Special Technology Session. The first summarizes the lessons learned applying the Chimera method [9], the second values the benefit of time resolving methods [10], and the last the experience with applying a completely different simulation approach with Lattice Boltzmann methods [11].

Just as a teaser here for the comparison of experimental and numerical simulation data, Fig. 10.11 presents the PIV images obtained from static measurements with retracted and deployed Krueger flap within the first ONERA L1 wind tunnel test overlaid on the flow fields obtained by the corresponding CFD simulations. This first and rough comparison raises high expectations both on the quality of experimental data and on the maturity of simulation methods to handle the designated flow.

10.5 Conclusion

The project UHURA aimed at validating numerical simulation methods for the aerodynamics of moving high-lift devices. These validated methods shall further on help mitigating critical flow states at innovative high-lift systems, especially Krueger flaps to be implemented at laminar technology wings.

The project created a unique and comprehensive database for such flows by conducting 5 wind tunnel campaigns in three different tunnels using two different models of same shape but different scale. It cannot be emphasized enough that this was a very special journey since the COVID-19 pandemic mostly prevented travelling and thus all the tests performed in a cooperative way had to be conducted with remote support only.

In total, 34 GB of time resolved data have been collected, not including the PIV and SPR images. This database is now ready for supporting validation activities of numerical simulation methods.

Acknowledgements The project leading to this publication has received funding from the European Union's Horizon 2020 Research and Innovation Framework Programme under grant agreement No 769088. Further on, special thanks are attributed to all the contributing staff not explicitly named here, mainly all the colleagues contributing to the project—whether in simulations or experiment, all the colleagues at the testing facilities and all the colleagues running the highly innovative measurement technologies.

References

1. Krüger W (1943) Über eine neue Möglichkeit der Steigerung des Höchstauftriebes von Hochgeschwindigkeitsprofilen, AVA-Bericht 43/W/64
2. Rudolph PKC (1996) High-lift systems on commercial subsonic airliners, NASA-CR-4746
3. Boeing Commercial Airplane Group (1999) High Reynolds number hybrid laminar flow control (HLFC) flight experiment—II. Aerodynamic design, NASA CR-1999-209324
4. Strüber H, Wild J (2014) Aerodynamic design of a high-lift system compatible with a natural laminar flow wing within the DeSiReH project. In: 29th congress of the international council of the aeronautical sciences, Paper-ID ICAS 2014-0300
5. Iuliano E, Quagliarella D, Wild J (2022) Krueger high-lift system design optimization. In: The 8th European congress on computational methods in applied sciences and engineering—ECCOMAS congress 2022, Paper-ID 2241. https://doi.org/10.23967/eccomas.2022.159
6. Wild J (2013) Mach and Reynolds number dependencies of the stall behavior of high-lift wing sections. J Aircr 50(4):1202–1216. https://doi.org/10.2514/1.C032138
7. Ciobaca V, Dandois J (2017) High Reynolds number high-lift airfoil testing with flow control. In: 35th AIAA applied aerodynamics conference, Paper-ID AIAA 2017-3245. https://doi.org/10.2514/6.2017-3245
8. Tanguy G, Monnier JC, Verbeke C, Wild J (2022) Experimental aerodynamic investigation of a Krueger flap device using particle image velocimetry. In: 9th European conference for aeronautics and space sciences (EUCASS), Paper-ID 6067. https://doi.org/10.13009/EUCASS2022-6067
9. Hasabnis A, Maseland H, Moens F, Prachař A, Wild J (2022) Lessons learnt from Chimera method application to a deploying Krueger device. In: The 8th European congress on computational methods in applied sciences and engineering—ECCOMAS congress 2022, Paper-ID 2246. https://doi.org/10.23967/eccomas.2022.167
10. Wallin S, Montecchia M, Eliasson P, Prachař A (2022) Scale-resolved simulations of the deployment and retraction of a Krueger high-lift device. In: The 8th European congress on computational methods in applied sciences and engineering—ECCOMAS congress 2022, Paper-ID 2245
11. Ponsin J, Lozano C (2022) Lattice Boltzmann simulation of a deploying Krueger device. In: The 8th European congress on computational methods in applied sciences and engineering—ECCOMAS congress 2022, Paper-ID 2213. https://doi.org/10.23967/eccomas.2022.163

Chapter 11
Morphing Effects on the Aerodynamic Performances of a Supercritical Wing's Prototype in Transonic Flow Conditions

C. Jimenez-Navarro, J. B. Tô, C. Rouaix, P. Doerffer, A. Marouf, R. El Akoury, and M. Braza

Abstract A detailed parametric study for an Airbus A320 morphing wing, the tRS: Transonic Reduced Scale prototype of the H2020 N° 723402 SMS project [1, 2], has been carried out through numerical simulations at cruise conditions. The present work is a continuation of the studies performed under the SMS project an also in the framework of the H2020 TEAMAERO "Towards Effective Flow Control and Mitigation of Shock Effects in Aeronautical Applications" [3] European research project. The chord's length, 15 cm, was selected to adapt the prototype in the dimensions of the wind tunnel of IMP-PAN in Gdansk, partner of the SMS project. The trailing edge of the profile is actuated by electric piezoactuators whose motion has been modelled numerically using the High-Fidelity CFD solver Navier-Stokes Multi Block (NSMB) code. The simulations have provided more physical insights on the interaction between the shock wave, the shear-layers past the separation and the vortex dynamics in the wake. It was found that trailing edge actuations, by means of vibrations in the order of (100–750) Hz and slight deformations in the order of 0.2 mm, are able to modify the transonic interaction and effectively control the shock's position. A lock-in effect has been obtained between the shock motion and the actuation frequency leading to considerable benefits. A good comparison of the shock's motion between experiments carried out by IMP-PAN institute and the present simulations is obtained. Furthermore, an increase of the lift and the aerodynamic performance (lift-to-drag ratio) has been obtained in the order of 1%, for actuation frequencies 510 and 720 Hz, together with a reduction of the *rms* of the forces, indicating an attenuation of the transonic buffet. Therefore, the present morphing concept is to be considered as a promising way towards disruptive future wing design.

Keywords Morphing effects · Supercritical wing · Transonic flow

---

C. Jimenez-Navarro (✉) · J. B. Tô · C. Rouaix · P. Doerffer · A. Marouf · R. El Akoury · M. Braza
Institut de Mécanique des Fluides de Toulouse (IMFT) UMR 5502 CNRS-INPT-UT3, 2 Allée du Prof. Camille Soula, 31400 Toulouse, France
e-mail: cesar.jimeneznavarro@imft.fr

T. Tuovinen et al. (eds.), *Advanced Computational Methods and Design for Greener Aviation*, Computational Methods in Applied Sciences 59,
https://doi.org/10.1007/978-3-031-61109-4_11

11.1 Introduction

The study of the wake dynamics over supercritical wings in transonic regime have been the object of numerous research studies concerning the Shock Boundary Layer Interaction (SBLI) and the buffet dynamics. The present article focuses on the role of near wake and shear layer unsteadiness to the buffet phenomenon. This, because given previous studies in our group, a strong relation between this unsteadiness and the shock motion was found since the last two decades. The Navier-Stokes simulations by Bouhadji et al. [4] and Bourdet et al. [5] had illustrated the strong dependency of the von Kármán mode on the SBLI and on the buffet development. Szubert et al. [6] demonstrated that suppression of the von Kármán mode through introduction of a splitter plate was able to suppress the buffet instability. By means of appropriate turbulence modelling closures used in the present article, Szubert et al. [7] quantified the interaction among the buffet, the shear layer and the von Kármán instabilities, as well as Simiriotis, 2020 [8].

Based on these studies, the idea of using different morphing concepts in the rear part of a wing was expected to produce significant feedback effects on the overall area of the wing, up to significant upstream distance of the SBLI and drastically modify the forces. This main idea has been fully exploited during the SMS and TEAMAERO EU research projects. The aerodynamic performance of a bio-inspired electroactive morphing concepts, first developed: by the multidisciplinary research team of the Laboratories IMFT and LAPLACE, thanks to the acquired knowledge through the research projects supported by the Foundation STAE—"Sciences et Technologies pour l'Aéronautique et l'Espace": EMMAV- Electroactive Morphing for Micro-Airvehicles, DYNAMORPH—DYNAmic regimes electroactive MORPHing and the platform SMARTWING [9]. These morphing concepts are able to manipulate different classes of turbulent structures which in turn produce optimal interactions with the actuations and create a suitable fluid-structure interaction system, able to increase the aerodynamic performance. The European research project SMS—"Smart Morphing and Sensing for aeronautical configurations" was a continuation of these research efforts and showed promising results in terms of improved aerodynamic performance even in large-scale aeronautical applications [2, 10–12].

In continuation of a first morphing study in cruise conditions by Tô et al. [13], the present article analyses in detail the lock-in mechanism of the shock's motion to the actuation frequency and the evaluation of the averaged and *rms* values of the aerodynamic forces in a wide range of vibration frequencies. The optimal ranges will be defined thanks to both increase of the lift-to-drag ratio and simultaneous minimisation of the *rms* of the forces at a Mach number and incidence provided by AIRBUS, endorser of the SMS project. It is essential to study the influence of buffet dynamics and how to diminish its harmful effects in these conditions since it can lead to undesired dynamic conditions (i.e. dip-flutter instability amplification with negative damping). Previous numerical works have evidenced the significant feedback effects of the near trailing edge wake dynamics in the SBLI and have shown that it is feasible to use the aforementioned morphing technology to produce a lock-in of the buffet's frequency with the actuation frequency and furthermore to reduce the separated region and wake's width [13].

Earlier attempts to gain insight on the effects of flapping on transonic buffet were carried out by Lee [14] and Tijdeman [15] where the airfoil flap was actuated at a frequency in the order of the buffet natural frequency ($f_a \approx 100Hz$). These studies showed that it is possible to control the shock's location by modifying the downstream pressure caused by the flap deflection and therefore they constitute a valuable heritage for flow control techniques research. Recent researches by Raveh [16] and Iovnovich [17] performed numerical studies on large amplitude flapping motions with a frequency up to the double of the buffet natural frequency ($f_a < 2f_b$) and reported the existence of a synchronization of the buffet and the flapping motions. The work presented in this paper goes beyond the state of art by studying small trailing edge deformations at relatively higher frequencies, in the range of 100 to 800 Hz, that are possible thanks to the usage of new generation electroactive materials.

The present study provides a detailed parametric space for the vibration frequency effects on the buffet and on the aerodynamic performances. The paper is structured in the following way: In the first section the numerical methodology is explained. Secondly, the problematic of buffet is analyzed from the numerical point of view and how morphing can play an essential role on buffet attenuation. Third, the results from the parametric study on frequency are presented and discussed and finally the conclusions are gathered in the last section of the paper.

11.2 Numerical Method

The numerical simulations have been undertaken with the Navier-Stokes Multi-Block (NSMB) code which is a compressible solver developed by the NSMB European Consortium [18]. The unsteady compressible regime Navier-Stokes equations in NSMB are formulated in conservative from as follows

$$\frac{\partial}{\partial t}(W) + \frac{\partial}{\partial x}(f - f_v) + \frac{\partial}{\partial y}(g - g_v) + \frac{\partial}{\partial z}(h - h_v) = 0 \qquad (11.1)$$

The state vector is expressed as

$$W = \begin{bmatrix} \rho \\ \rho u \\ \rho v \\ \rho w \\ \rho E \end{bmatrix}$$

The convective fluxes are given by

$$f = \begin{bmatrix} \rho u \\ \rho u^2 + p \\ \rho uv \\ \rho uw \\ u(\rho E + p) \end{bmatrix}, g = \begin{bmatrix} \rho v \\ \rho vu \\ \rho v^2 + p \\ \rho vw \\ v(\rho E + p) \end{bmatrix}, h = \begin{bmatrix} \rho w \\ \rho wu \\ \rho wv \\ \rho w^2 + p \\ w(\rho E + p) \end{bmatrix}$$

The viscous fluxes are expressed as

$$f_v = \begin{bmatrix} 0 \\ \tau_{xx} \\ \tau_{xy} \\ \tau_{xz} \\ (\tau U)_x - q_x \end{bmatrix}, g_v = \begin{bmatrix} 0 \\ \tau_{yx} \\ \tau_{yy} \\ \tau_{yz} \\ (\tau U)_y - q_y \end{bmatrix}, h_v = \begin{bmatrix} 0 \\ \tau_{zx} \\ \tau_{zy} \\ \tau_{zz} \\ (\tau U)_z - q_z \end{bmatrix}$$

where,

$$\tau_{ij} = \mu(\frac{\partial u_j}{\partial x_i} + \frac{\partial u_i}{\partial x_j}) - \frac{2}{3}\mu\frac{\partial u_k}{\partial x_k}\delta_{ij}$$

and q_i is the heat flux due to conduction computed according to Fourier's law.

For solving the unsteady Navier Stokes equations, an implicit dual time-stepping second-order accurate method was used with an outer time step of $dt = 5 \cdot 10^{-6}$ and the number of inner iterations set to $N_{inner} = 150$. These values were determined after a detailed study, to be the most suitable in terms of speed and accuracy by previous numerical studies [13]. Regarding the calculation of fluxes, Roe's upwind scheme was utilized together with the 3rd order Monotonic Upwind Scheme for Conservation Laws (MUSCL) [19]. The statistics were carried out around 1 s of physical time where at least 100 buffet cycles were captured in the static (e.g. non morphing) case.

11.3 Turbulence Modelling

The turbulence model chosen for the study is the Organized Eddy Simulation (OES) [7, 20, 21]. This model allows respecting the dual character of turbulence where the chaotic turbulence is embedded in the coherent structures [20, 22]. By using the ensemble average, this model is able to solve the coherent structures in the frequency spectrum and model the chaotic turbulence by means of a modified modelling for the turbulence stresses to take into account the energy subtracted by the coherent eddies. The OES approach has been largely used in previous research activities and showed remarkable efficiency in the prediction of high Reynolds number and highly detached flows, by providing a physically correct development of the flow instabilities and related coherent structures, even in two dimensions, because the approach is not intrinsically 3D. In the present study, due to main physical characteristics of the coherent structures that can be assessed by a two-dimensional approximation, it was possible to perform a large parametric study in 2D concerning the vibration effects and to detect optimal intervals in which a narrower space of these parameters will be used in a 3D study afterwards.

11.4 Physical Parameters

The A320 wing prototype has been used in this study to continue in the same line as the work carried out by multiple laboratories and teams of researchers, partners of the European project SMS - Smart Morphing and Sensing for aeronautical configurations [2]. It consists of a 15cm A320 wing prototype that is still being used in experiments at IMP-PAN Academy of Science in Gdansk, Poland (Fig. 11.1). The boundary conditions selected for the numerical simulations have been taken in accordance with the wind tunnel experiments flow parameters, using an angle of attack of $\alpha = 1.8°$, a free-stream Mach number of 0.78 with the values of static pressure and static temperature set to 67574 Pa and 261.22 K respectively, leading to a chord-based Reynolds number of $2.06 \cdot 10^6$. The upstream (inlet) freestream turbulence intensity of $3.75 \cdot 10^{-5}\ m^2 s^{-2}$.

11.5 Grid and Numerical Parameters

The domain and the geometry have been discretized in a 2D structured multi-block C type mesh with 64 blocks and 194000 elements. The values of y^+ are of the order of $1 \cdot 10^{-3}$ with around 70 layers in the boundary layer. The domain, geometry and main dimensions are presented in Fig. 11.2.

Since structural morphing needs a moving grid, the cells associated to the moving trailing edge region and the cells in the nearby region are moved by and Arbitrary Lagrangian Eulerian (ALE) [23] method with a Transfinite 2D interpolation (TFI2D). The displacement of the trailing edge is modelled by using a quadratic function given by Eq. 11.2 where the parameters that govern the trailing edge motion are the actuation frequency (f_a), the semi-amplitude (h_p) and the length of the deformed area (L_p). Figure 11.3 is provided to illustrate the numerical implementation.

$$z(x,t) = \left(\frac{2h_p}{3L_p^2}(x - x_0)^2 + \frac{h_p}{3L_p}(x - x_0) \right) sin(\omega t) \tag{11.2}$$

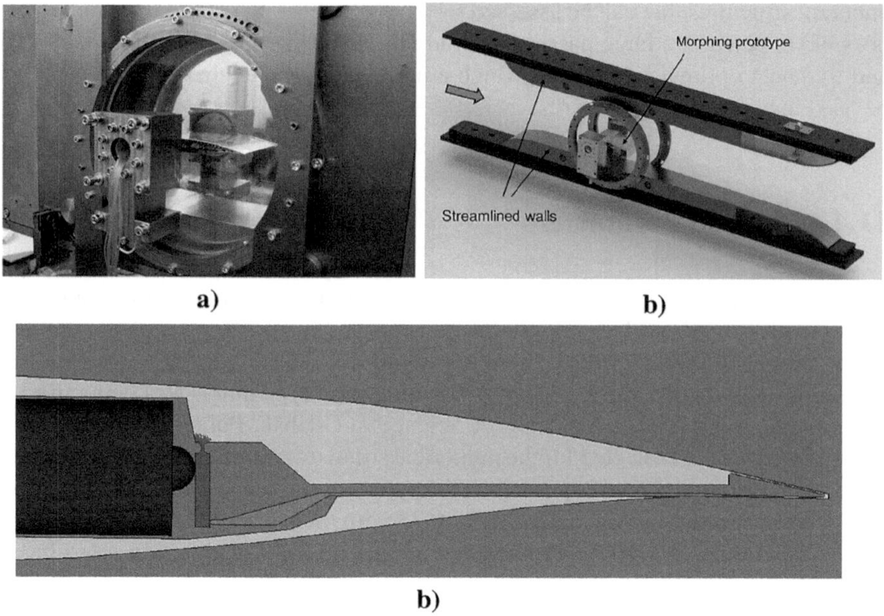

Fig. 11.1 The tRS A320 prototype in **a** the IMP-PAN's transonic wind tunnel, **b** wind tunnel streamlined walls and **c** Internal schematics of tRS prototype constructed by the LAPLACE laboratory with the embedded piezoelectric actuator (blue) and the force transmission chain (red and green)

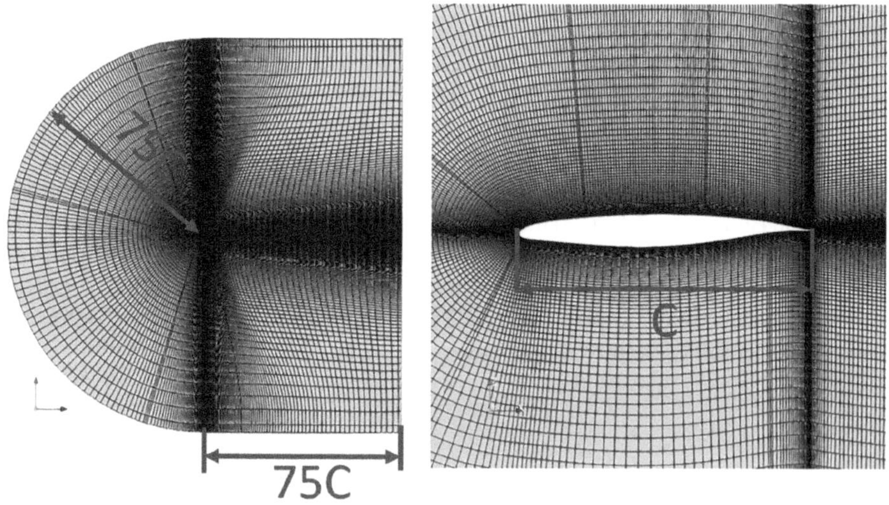

Fig. 11.2 View of the mesh and the computational domain dimensions

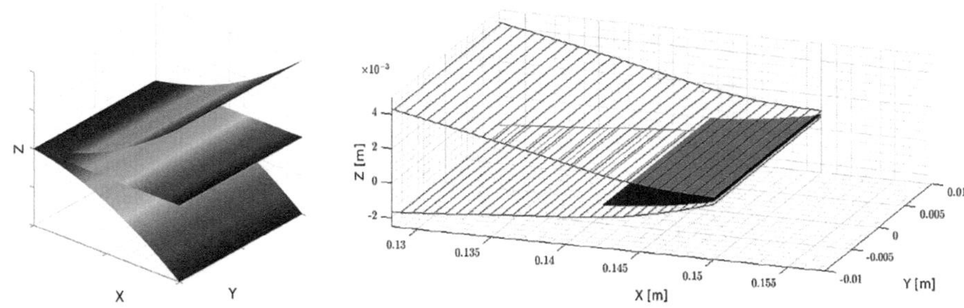

a) Deformation of the near trailing edge area in different stages

b) Instantaneous upward mesh deflection

Fig. 11.3 Schematic description of the displacement by means of the second order polynomial deformation in two and three-dimension

where,

- h_p is the patch semi-amplitude.
- L_p is the patch length.
- x_p is the initial coordinate of the patch in the x direction.
- ω is the angular frequency ($\omega = \frac{2\pi}{f_a}$) obtained as a function of the actuation frequency (f_a) .

11.6 Physical Analysis of the Flow Structures in the Unactuated Case

In the present Mach number regime and angle of incidence, buffet instability may occur. It consists of large amplitude shock oscillations that typically occur on the suction side but it can also take place on the pressure side in case for example of symmetric aerofoil's section. In the present study, investigation of the morphing effects on the buffet will be discussed. Figure 11.4 is a 3D numerical simulation of the flow by using the 2D grid extruded over 7cm along the span. The λ-shock structure can be seen as well as the separated region and different classes of vortices due to the shear layers and wake unsteadiness. These vortices undergo specific spanwise undulations along the span, due to development of secondary instabilities. The larger undulated vortices are the 3D modification of the von Kármán vortices. The Kelvin-Helmholtz and von Kármán instabilities have been obtained. The Kelvin-Helmholtz instability is formed right after the flow separation due to the intense shear between the high-speed flow coming from the supersonic region and the low-speed recirculating flow. On the other hand, the von Kármán instability is created due to the shearing of the velocity profiles much downstream of the trailing edge in the wake which creates the alternating vortices afterwards, due to the presence of two inflexion points in the

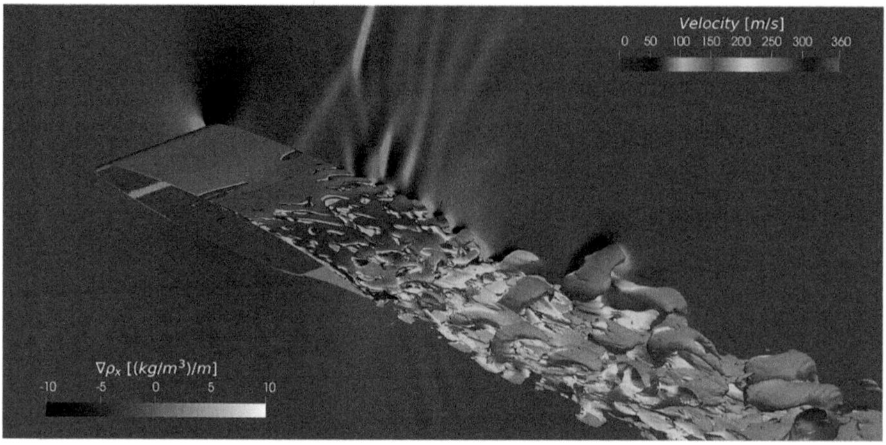

Fig. 11.4 Flow over the A320 TRs prototype: Q-criterion contours coloured with velocity magnitude and density gradients displayed at the symmetry plane

velocity profiles in that area. The development of these instabilities have significant effects on the wing's aerodynamic performance. For instance, they can cause drag and lift fluctuations, which can lead to structural vibrations, noise generation and the stability and control of the aircraft.

The flow conditions considered in the present study lead to a weak form of buffet where the mean position of the shock is $\frac{\overline{x}}{c} = 0.659$ and the root mean square (*rms*) of the shock wave position is $\frac{x_{rms}}{c} = 0.6 \cdot 10^{-3}$. In Fig. 11.5, the contour of time-averaged Mach number is presented which also shows the mean shock's location. To study the buffet dynamics, a Power Spectral Density (PSD) analysis was used to investigate the turbulence spectrum and the influence of the main flow instability modes. Figure 11.6 shows the PSD analysis on the unactuated case with the full signal window and a window containing a small portion of the window's length to clearly visualize the low and high frequency modes in all the spectrum. As shown in Fig. 11.6a, the PSD shows a prominent peak for the most energetic mode in the signal which corresponds to the buffet frequency ($f_b = 106Hz$, $St = 0.063$), which will be taken as natural frequency since it corresponds to the buffet frequency without morphing actuations. In Fig. 11.6b, the high frequency events associated to the shear layer mode (or KH vortices) around 20 kHz are clearly identified. The von Kármán instability is in the order of 6.8 kHz, whose frequency bump is present in the PSD from Fig. 11.6a and was also confirmed by tracking the von Kármán vortices from flow visualizations. Therefore, these two figures highlight that the three main modes that play an important role on the evolution of the forces are the buffet, the von Kármán and Kelvin-Helmholtz modes.

Fig. 11.5 Time-averaged Mach number contours for the unactuated A320 airfoil

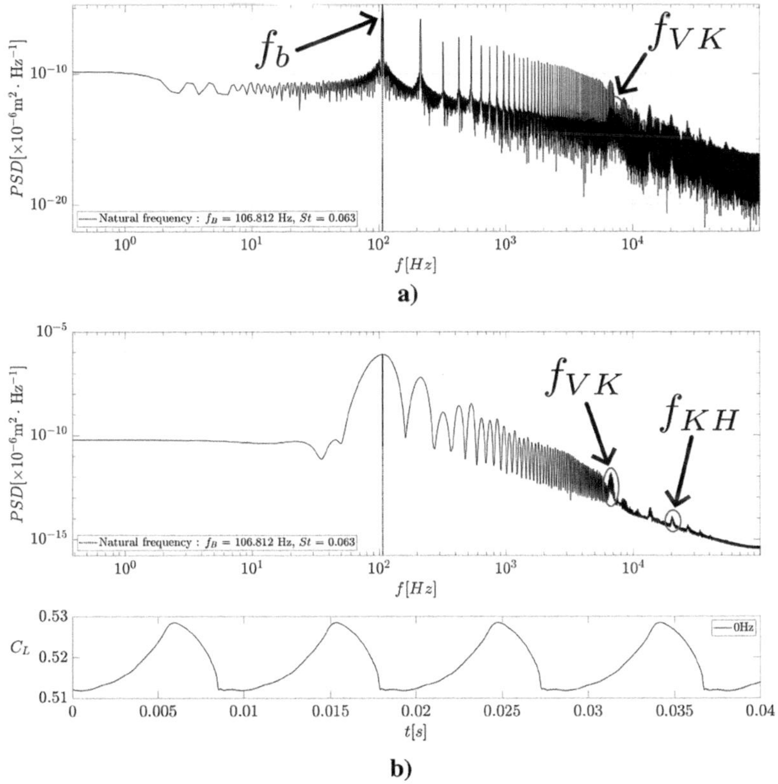

Fig. 11.6 PSD analysis on the lift coefficient showing **a** the buffet frequency and its harmonics and **b** a zoom on the buffet frequency, the von Kármán (VK) and Kelvin Helmoltz (KH) modes

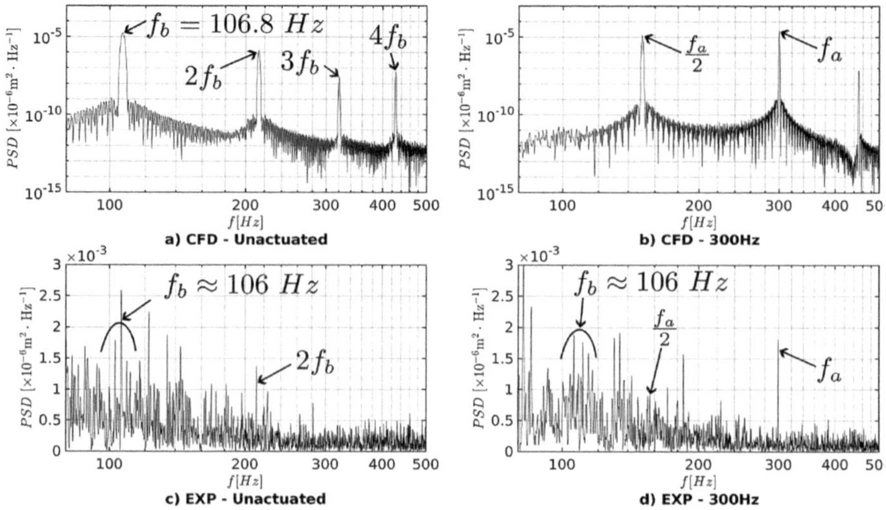

Fig. 11.7 PSDs on **a** C_L signal for the numerical unactuated case, **b** C_L signal for the numerical case actuated at 300 Hz, **c** shock wave's location for the experimental unactuated case and **d** shock wave's location for the experimental case actuated at 300 Hz

The numerical results have been compared with the experimental data provided by IMP-PAN. PSDs were computed for the lift coefficient signal for the numerical simulations and for the shock's location evolution from Schlieren image processing of the experimental data by IMP-PAN. The results are displayed in Fig. 11.7. The shock oscillates with a frequency of $f_b = 106$ Hz for the numerical unactuated case (Fig. 11.7a). In the experimental case, the PSD displays a broadband peak centered in a similar region as in the numerical case (Fig. 11.7c). The formation of this bump is due to the stronger effect of the smaller chaotic turbulence structures than in the 2D simulations. The harmonics higher than $2f_b$ are significantly attenuated in the experiments. The actuation frequency is visible with a clear peak at $f_a = 300$ Hz in both cases of simulations and experiments, in Fig. 11.7b and d. Figure 11.8 is presented to have a better understanding the frequency named here $\frac{f_a}{2}$ which is attributed by the authors to a modulation mode of the shock wave oscillation's amplitude. Since the variation of the lift coefficient signal changes every 2 actuation periods, it is easy to understand that the amplitude changes with a frequency on the order of half of the actuation frequency.

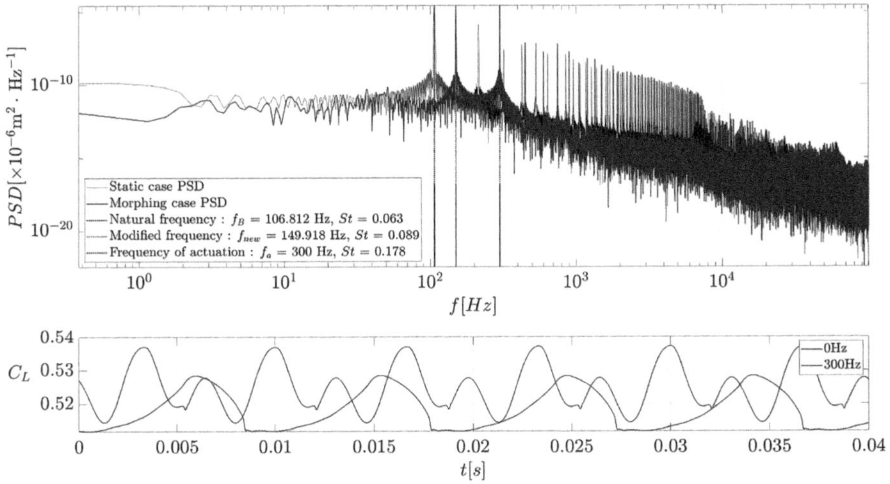

Fig. 11.8 PSD analysis on the lift coefficient for the case $f_a = 300$ Hz compared with the unactuated case $f_a = 0$ Hz

11.7 Parametric Study on the Vibration Frequency with Amplitude Set to $\theta = \pm 1°$

In this section, the results for the parametric study regarding the actuation frequency at constant amplitude of the high frequency low amplitude flapping trailing edge morphing will be presented and discussed. The actuated length was considered as a constant parameter in all the simulations and it value was set to $p_l = 7.5 \cdot 10^{-3} m$ (5% of the airfoil chord) for being a realistic from the manufacturing point of view. The frequencies considered in this study are contained in the range $f_a = [100 - 2000]$ Hz. The force coefficients from the first parametric study with the amplitude set to 0.2 mm ($\theta = \pm 1°$) are summarized in Figs. 11.9 and 11.10.

In general terms, the trailing edge vibration leads to an increase in lift and drag but not always in the same proportion, as shown in the lift to drag ratio evolution. From this set of results interesting areas can be noticed, the range of frequencies $f_a = [250 - 390]$Hz seems to be a promising area of actuation in terms of *rms* reduction with respect to the static case. A synchronization of the shock's motion and the actuation frequency was observed between $f_a = [365 - 510]$ Hz, where a significant increase of C_L/C_D as a function of the vibration frequency is perceived, and at $f_a = 720$ Hz which is the most optimal frequency in terms of lift to drag ratio. This phenomenon will be referred to as Lock-In Effect [13] (Fig. 11.11). The PSD on pressure fluctuations at monitor 4, which is located at the shock's foot is presented as an example of lock-in effect in Fig. 11.12 where a modified buffet frequency appears exactly at the actuation frequency.

To study this phenomenon, the shock's position is tracked during the simulation. This is done by means of density gradients recorded on a line placed across the

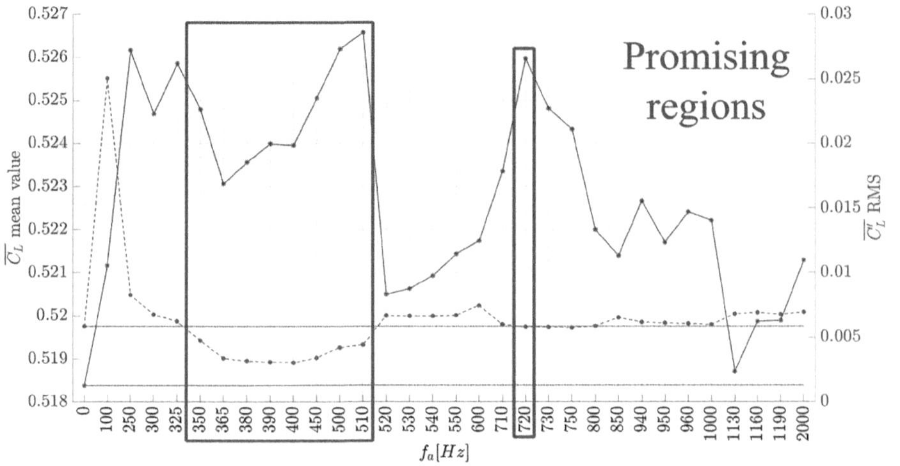

Fig. 11.9 Mean (blue) and *rms* (orange) lift coefficient evolution for the studied frequencies with $\theta = \pm 1°$. Two horizontal lines are presented for each set of results in order to show the relative evolution with respect to the unactuated case ($f_a = 0$Hz)

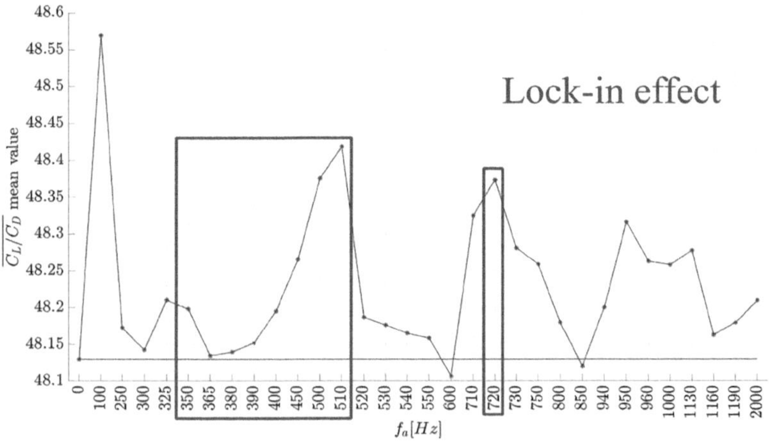

Fig. 11.10 Mean (blue) lift to drag ratio coefficient evolution for the studied frequencies with $\theta = \pm 1°$. A horizontal line is presented to show the relative evolution with respect to the unactuated case ($f_a = 0$Hz)

shock wave, as shown in Figs. 11.13 and 11.14. The position is plotted as a function of time in Fig. 11.15 for different interesting frequencies. By performing the mean and *rms* calculations of these signals, one can obtain the results shown in Fig. 11.15. These results are quite meaningful because they highlight that by acting at the right frequencies it is possible to push the shock downstream and reduce its level of fluctuation, leading to a significant buffet attenuation and lift increase. The frequencies that lead to an optimal buffet attenuation found in this study are between the range

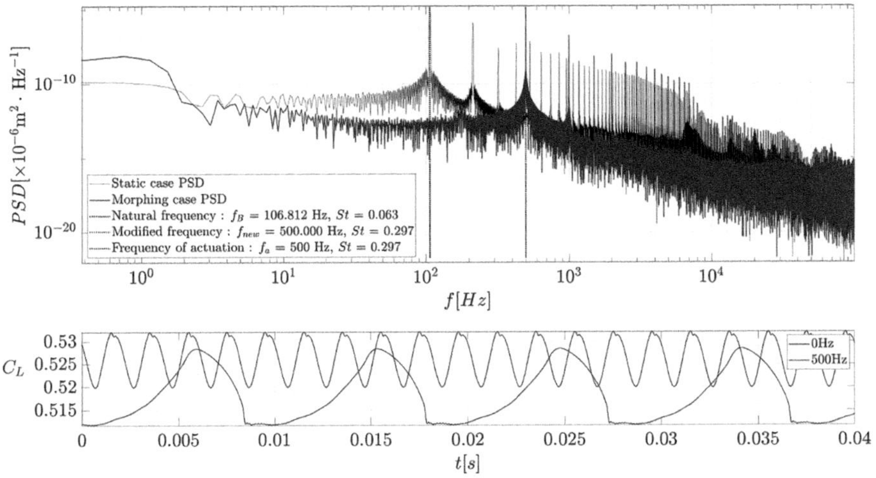

Fig. 11.11 PSD analysis on the lift coefficient for the case $f_a = 500$Hz compared with the unactuated case $f_a = 0$ Hz

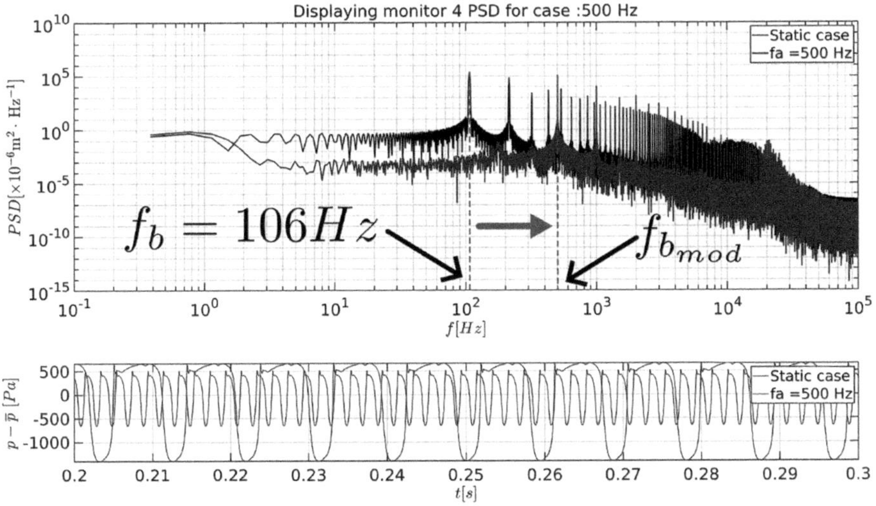

Fig. 11.12 PSD analysis on the pressure fluctuations at the monitor point number 4 for the case $f_a = 500$ Hz compared with the unactuated case $f_a = 0$ Hz

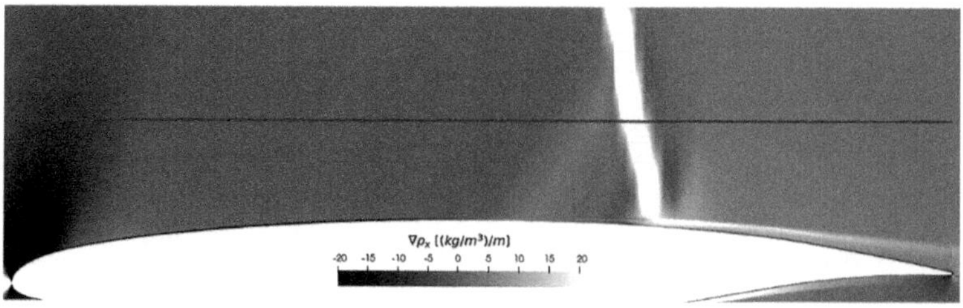

Fig. 11.13 Instantaneous density gradients in X direction contour with the line used to track the shock wave's position displayed in red

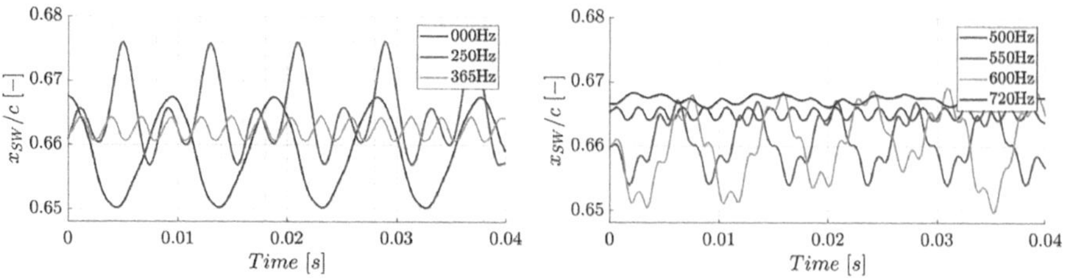

Fig. 11.14 Shock wave's position evolution for different interesting frequencies with $\theta = \pm 1°$

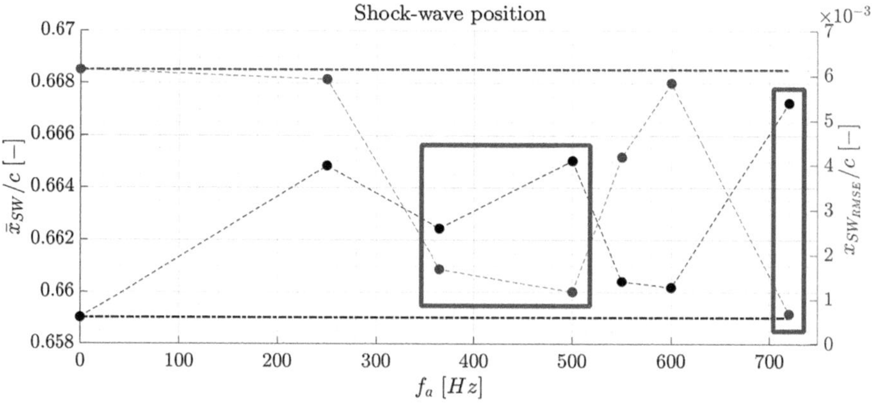

Fig. 11.15 Mean (black) and *rms* (orange) shock's position evolution for interesting frequencies with $\theta = \pm 1°$. Two horizontal lines are presented for each set of results to show the relative evolution with respect the unactuated case ($f_a = 0$Hz)

$f_a = [365 - 510]$ Hz and around $f_a = 720$ Hz, which are exactly the frequencies where the lock-in effect occurs. When the trailing edge moves downwards, it induces an increase of the effective angle of attack due to an increase of the airfoil's curvature that leads to a reduction of the local pressure (increase of the adverse pressure gradient). On the other hand, the opposite takes place when it moves upwards and a pressure wave is generated that travels upstream. Therefore it can be stated that an optimal periodic pressure fluctuation at the trailing edge can help to move downstream the formation of the shock by limiting the amplification of the buffet instability and providing increase of lift.

11.8 Conclusions

A parametric study on the frequency and amplitude of a flapping trailing edge applied in a specified area near the trailing edge has been carried out for the tRS A320 prototype of the SMS project which has been tested experimentally at IMP-PAN. It has been shown that the shock wave dynamics are mostly controlled by means of pressure fluctuations generated by the trailing edge actuations, that are strongly affected by the near wake instabilities and coherent vortices. A lock-in effect of the buffet frequency at the actuation frequency is induced for several vlaues of the actuation frequency which permits to control the shock oscillation frequency. The oscillation's amplitude is reduced with the increase in actuation frequency until a limit where the shock's motion does not synchronize with the actuations anymore. The most promising parametric configurations take place when lock-in occurs at $f_a = [365 - 510]$ Hz and $f_a = 720$ Hz, acting with 0.2 mm of amplitude ($\theta = \pm 1°$). They lead to a lift increase in the order of 1.5% and lift-to-drag increase in the order of 1%.

Acknowledgements The authors are grateful to the European Commission for the attribution of the EU projects H2020 N° 723402 SMS—"Smart Morphing and Sensing for aeronautical configurations", https://cordis.europa.eu/project/id/723402 as well as TEAMAERO—"Towards Effective Flow Control and Mitigation of Shock Effects in Aeronautical Applications", https://cordis.europa.eu/project/id/860909/fr. They are grateful to the EU—PRACE allocation, project N°2021250049, BEAMW—"Bioinspired Electroactive Aeronautics Morphing Wings" as well as to the French Supercomputing Centres TGCC and CINES.

References

1. Smart morphing and sensing for aeronautical configurations, https://cordis.europa.eu/project/id/723402, www.smartwing.org/SMS/EU
2. Braza M, Rouchon JF, Tzabiras G, Auteri F, Flaszynski P (2023) Smart morphing and sensing for aeronautical configurations; prototypes, experimental and numerical findings from the H2020 N° 723402 SMS EU Project. Notes on numerical fluid mechanics and multidisciplinary design, vol 153. Springer
3. TEAMAero, https://cordis.europa.eu/project/id/860909/
4. Bouhadji A, Braza M (2003) Organised modes and shock-vortex interaction in unsteady viscous transonic flows around an Aerofoil: part i: Mach number effect. Comput Fluids 32(9):1233–60
5. Bourdet S, Bouhadji A, Braza M, Thiele F (2003) Direct numerical simulation of the three-dimensional transition to turbulence in the transonic flow around a wing. Flow Turbul Combust 71(1):203–20

6. Szubert D, Grossi F, Jimenez-Garcia A, Guibert V, Hoarau Y, Saintlos S. , Hunt J., Braza M (2012) Feedback effects and stochastic forcing response of the transonic buffet around an airfoil including trailing-edge-plate at high Reynolds. ECCOMAS Conference, Viena
7. Szubert D, Grossi F, Jimenez-Garcia A, Hoarau Y, Hunt JC, Braza M (2015) Shock-vortex shear-layer interaction in the transonic flow around a supercritical airfoil at high Reynolds number in buffet conditions. J Fluids Struct 55:276–302
8. Simiriotis N (2020) Numerical study and physical analysis of electroactive morphing wings and hydrodynamic profiles at high Reynolds number turbulent flows, PhD. http://www.theses.fr/2020INPT0041
9. The Smart Morphing Centre, http://smartwing.org
10. Marouf A, Hoarau Y, Vos JB, Charbonnier D, Bmegaptche-Tekap Y, Braza M (2019) Evaluation of the aerodynamic performance increase thanks to a morphing A320 wing with high-lift flap by means of CFD Hi-Fi approaches. In: AIAA aviation forum, p 2912
11. Auteri F, Savino A, Zanotti A, Gibertini G, Zagaglia D, Bmegaptche-Tekap Y, Braza M (2022) Experimental evaluation of the aerodynamic performance of a large-scale high-lift morphing wing. Aerosp Sci Technol 124:107515
12. Marouf A, Simiriotis N, Tô JB, Bmegaptche-Tekap Y, Hoarau Y, Braza M (2020) DDES and OES simulations of a morphing airbus A320 wing and flap in different scales at high Reynolds. In: Progress in hybrid RANS-LES modelling, vol. 143. Springer, pp 249–58
13. Tô JB, Simiriotis N, Marouf A, Szubert D, Asproulias I, Zilli DM, Hoarau Y, Hunt JC, Braza M (2019) Effects of vibrating and deformed trailing edge of a morphing supercritical airfoil in transonic regime by numerical simulation at high Reynolds number. J Fluids Struct 91:102595
14. Lee B (1989) Oscillatory shock motion caused by transonic shock boundary-layer interaction. AIAA J 28(5):942–4
15. Tijdeman H (1977) Investigations of the transonic flow around oscillating airfoils, NLR-TR 77090 U
16. Raveh D, Dowell E (2011) Frequency lock-in phenomenon for oscillating airfoils in buffeting flows. J Fluids Struct 27(1):89–104
17. Iovnovich M, Raveh D (2012) Transonic unsteady aerodynamics in the vicinity of shock-buffet instability. J Fluids Struct 29:131–42
18. Hoarau Y, Pena D, Vos JB, Charbonier D, Gehri A, Braza M, Deloze T, Laurendeau E (2016) Recent developments of the Navier Stokes multi block (NSMB) CFD solver. In: 54th AIAA aerospace sciences meeting, p 2056
19. Van Leer B (1979) Towards the ultimate conservative difference scheme V a second-order sequel to Godunov's method. J Comput Phys 32(1):101–36
20. Braza M, Perrin R, Hoarau Y (2006) Turbulence properties in the cylinder wake at high Reynolds numbers. J Fluids Struct 22(6–7):757–71
21. Bourguet R, Braza M, Harran G, El Akoury R (2008) Anisotropic Organised Eddy Simulation for the prediction of non-equilibrium turbulent flows around bodies. J Fluids Struct 24(8):1240–51
22. Perrin R, Cid E, Cazin S, Sevrain A, Braza M, Moradei F, Harran G (2007) Phase-averaged measurements of the turbulence properties in the near wake of a circular cylinder at high Reynolds number by 2C-PIV and 3C-PIV. Exp Fluids 42(1):93–109
23. Donea J, Giuliani S, Halleux JP (1982) An arbitrary Lagrangian-Eulerian finite element method for transient dynamic fluid-structure interactions. Comput Methods Appl Mech Eng 33:689–723

Chapter 12
Numerical Simulation of an A320 Morphing Wing Through Frequency Modulation in the Vicinity of the Trailing Edge in Subsonic Regimes

C. Rouaix, C. Jiménez-Navarro, M. Carvalho, A. Marouf, Y. Hoarau, R. El Akoury, C. Raibaudo, G. Harran, H. Hangan, J. F. Rouchon, and M. Braza

Abstract Bioinspired electroactive morphing is investigated around an A320 wing at high Reynolds number in subsonic regime by applying slight deformations and vibrations in the near trailing edge region. The effects of morphing on the aerodynamic performance are studied numerically using the Navier Stokes Multi-Block (NSMB) code on the 70 cm chord Reduced Scale (RS) prototype of the European Union project "Smart Morphing and Sensing for aeronautical configurations"—SMS (https://www.smartwing.org/SMS/EU), at 10° of incidence and for a Reynolds number of 1 million. This study focuses on the effects of a constant actuation frequency and on the time modulation of this frequency following a linear evolution, called wobulation, which to the authors' knowledge, is new in the state-of-the-art with respect to morphing wings. An extensive 2D numerical parametric study has been performed, which allows the evaluation of optimal morphing parametric

C. Rouaix (✉) · C. Jiménez-Navarro · M. Carvalho · A. Marouf · R. El Akoury · G. Harran · M. Braza
IMFT—Institut de Mécanique des Fluides de Toulouse UMR 5502 CNRS-INPT-UT3, 2 Allée du Professeur Camille Soula, 31400 Toulouse, France
e-mail: clement.rouaix@imft.fr; clement.rouaix@ontariotechu.ca

C. Rouaix · M. Carvalho · H. Hangan
Faculty of Engineering and Applied Science (FEAS), Ontario Tech University, 2000 Simcoe St. N., Oshawa, ON L1G0C5, Canada

A. Marouf · Y. Hoarau
ICUBE—Laboratoire des sciences de l'ingénieur, de l'informatique et de l'imagerie, UMR 7357, CNRS-Univ.Strasbourg-ENGEES-INSA, Strasbourg, France

M. Carvalho · J. F. Rouchon
LAPLACE—Laboratoire Plasma et Conversion d'Énergie UMR 5213 CNRS-INPT-UT3, Toulouse, France

C. Raibaudo
Université d'Orléans, INSA-CVL, PRISME, 4229 Orléans, EA, France

T. Tuovinen et al. (eds.), *Advanced Computational Methods and Design for Greener Aviation*, Computational Methods in Applied Sciences 59,
https://doi.org/10.1007/978-3-031-61109-4_12

ranges. The studied interval of constant actuation frequency is $f_a = [10 - 600]$ Hz and the amplitude of vibrations is $a_p = 0.7$ mm. The High Fidelity CFD simulations used the Organised Eddy Simulation (OES) turbulence modelling approach, sensitized to coherent structures development. A thorough understanding of the physical phenomena that occur with both constant actuation frequency and wobulation has been undertaken. By beneficially manipulating the surrounding turbulence, it is shown that the optimal constant actuation frequency ranges increase the mean lift by an order of 2% and reduce the mean drag by an order of 1%. It has been shown that wobulation allows the detection of optimal intervals for constant actuation frequency, resulting in an overall increase in aerodynamic performance.

12.1 Introduction

Previous research projects of the multi-disciplinary research platform involving the French laboratories IMFT and LAPLACE, such as Electroactive Morphing for Micro-Airvehicles–EMMAV or Dynamic Regime Electroactive Morphing–DYNAMORPH, have investigated the bio-inspired concept of electroactive morphing to increase aerodynamic performance of wings. [1] and [2] experimentally investigated the so-called *hybrid electroactive morphing*, applied to the near trailing edge region of a deformable flat plate and a NACA0012 wing, respectively, for an incidence of $\alpha = 10°$ and with a Reynolds number of $Re = 2 \times 10^5$. This hybrid morphing, capable of operating on different time and length scales, was implemented using Shape Memory Alloys (SMA) and "push-push" ceramic piezoelectric actuators. As a result, the morphing wing was able to combine large displacement actuation (in the order of a few *cm*) at low frequency (in the order of 1Hz) and relatively small deformations (in the order of a few *mm*) at Higher Frequency Vibrating Trailing Edge (HFVTE) (in the order of kHz), [3–5]. In these studies, the trailing edge shape modification was shown to have a significant impact over the main flow instability modes such as the Von-Kármán, the shear layers associated with Kelvin–Helmholtz vortices, and the smaller scale vortices due to chaotic turbulence.

In the context of the SMS project, [3, 6] and [7] experimentally and numerically investigated morphing effects by means of trailing edge slight deformations in the frequency range of $f_a = [100 - 400]$ Hz (previously referred to as HFVTE) over the Reduced Scale (RS) A320 prototype in the low-subsonic regime, at an incidence of $\alpha = 10°$ and a number of $Re = 1 \times 10^6$. An extensive parametric study was performed on the trailing edge actuation amplitude and frequency. The constant actuation frequency was identified as the main driving factor for the manipulation of the wake dynamics. Indeed, the wake structures were significantly affected by morphing when actuating at frequencies close or above the natural frequencies of the separated shear layers. More recently, for the same configuration ($\alpha = 10°$ and $Re = 1 \times 10^6$) and for a constant actuation frequency of $f_a = 300$ Hz, which was shown to be optimal by the aforementioned 2D simulations, a three-dimensional numerical simulation was performed around the RS wing prototype, [8]. This study showed that under morphing actuation, the three-dimensional flow behaviour is modified with an attenuation of the secondary instability and a formation of more organised vortex rows producing a quasi-two-dimensional wake. It was found that the vortex breakdown suppresses the harmful vortices and enhances those that contribute to the increase in circulation, similar to the movement of large-span hunting bird feathers, with a resulting increase in lift, [3] and [8]. Up to now, HFVTE morphing has only been studied for constant actuation frequency and amplitude.

The time modulation of the frequency is characterized by a continuous sweep over a range, and is referred to as wobulation in the literature. This term has been widely employed in image display systems, acoustic, radar and radio technologies ([9, 10] and [11]). The wobulation of the trailing edge actuation frequency has been experimentally performed for the first time in literature in the context of the SMS project, [7]. The objective was to design a closed-loop feedback controller in real time, able to reduce the variance of the wall pressure and of the forces and to increase the aerodynamic performance, as well as simultaneously reduce the noise sources. A linear time variation of the frequency was applied on the RS A320 prototype that allowed for the characterization of the dynamic system in open-loop.

The present work is a continuation of the aforementioned studies of morphing effects by constant actuation frequency and wobulation, with the aim of numerically investigating the effects of these two types of morphing and discussing a comparison and interrelation between them. A 2D approximation is adopted in this study due to the fact that the dynamics of the main instabilities and coherent structures present in the flow have a 2D origin. This allows to investigate wobulation over a wide range of parameters, allowing the identification of optimal morphing trailing edge actuation frequency and amplitude for further selective studies in 3D.

12.2 Numerical Method

12.2.1 Governing Equations

The Navier–Stokes Multi Block (NSMB) solver [12], which solves equations under the following form, was used for the numerical simulations:

$$\frac{\partial}{\partial t}(W) + \frac{\partial}{\partial x}(f - f_v) + \frac{\partial}{\partial y}(g - g_v) + \frac{\partial}{\partial z}(h - h_v) = 0 \tag{12.1}$$

where x, y, z and t denote space cartesian coordinates and time, respectively. The state vector W, the convective fluxes f, g, h and the viscous fluxes f_v, g_v, h_v are defined as follows:

$$W = \begin{Bmatrix} \rho \\ \rho u \\ \rho v \\ \rho w \\ \rho E \end{Bmatrix},$$

$$f = \begin{Bmatrix} \rho u \\ \rho u^2 + p \\ \rho u v \\ \rho u w \\ u(\rho E + p) \end{Bmatrix}, g = \begin{Bmatrix} \rho v \\ \rho v u \\ \rho v^2 + p \\ \rho v w \\ v(\rho E + p) \end{Bmatrix}, h = \begin{Bmatrix} \rho w \\ \rho w u \\ \rho w v \\ \rho w^2 + p \\ w(\rho E + p) \end{Bmatrix}$$

$$f_v = \begin{Bmatrix} 0 \\ \tau_{xx} \\ \tau_{xy} \\ \tau_{xz} \\ (\tau U)_x - q_x \end{Bmatrix}, g_v = \begin{Bmatrix} 0 \\ \tau_{yx} \\ \tau_{yy} \\ \tau_{yz} \\ (\tau U)_y - q_y \end{Bmatrix}, h_v = \begin{Bmatrix} 0 \\ \tau_{zx} \\ \tau_{zy} \\ \tau_{zz} \\ (\tau U)_z - q_z \end{Bmatrix}$$

where u, v and w are the cartesian velocity components, p is the pressure, ρ is the density, E is the total energy, q_* is the heat flux and τ is the shear tensor. The viscous dissipation is calculated as: $(\tau U)_* = \tau_{*x}u + \tau_{*y}v + \tau_{*z}w$.

12.2.2 Numerical Parameters

An implicit dual time-stepping with a second-order implicit backward difference scheme was used for the temporal discretization. The outer time step was determined after an extensive numerical study and was chosen to be equal to $\Delta t = 1 \times 10^{-5}$ s, giving a CFL number of 50, together with a number of inner loop iterations of 60 to ensure proper convergence, [6]. Diffusion terms were spatially discretized employing a second-order central scheme. Advection terms were spatially discretized using a fourth-order standard central skew-symmetric scheme with artificial second and fourth-order dissipation terms. Several preconditioning methods were investigated to improve the accuracy for the low-subsonic flow (Mach number of $Ma = 0.063$), and the artificial compressibility using LU-SGS method was selected after giving better agreement of the velocity deficit with the experimental results [13, 14].

The Arbitrary Lagrangian Eulerian (ALE) approach [15] was used to account for the deformation and vibration of the near trailing-edge region, similar to the Macro-Fiber Composite (MFC) ceramic piezoelectric actuator patches implemented to actuate the Reduced Scale (RS) prototype in the experiments. The computation of the new deformed grid was ensured using remeshing methods through Volume Spline Interpolation (VSI) and Trans-Finite Interpolation (TFI) [16]. The deformed shape of the patches follows a second-order polynomial expression that mimics the

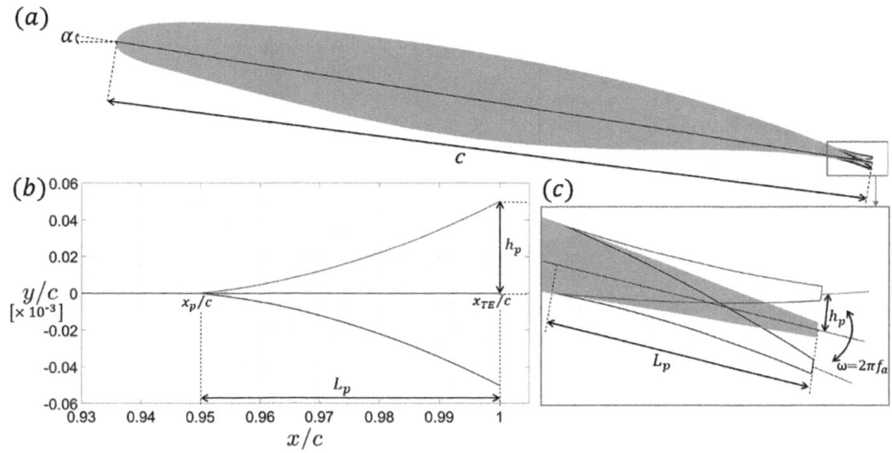

Fig. 12.1 **a** Supercritical A320 airfoil with a chord of $c = 70$ cm and an incidence of $\alpha = 10°$. **b** Vertical displacement y in the near trailing edge region following the second-order polynomial expression (12.2). **c** Zoom of the morphed trailing edge

deformation of the MFC piezoelectric actuators in experiments [3], as shown in Fig. 12.1:

$$y(x,t) = \left(\frac{2h_p}{3L_p^2}\left(x - x_p\right)^2 + \frac{h_p}{3L_p}\left(x - x_p\right) \right) sin(\omega_a t) \tag{12.2}$$

where y is the vertical displacement, x is the chordwise direction, h_p is the semi-amplitude at the trailing edge location x_{TE}, $L_p = 3.5$ cm is the patch length on which the displacement is achieved, x_p is the origin of the displacement defined by $x_p = c - L_p$. The total amplitude of the trailing edge is equal to $a = 2h_p = 0.7$ mm. ω_a is the angular frequency related to the trailing edge actuation frequency such as $\omega_a = 2\pi f_a$. In this study, the trailing edge actuation frequency can be constant (f_a) or time modulated with wobulation (f_w). The effect of the amplitude has been previously studied and showed an optimal behavior for this order of magnitude, while amplitudes larger than 3mm led to a significant increase in drag [17].

12.2.3 Computational Domain and Boundary Conditions

The computational domain used a rectangular shape with dimensions representative of those of the S4 subsonic wind tunnel of IMFT used in experiments. The 2D structured mesh with 3×10^5 cells illustrated in Fig. 12.2 was chosen after a mesh convergence study, which showed a reasonable agreement with the experimental results [13, 17]. The height of the first cell layer was set to obtain a y^+ of 0.1. Slip

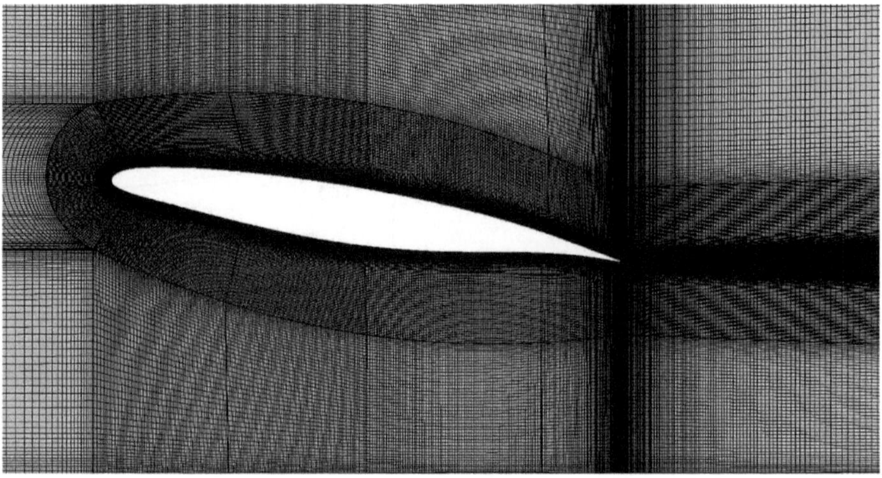

Fig. 12.2 Zoom around the A320 airfoil of the 2D multi-block structured mesh, using 64 blocks and approximately 3×10^5 finite volume cells with quadrilateral shape

conditions were implemented for the upper and lower walls, while no-slip conditions were used on the airfoil's surface. A Dirichlet boundary condition on the velocity was imposed at the inlet with a freestream velocity of $u_\infty = 21.5\,\mathrm{m/s}$, together with a turbulence intensity level of 0.1% of the freestream velocity. In addition, due to the significant decay of the turbulent kinetic energy from the inlet up to the wing caused by the dissipation rate, the ambient sustaining terms have been activated [18, 19]. As a result, the turbulent intensity around the airfoil is physically well preserved despite the turbulent decay observed from the inlet region to the airfoil. Non-reflective conditions for the outlet boundary have been imposed [20], based on a more general non-reflective formulation [21] that is suitable for incompressible flow regimes and ensures mitigation of spurious feedback effects from the outlet boundary pressure.

12.2.4 Turbulence Modelling

The Organised Eddy Simulation (OES) approach was used [19, 22, 23]. This turbulence modelling, based on the phase averaging of the Navier–Stokes equations, is based on the splitting of the turbulence spectrum into a part containing the organized physical processes (the resolved turbulence) and a second part regrouping the chaotic processes. The distinction between the structures to be resolved and those to be modeled is based on their physical character and not their size. Therefore, this approach is sensitized to capture the flow instabilities and the development of coherent structures, while modelling the chaotic turbulence effects, thus reproducing the dual character of turbulence, organised and chaotic [24]. Due to the non-linear interaction among the multitude of coherent modes and chaotic processes, the turbulence spectrum has a modified form that deviates from the one corresponding to the statistical equilibrium, as has been demonstrated experimentally by [22, 25, 26] and [27]. The OES approach is particularly suitable in the present case because it is able to capture the near wall turbulence stress anisotropy in the detached flow, which is characterized by the non-equilibrium turbulence region produced by the relatively high angle of attack and high Reynolds number involved. Because the OES approach is non-inherently 3D, it allows the investigation of morphing effects through an extensive 2D parametric study.

12.3 Flow Physics of the Static (Non-actuated) Case

Figure 12.3a shows an instantaneous 2D snapshot of the flow field obtained for the static reference case by means of streaklines visualisation. The flow comes from the left towards the right and the color of each particle represents its initial position.

As shown in Fig. 12.3a, the turbulent wake is formed by two interfacial shear layers (upper and lower) containing Kelvin–Helmholtz vortices. These shear layers interact non-linearly with each other in the near trailing edge region and form a clear Von-Kármán alternating eddy pattern (vortex shedding) further downstream in the wake. Due to the relatively high angle of attack and because of the supercritical shape of the A320 airfoil section involved in the configuration studied, the effective shear along the lower shear layer is significantly higher in comparison with the upper shear layer formed on the suction side.

The separated shear layers and vortex shedding natural frequencies were first assessed by tracking their patterns visualised by means of streakline snapshots, similar to the one shown in Fig. 12.3a. These values were then confirmed by spectral analysis of the vertical velocity component for monitor points located at strategic positions (Figs. 12.3b and c). Table 12.1 summarizes the main natural frequencies in the flow for the static case together with their respective nominal Strouhal number based on the chord of the airfoil.

12.4 Morphing: Parametric Study on the Constant Actuation Frequency

Figure 12.4 shows streakline visualization of the wake for the static case and for three other constant actuation frequencies. As can be seen in Fig. 12.4b, actuating at (or close to) the Von-Kármán natural frequency strongly reinforced the alternating vortex shedding. This increases the circulation, which in turn increases the mean lift coefficient, as shown in Table 12.2. However, the drag is drastically increased due to the thickening of the wake and to the *feedback* effect on the separation area, inducing

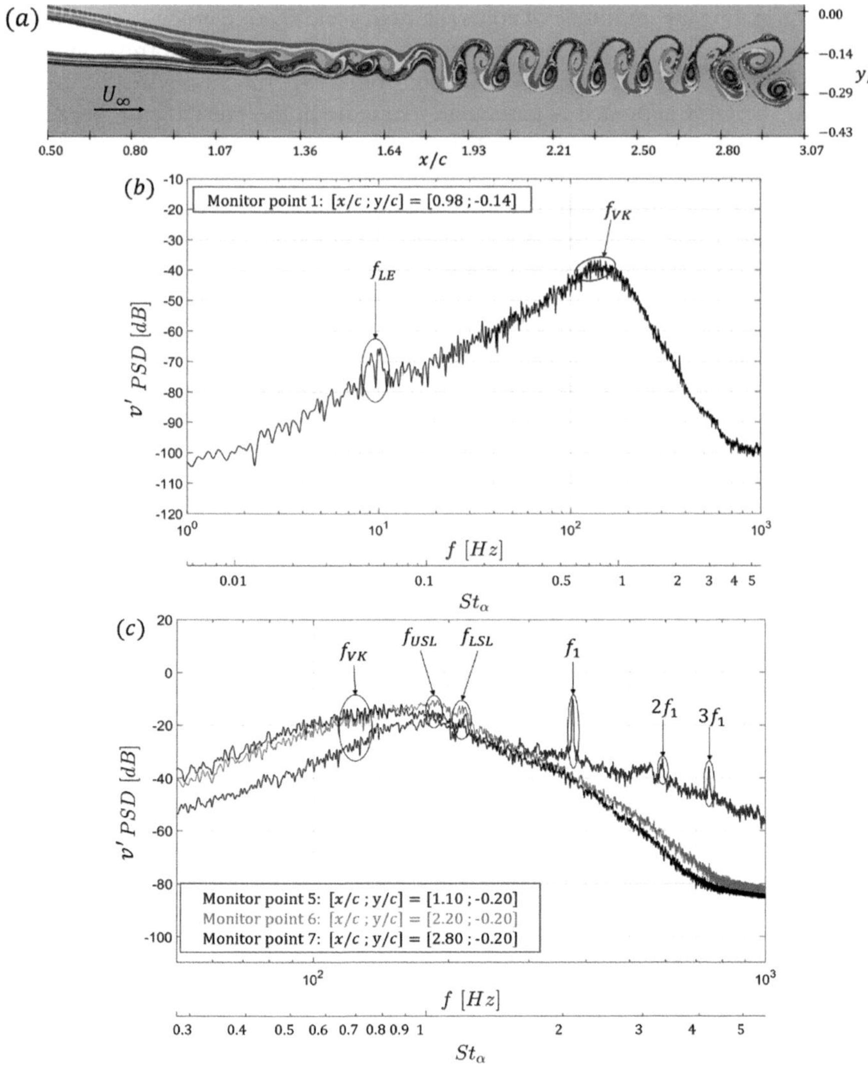

Fig. 12.3 **a** Streaklines visualisation of the flow field for the static reference case ($Re = 1 \times 10^6$ and $\alpha = 10°$). **b** PSD of the vertical component of the velocity fluctuations $v\prime$ for monitor point 1 located in the recirculation region and **c** for monitor points 5, 6 and 7 located along a horizontal line in the wake

a decrease in the mean lift-to-drag ratio. A significant increase in the *rms* value of the aerodynamic forces is also observed, showing therefore that actuation close to the vortex shedding natural frequency is detrimental to the aerodynamic performance.

Actuating at (or close to) the lower shear layer natural frequency strongly reinforced the lower shear layer vortices in comparison with the static case as illustrated

Table 12.1 Natural frequencies for the static case: f_{LE}, f_{VK}, f_{USL} and f_{LSL} respectively correspond to the leading edge small separation bubble, Von-Kármán, upper shear layer and lower shear layer natural frequencies. f_1 corresponds to an interaction mode which is an n-integer combination of shear layers instabilities

	$f[Hz]$	$\text{St} = \frac{fc}{U_\infty}$
Leading edge: f_{LE}	8–12	0.26–0.39
Von-Kármán: f_{VK}	130–140	4.23–4.56
Upper shear layer: f_{USL}	180–190	5.86–6.19
Lower shear layer: f_{LSL}	205–215	6.67–7.00
Interaction mode: f_1	370–375	12.05–12.21

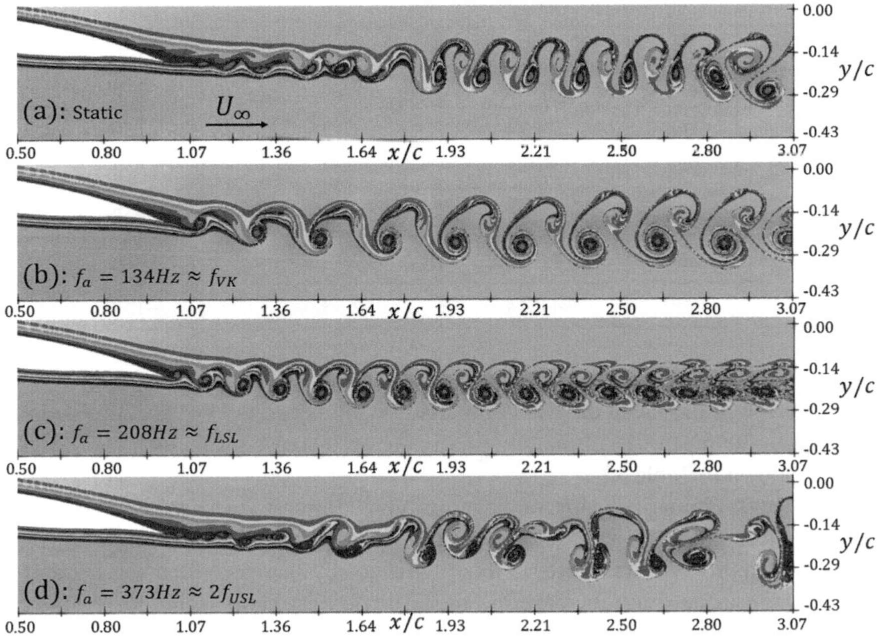

Fig. 12.4 Streakline visualisation in the wake for the static and constant frequency actuation cases– **a** static case, **b** $f_a = 134$ Hz, **c** $f_a = 208$ Hz and **d** $f_a = 373$ Hz

in Fig. 12.4c. This produces as a main effect a significant increase in the local circulation around the airfoil, providing the highest increase in mean lift (in the order of 8% as shown in Table 12.2) compared to all the cases investigated in this study. Although the mean drag is also increased, the lift increase is more important, resulting in an increase in the mean lift-to-drag ratio of the order of 2%.

As shown in Fig. 12.4d, actuating at (or close to) the upper shear layer natural frequency or its first harmonics delays the appearance of Von-Kármán alternated eddies, which are "pushed" further downstream in the wake. This actuation results

Table 12.2 Aerodynamic performance for different constant actuation frequency cases. Results are shown in relative change of the morphing cases in comparison with the static case. Any beneficial effect on aerodynamic performance is highlighted in green, any detrimental effect is shown in orange

	f_a [Hz]	$\left[\frac{(\overline{C_D})_c-(\overline{C_D})_s}{(\overline{C_D})_s}\right]$ × 100 [%]	$\left[\frac{(\overline{C_L})_c-(\overline{C_L})_s}{(\overline{C_L})_s}\right]$ × 100 [%]	$\left[\frac{(\overline{C_L/C_D})_c-(\overline{C_L/C_D})_s}{(\overline{C_L/C_D})_s}\right]$ × 100 [%]
-	15	-1.09	+0.02	+1.12
$\sim f_{VK}/2$	69	-1.00	+0.02	+1.03
f_{VK}	134	+11.3	+4.92	-5.69
$\sim f_{USL}$	185	-0.39	+0.37	+0.76
	192	-0.83	+0.28	+1.12
$\sim f_{LSL}$	208	+5.25	+7.32	+1.97
	210	+6.52	+7.98	+1.37
$\sim 2f_{VK}$	260	+4.74	+3.29	-1.39
	274	+2.48	+3.07	+0.58
$f_1 \sim 2f_{USL}$	373	-0.36	+0.62	+0.98
$\sim 3f_{USL}$	550	+0.69	+2.18	+1.48
-	590	+0.44	+2.33	+1.88

in a decrease in the mean drag in comparison with the static case, together with an increase in the mean lift, ultimately producing an increase in the mean lift-to-drag ratio as shown in Table 12.2.

Table 12.2 summarizes some of the more interesting cases in terms of aerodynamic performance. These parameters are used in ongoing studies for a multi-objective optimization problem with criteria to simultaneously increase the mean lift and or decrease the mean drag, increase the mean lift-to-drag ratio while decreasing the aerodynamic forces *rms* (or keeping the lowest increase of *rms*) with morphing compared to the static case. As discussed in the next section, wobulation allows to accurately and efficiently detect optimal ranges of constant actuation frequency, both in terms of mean aerodynamic forces and *rms*.

12.5 Time Modulation of the Actuation Frequency

As mentioned above, the modulation of the trailing edge actuation frequency (or wobulation) follows a linear time variation, that is defined by an initial frequency f_i and a constant frequency increment Δf_w (or wobulation rate of sweep). For the sake of clarity and without loss of generality, the following discussion focuses only on the lift *rms*, but remains valid for mean drag, drag *rms* and mean lift.

When the lift *rms* signal obtained with a large number of constant actuation frequencies is superimposed on the one obtained with wobulation by a single continuous sweep over a range of actuation frequencies, the similarity between the two signals is striking as shown in Fig. 12.5d. By correlating these two signals, it is possible to find a relationship between wobulation and constant actuation frequencies. Those similarities between wobulation and constant actuation frequencies are also observed for mean drag, drag *rms* and mean lift as shown in Fig. 12.5a, b and c respectively. Pearson's correlation coefficient analysis, which indicates the degree of similarity between two signals, provided a relatively high similarity of 0.78, 0.97, 0.88 and 0.96 for mean drag, drag *rms*, mean lift and lift *rms* respectively.

Therefore, wobulation can be thought of as a continuous sweep of an infinite number of successive constant actuation frequencies. However, this sweep also contains memory effects of the previous states. The variation of aerodynamic forces obtained during a wobulation sweep is due to the trailing edge actuation frequency successive lock-in on the natural frequencies of the dynamic system, such as for instance the Von-Kármán vortex shedding, or the two shear layers, their harmonics and subharmonics. In one sweep over a range of actuation frequencies, the present wobulation is able to accurately and efficiently detect optimal constant actuation frequencies in terms of aerodynamic forces mean and *rms*, beneficial for the aerodynamic performance increase.

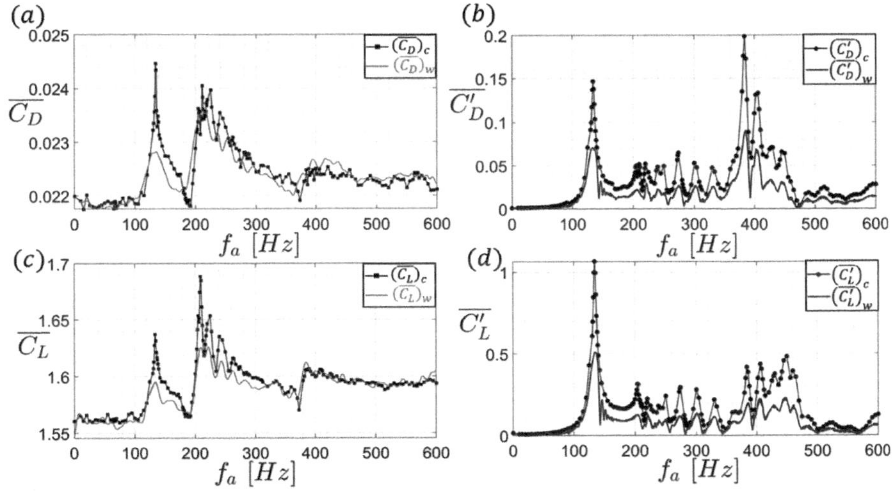

Fig. 12.5 Aerodynamic forces mean and *rms* plotted versus the actuation frequency for constant actuation frequency in black and for wobulation ($f_i = 0$ Hz and $\Delta f_w = 50$ Hz) in color (green and blue). **a** mean drag $\overline{C_D}$, **b** drag *rms* $\overline{C_{D'}}$, **c** mean lift $\overline{C_L}$, and **d** lift *rms* $\overline{C_{L'}}$

12.6 Conclusions

In this paper, the wobulation of the trailing edge actuation frequency has been introduced in the context of morphing. In a first step, the physical analysis of the flow for the static reference case (non-actuated) was performed, with the assessment of the natural frequencies of the shear-layers associated with Kelvin–Helmholtz vortices and of the vortex shedding related to the Von-Kármán instability. Then, the effects of electroactive morphing using a constant trailing edge actuation frequency and a constant vibration amplitude of 0.7 mm have been discussed for frequencies included in the range $f_a = [10 - 600]$ Hz. It has been shown that actuating at (or close to) the upper and lower shear layers natural frequencies and near their harmonics is beneficial for aerodynamic performance. Actuating close to the upper shear layer natural frequency or its first harmonics leads to approximately a 1% mean drag reduction and a 1% mean lift-to-drag ratio increase. Actuating close to the lower shear layer natural frequency results in an increase in mean lift of 8% and a mean lift-to-drag ratio increase of 2%. Finally, wobulation has proven to be an accurate and efficient method to detect optimal constant actuation frequencies beneficial for aerodynamic performance. This numerical study is a first step for experiments toward designing a closed-loop feedback controller in real-time over the A320 Reduced Scale morphing prototype.

Acknowledgements The authors are grateful to the European Commission for the attribution of the EU projects H2020 N° 723402 SMS—"Smart Morphing and Sensing for aeronautical configurations", https://cordis.europa.eu/project/id/723402, as well as to the French project N° CE05-000 supported by the "Agence Nationale de Recherche"—ANR. The authors are grateful to the EU-PRACE allocation, project N°2021250049 BEAMW—"Bioinspired Electroactive Aeronautics Morphing Wings" and to the French Supercomputing Centres TGCC and CINES.

References

1. Chinaud M, Rouchon JF, Duhayon E, Scheller J, Cazin S, Marchal M, Braza M (2014) Trailing-edge dynamics and morphing of a deformable flat plate at high Reynolds number by time-resolved PIV. J Fluids Struct 47:41–54. https://doi.org/10.1016/j.jfluidstructs.2014.02.007
2. Scheller J, Chinaud M, Rouchon JF, Duhayon E, Cazin S, Marchal M, Braza M (2015) Trailing-edge dynamics of a morphing NACA0012 aileron at high Reynolds number by high-speed PIV. J Fluids Struct 55:42–51. https://doi.org/10.1016/j.jfluidstructs.2014.12.012
3. Jodin G, Motta V, Scheller J, Duhayon E, Döll C, Rouchon JF, Braza M (2017) Dynamics of a hybrid morphing wing with active open loop vibrating trailing edge by time-resolved PIV and force measures. J Fluids Struct 74:263–290. https://doi.org/10.1016/j.jfluidstructs.2017.06.015

4. Jodin G, Scheller J, Duhayon E, Rouchon JF, Braza M (2017) Implementation of a hybrid electro-active actuated morphing wing in wind tunnel. Solid State Phenomena, Trans Tech Publications 2017, vol 260, pp 85–91. https://doi.org/10.4028/www.scientific.net/SSP.260.85
5. Scheller J (2015) Electroactive morphing for the aerodynamic performance improvement of next generation airvehicles. PhD Thesis, Toulouse INPT-IMFT. https://oatao.univ-toulouse.fr/14479/
6. Simiriotis N, Jodin G, Marouf A, Elyakime P, Hoarau Y, Hunt JCR, Rouchon JF, Braza M (2019) Morphing of a supercritical wing by means of trailing edge deformation and vibration at high Reynolds numbers: experimental and numerical investigation. J Fluids Struct 91:102676. https://doi.org/10.1016/j.jfluidstructs.2019.06.016
7. Braza M, Rouchon JF, Tzabiras G, Auteri F, Flaszynski P (2023) Smart morphing and sensing for aeronautical configurations-prototypes, experimental and numerical findings from the H2020 N° 723402 SMS EU project, vol 153. Notes on numerical fluid mechanics and multidisciplinary design, Publisher Springer. https://doi.org/10.1007/978-3-031-22580-2
8. Marouf A, Hoarau Y, Rouchon JF, Braza M (2023) Three-dimensional simulation effects of trailing-edge actuation on a morphing A320 wing by means of hybrid turbulence modelling. Int J Numer Methods Heat Fluid Flows 33(4):1436–1457. https://doi.org/10.1108/HFF-09-2022-0559
9. Kim K, Han S, Yoon J, Kwon S, Park HK, Park W (2016) Lithographic resolution enhancement of a maskless lithography system based on a wobulation technique for flow lithography. Appl Phys Lett 109(23):234101. https://doi.org/10.1063/1.4967373
10. Ferrand G, Luong M, Huber G, Desvaux H (2015) A more accurate tuning-matching technique for NMR probes using wobulation and variable phase shifter. Concepts Magn Reson Part B Magn Reson Eng 45(2):59–68. https://doi.org/10.1002/cmr.b.21277
11. Zaitseva KA, Korolev VI, Akhi AV, Butyrskii EY (2017) Perception of noises with temporally unstable frequency modulation of tonal components by the dolphin Tursiops truncates. J Evolut Biochim Physiol 53(3):241–244. https://doi.org/10.1134/S0022093017030097
12. Hoarau Y, Pena D, Vos JB, Charbonier D, Gehri A, Braza M, Deloze T, Laurendeau E (2016) Recent developments of the Navier Stokes Multi Block (NSMB) CFD solver. In proceedings of the 54th AIAA aerospace sciences meeting. San Diego, California, USA, https://arc.aiaa.org/doi/abs/10.2514/6.2016-2056
13. Marouf A (2020) Physical analysis of electroactive morphing concepts for the aerodynamic performance increase of future wing design through high-fidelity numerical simulation and turbulence modelling in high Reynolds number. PhD thesis, University of Strasbourg & Toulouse INPT-IMFT. https://theses.hal.science/tel-03855969/
14. Chorin AJ (1967) A numerical method for solving incompressible viscous flow problems. J Comput Phys 2(1):12–26. https://doi.org/10.1016/0021-9991(67)90037-x
15. Donea J, Giuliani S, Halleux JP (1982) An arbitrary lagrangian-eulerian finite element method for transient dynamic fluid–structure interactions. Comput Methods Appl Mech Eng 33(1/3):689–723. https://doi.org/10.1016/0045-7825(82)90128-1
16. Guillaume M, Gehri A, Stephani P, Vos J, Mandanis G (2011) F/A-18 vertical buffeting calculations using unsteady fluid structure interaction. Aeronaut J 115(1167):285–294. https://doi.org/10.1017/S0001924000005777
17. Simiriotis N (2020) Numerical study and physical analysis of electroactive morphing wings and hydrodynamic profiles at high Reynolds number turbulent flows. PhD Thesis, Toulouse INPT-IMFT. https://theses.hal.science/tel-04164493/document
18. Spalart PR, Rumsey CL (2007) Effective inflow conditions for turbulence models in aerodynamic calculations. AIAA J 45(10):2544–2553. https://doi.org/10.2514/1.29373
19. Szubert D, Grossi F, Garcia AJ, Hoarau Y, Hunt JCR, Braza M (2015) Shock-vortex shear-layer interaction in the transonic flow around a supercritical airfoil at high Reynolds number in buffet conditions. J Fluids Struct 55:276–302. https://doi.org/10.1016/j.jfluidstructs.2015.03.005
20. Poinsot TJ, Lele SK (1992) Boundary conditions for direct simulations of compressible viscous flows. J Comput Phys 101(1):104–129. https://doi.org/10.1016/0021-9991(92)90046-2

21. Jin GY, Braza M (1993) A nonreflecting outlet boundary condition for incompressible unsteady Navier-Stokes calculations. J Comput Phys 107(2):239–253. https://doi.org/10.1006/jcph.1993.1140
22. Bourguet R, Braza M, Harran G, El Akoury R (2008) Anisotropic organised eddy simulation for the prediction of non-equilibrium turbulent flows around bodies. J Fluids Struct 24(8):1240–1251. https://doi.org/10.1016/j.jfluidstructs.2008.07.004
23. Braza M, Perrin R, Hoarau Y (2006) Turbulence properties in the cylinder wake at high Reynolds numbers. J Fluids Struct 22(6/7):757–771. https://doi.org/10.1016/j.jfluidstructs.2006.04.021
24. Brown GL, Roshko A (1974) On density effects and large structure in turbulent mixing layers. J Fluid Mech 64(4):775–816. https://doi.org/10.1017/S002211207400190X
25. Perrin R, Braza M, Cid E, Cazin S, Chassaing P, Mockett C, Reimann T, Thiele F (2008) Coherent and turbulent process analysis in the flow past a circular cylinder at high Reynolds number. J Fluids Struct 24(8):1313–1325. https://doi.org/10.1016/j.jfluidstructs.2008.08.005
26. Perrin R, Braza M, Cid E, Cazin S, Barthet A, Sevrain A, Mockett C, Thiele F (2007) Obtaining phase averaged turbulence properties in the near wake of a circular cylinder at high Reynolds number using POD. Exp Fluids 43:341–355. https://doi.org/10.1007/s00348-007-0347-6
27. Perrin R, Cid E, Cazin S, Sevrain A, Braza M, Moradei F, Harran G (2007) Phase-averaged measurements of the turbulence properties in the near wake of a circular cylinder at high Reynolds number by 2C-PIV and 3C-PIV. Exp Fluids 42:93–109. https://doi.org/10.1007/s00348-006-0223-9

Part IV
Simulation and Application for Transport and Manufacturing

Chapter 13
Design Optimization Based on Multi-fidelity Metamodels

Alberto Clarich, Luca Battaglia, Carlo Poloni, Lucia Parussini, Haysam Telib, and Angela Scardigli

Abstract This paper illustrates how multi-fidelity metamodels can be efficiently applied to save time and costs in parametric design optimization, which normally requires simulating numerically a large number of designs. Datasets of high fidelity (HF) and middle-low fidelity (LF) simulations, obtained for instance from computational models solved by grid discretizations of different accuracy, can be used together to feed the surrogate model, improving the accuracy of the response function prediction and reducing the overall computational cost at the same time. The methodologies proposed in this paper include adaptive Design of Experiments algorithms to define the optimal dataset of design simulations, and efficient multi-fidelity surrogate methods for scalar fields (Cokriging) and vector fields (Reduced Order Models). All the methods are tested and applied to CFD test cases.

Keywords Design optimization · Multi-fidelity metamodels · ModeFRONTIER

13.1 Introduction

Design optimization is a strategic key in industrial design: it is an essential component to produce more efficient, better performing and more sustainable products.

At this end, it is normally necessary to computationally evaluate the objective functions of a large number of candidate solutions, at the variation of the design parameters, until the optimal criteria are satisfied. If the availability of process automation platforms and of increasing computational resources help today to achieve solutions

A. Clarich (✉) · L. Battaglia
ESTECO SpA, Trieste, Italy
e-mail: clarich@esteco.com

C. Poloni · L. Parussini
University of Trieste, Trieste, Italy

H. Telib · A. Scardigli
Optimad, Torino, Italy

T. Tuovinen et al. (eds.), *Advanced Computational Methods and Design for Greener Aviation*, Computational Methods in Applied Sciences 59,
https://doi.org/10.1007/978-3-031-61109-4_13

that were not feasible few years ago, it is still particularly important to elaborate efficient optimization methodologies to obtain the optimal results with the lowest number of design evaluations, to save costs and time.

In that sense, the adoption of meta-models, i.e. mathematical or statistical models able to approximate the response of the system on the basis of a dataset of design simulations, help the designers to reach this purpose, since the overall cost is generally lower than the one necessary to evaluate each design proposed by the optimization algorithm by a numerical simulation. By the way, the impact of the design simulations number for the meta-model training can still be relevant in many cases, in particular if the simulation model is a high-fidelity (HF). How it is then possible to train efficient meta-models with a lower computational cost? And in particular, how can low-fidelity (LF) models be introduced in the training process, to reduce the overall computational time?

In this paper, we present two different multi-fidelity methodologies to answer to these questions, all of them perfectly compatible with a process automation platform and therefore directly applicable to industrial design. The first method is based on Cokriging algorithm filled by a nested DOE sampling (Dataset Reducer), that aims to build a scalar field meta-model combining two or more series of samples of different fidelity, defined in two consecutive steps (LF samples are used to determine the optimal position of the HF samples).

The second method, based on a multi-fidelity ROM (Reduced-Order-Models), allows instead to create a vector field meta-model, which approximate the original CFD model by a linear combination of the principal components of the original field.

13.2 Multi-fidelity Regression Models for Scalar Outputs

13.2.1 Cokriging Algorithm: Basic Definition

Cokriging is a multi-fidelity surrogate model based on Bayesian theory. It was originally proposed for enhancing the accuracy of a high-fidelity surrogate using supplementary observations of low-fidelity models. In essence, it aims at exploiting the cross-correlation between two or more Gaussian process surrogates through a stochastic auto-regressive scheme, and it yields a predictive posterior distribution for the high-fidelity model output that encodes the contribution of lower fidelity levels with quantified uncertainty (Fig. 13.1).

Cokriging was first proposed by Kennedy and O'Hagan in the landmark paper [1], and since then it has been used in many different applications, such as surrogate-based optimization [2]. The algorithm described in Kennedy and O'Hagan [1] has an unfavorable cubic scaling with the number of observed data and the number of surrogate models.

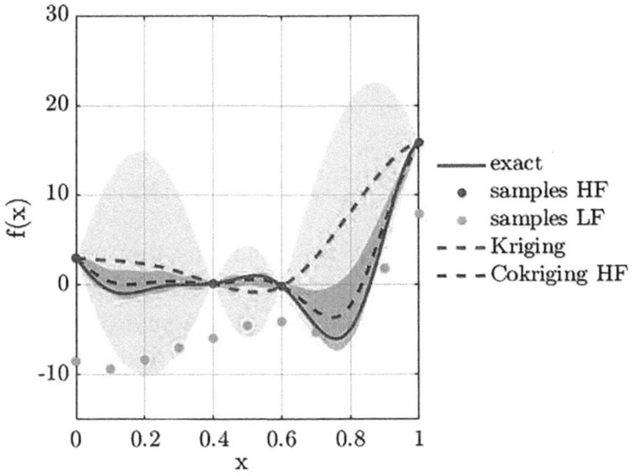

Fig. 13.1 Implementation of Cokriging by nested sampling points

To overcome this issue, Le Gratièt [3] proposed a recursive version of the Cokriging algorithm, which allows to break up the computation of posterior distribution into a sequence of distinct inferences of smaller dimensions. The advantage of this approach is clear: the inversion of the small covariance matrices associated with the sequence of inferences is less expensive and better conditioned than the inversion of the large covariance matrix associated with the full inference problem.

To describe the recursive Cokriging approach, suppose we have available multiple models, say s-models $(z_t(\mathbf{x}))_{t=1,\dots,s}$, providing an estimate of the real-valued scalar field.

$$z_t\colon \mathcal{X} \in \mathbb{R}^d \to \mathbb{R} \tag{13.1}$$

$$\mathbf{x} \to z_t(\mathbf{x}) \tag{13.2}$$

for $t = 1, \dots, s$, sorted by increasing order of fidelity, being $z_1(\mathbf{x})$ the model with the lowest level of fidelity, and modeled by Gaussian processes $(Z_t(\mathbf{x}))_{t=1,\dots,s}$.

Such Gaussian processes are related to each other by the autoregressive model:

$$\begin{cases} Z_t(\mathbf{x}) = \rho_{t-1}(\mathbf{x})\tilde{Z}_{t-1}(\mathbf{x}) + \delta_t(\mathbf{x}) \\ \tilde{Z}_{t-1}(\mathbf{x}) \perp \delta_t(\mathbf{x}) \\ \rho_{t-1}(\mathbf{x}) = \mathbf{g}_{t-1}^T(\mathbf{x})\boldsymbol{\beta}_{\rho_{t-1}} \end{cases} \tag{13.3}$$

where $\tilde{Z}_{t-1}(\mathbf{x})$ is a Gaussian process with distribution $\left(Z_{t-1}(\mathbf{x})|\mathbf{z}_{t-1}, \boldsymbol{\beta}_{t-1}, \boldsymbol{\beta}_{\rho_{t-1}}, \boldsymbol{\theta}_{t-1}\right)$, $\delta_t(\mathbf{x}) \sim \mathcal{GP}\left(\mathbf{f}_t^T(\mathbf{x})\boldsymbol{\beta}_t, k_t(\mathbf{x}, \mathbf{x}'; \boldsymbol{\theta}_t)\right)$ and $Z_1(\mathbf{x}) \sim \mathcal{GP}\left(\mathbf{f}_1^T(\mathbf{x})\boldsymbol{\beta}_1, k_1(\mathbf{x}, \mathbf{x}'; \boldsymbol{\theta}_1)\right)$.

k_t is the covariance function depending on hyper-parameters $\boldsymbol{\theta}_t$, $\boldsymbol{\beta}_t$ and $\boldsymbol{\beta}_{\rho_{t-1}}$ are vectors of unknown coefficients, $\mathbf{f}_t(\mathbf{x})$ and $\mathbf{g}_{t-1}(\mathbf{x})$ are chosen functions.

It can be shown that $\left(Z_t(\mathbf{x})|\mathbf{z}_t, \boldsymbol{\beta}_t, \boldsymbol{\beta}_{\rho_t}, \boldsymbol{\theta}_t\right)$, that is the posterior distribution of $Z_t(\mathbf{x})$, conditional on the observations and parameters of all lower-fidelity models, is Gaussian with mean and variance given by:

$$\widehat{m}_{Z_t}(\mathbf{x}) = \rho_{t-1}(\mathbf{x})\widehat{m}_{Z_{t-1}}(\mathbf{x}) + A_t(\mathbf{x}) \tag{13.4}$$

$$\hat{s}^2_{Z_t}(\mathbf{x}) = \rho^2_{t-1}(\mathbf{x})\hat{s}^2_{Z_{t-1}}(\mathbf{x}) + B_t(\mathbf{x}) \tag{13.5}$$

where $A_t(\mathbf{x})$ and $B_t(\mathbf{x})$ are functions that can be explicitly computed based on univariate GPR results at fidelity levels t and $t - 1$ [1].

This is very convenient from a computational viewpoint. In Le Gratièt [3], a fully Bayesian formulation of recursive Cokriging which incorporates prior information in the maximum likelihood estimation of the model hyper-parameters is given. This allows to speed up the estimation process by leveraging on analytical expressions.

In the Cokriging modeling process, high-fidelity (HF) and low-fidelity (LF) sampling points must be nested to satisfy the Markov property. Thanks to the recursive formulation of the multi-fidelity Cokriging model, we can easily adapt the method when the design sets are not nested [3]. The nested property of the experimental design sets is not necessary to build the model, but it allows for a simple estimation of the model parameters. In this paper we adopt the choice of a nested sampling also for a convenience in the selection of the HS samples, as it will be explained in the next section.

In Fig. 13.1 there is an example of Cokriging models obtained by nested sampling points. The models are obtained with the same HF samples, but different LF samples. Shaded areas correspond to confidence intervals $\widehat{m}_{Z_t}(x) \pm 3\hat{s}_{Z_t}(x)$ with t = 1, 2 where Z_1 is the LF model and Z_2 is the HF model.

13.2.2 Nested Sampling DOE: Dataset Reducer

Since the LF samples require less simulation time and they can be used to explore more widely the variable space, it might be convenient to use the information gathered from the LF dataset, to decide where to evaluate the HF dataset points. This is the approach we have here adopted, implementing the Dataset Reducer (DR) DOE, available in modeFRONTIER [4] optimization software.

The purpose of the DR DOE is to identify the optimal subset of training points (with a size $k < n$, where n is the size of the LF DOE) that allows to train a RSM meta-model with the minimum regression error. The high-fidelity model is then used to evaluate the optimal DOE dataset of size k, and the meta-model is trained combining the LF and the nested HF dataset (Fig. 13.2).

More in detail, the algorithm is executed into two phases: the Monte Carlo phase and the Annealing phase. In the Monte Carlo phase, a global search is performed and RSMs are iteratively trained on randomly selected training sets (size $k < n$) and validated using the set defined by the remaining (n - k) points. The training set with the lowest error (that is, with the best RSM quality) is selected.

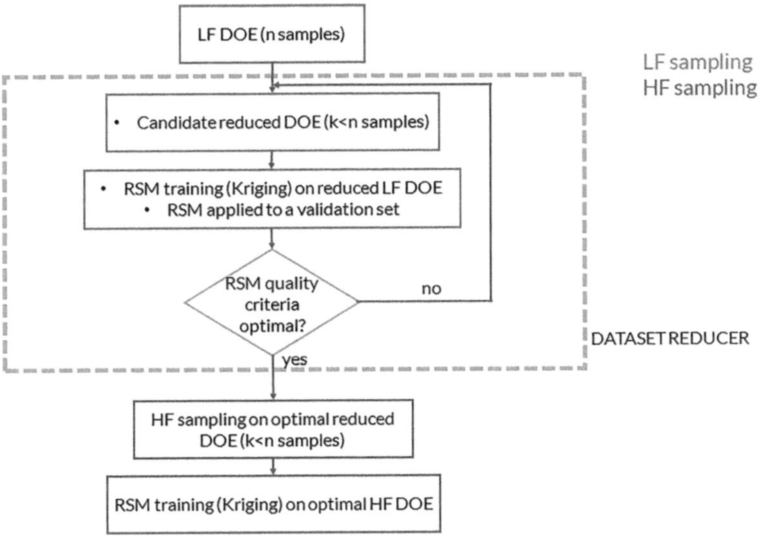

Fig. 13.2 Dataset reducer methodology for multi-fidelity meta-models training

In the Annealing phase, a local optimization search is performed starting from the best selected training set: small portions of it are iteratively substituted with the excluded points to progressively reduce the training error until the defined number of iterations is reached.

13.2.3 Application to an Aeronautical Test Case

To prove the efficiency of this methodology, it has been applied to an aeronautical test case.

A NACA0012 airfoil, in transonic conditions at Mach = 0.6, Re = 1.35E7 and angle of attack of 4°, has been simulated by ANSYS Fluent software using two grids of different quality (Fig. 13.3): the coarser grid (LF) has 30,000 elements while the refined one (HF) has 150,000 elements.

The aerodynamic coefficients obtained from the simulations for the baseline have been validated accordingly to literature results [5] for the selected flight conditions, so that the results with both grids differ from the experimental one reported in literature by less than 1% (Fig. 13.4).

The baseline airfoil has therefore been parameterized by adding two parametric Bezier curves of 4 control points on the upper and lower baseline profiles (Fig. 13.4): the goal of the study is to train a meta-model or Response Surface, to predict the aerodynamic coefficients in function of the 8 parameters (coordinates of the Bezier control points).

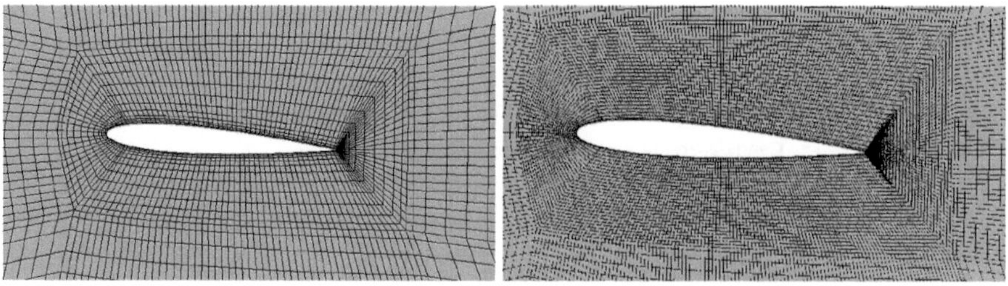

Fig. 13.3 NACA0012 airfoil. LF and HF grid (30,000 and 150,000 elements)

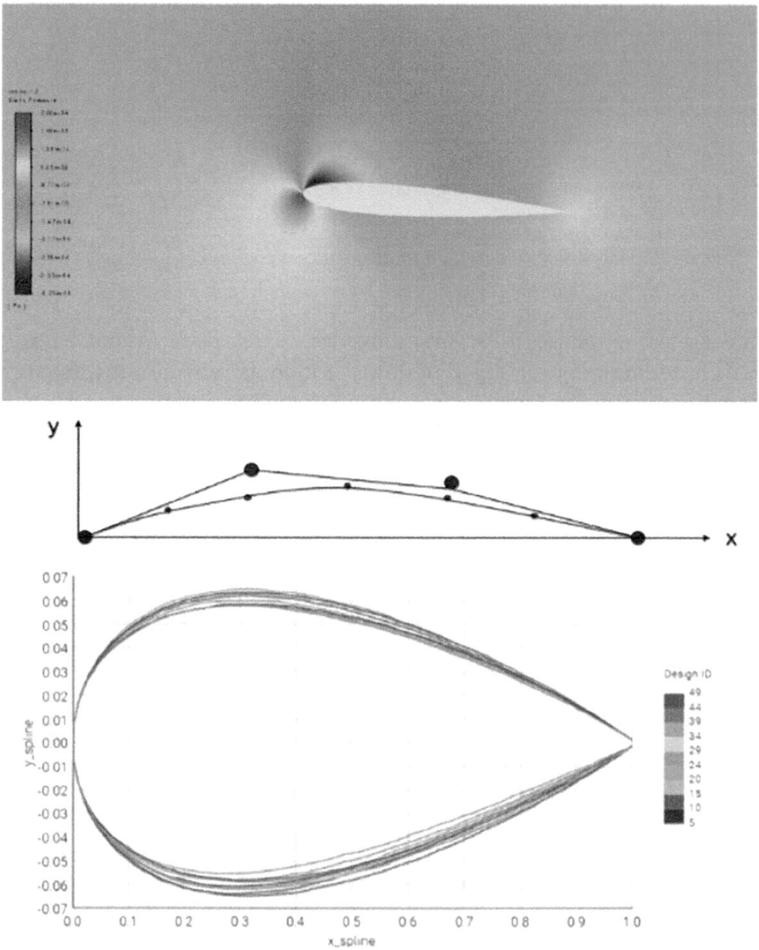

Fig. 13.4 Baseline simulation (M = 0.6, Re = 1.35E7, $\alpha = 4°$) and Bezier parameterization (below)

Table 13.1 Quality of RSM for Cd, Cl, Cm coefficients for Cokriging versus different training methods

Dataset for Cd	Mean relative error	R-squared
Cokriging (50 LF + 10 HF)	2.46E-3	0.950
10 HF samples by dataset reducer	2.25E-3	0.948
10 random HF samples	3.38E-3	0.881
Dataset for Cl	Mean relative error	R-squared
Cokriging (50 LF + 10 HF)	5.44E-3	0.980
10 HF samples by dataset reducer	1.13E-2	0.929
10 random HF samples	1.84E-2	0.845
Dataset for Cm	Mean relative error	R-squared
Cokriging (50 LF + 10 HF)	1.31E-1	0.984
10 HF samples by dataset reducer	1.71E-1	0.978
10 random HF samples	4.67E-1	0.891

As described in the previous paragraph, a uniform DOE of LF simulations has first been executed, for a total of 50 simulations. Selecting by the Dataset Reducer DOE the optimal subset of training points for each aerodynamic coefficient, 10 optimal sampling points have been re-evaluated by the HF grid.

The results of the Kriging RSM trained by the optimal set of 10 HF design points obtained by Dataset Reducer are reported in Table 13.1 for each coefficient, compared with the ones obtained with 10 fully random HF design simulations. The quality of the RSM is evaluated by using a validation set of 10 random points different from the training set, in terms of mean relative error and R-squared.

The accuracy of the Kriging RSM as expected is rather small by using as the training set a Random DOE (R-squared from 0.84 to 0.89), while it raises significantly using the optimal training subset found by Dataset Reducer (R-squared from 0.93 to 0.98).

Finally, as expected, the results obtained by Cokriging combining in a single model the LF dataset (50 points) and the optimal HF dataset for each coefficient (10 points), have been improved further (R-squared from 0.95 to 0.98), confirming the efficiency of the multi-fidelity approach.

13.3 Multi-fidelity Regression Models for ROM

The second approach presented in this paper is the one that combines models of different fidelity to train vector field meta-models, i.e. ROM (Reduced Order Models).

ROM models [6] can take advantage of the POD (Proper Orthogonal Decomposition) to represent a vector field (e.g., the velocity field around a vehicle) as a function of a reduced basis of the field eigenvectors.

We indicate a parameters set with $\Xi = [\boldsymbol{\mu}_1, \ldots, \boldsymbol{\mu}_\mathbf{N}] \epsilon \mathbb{R}^{P \times N}$, where $\boldsymbol{\mu}_i \epsilon \mathbb{R}^P$ with $\mathrm{i} = 1,..N$ are N different combinations of a number P of design parameters (i.e., N different designs defined by a DOE). Once the simulations are completed for each design, we indicate with $\boldsymbol{Y} = [\boldsymbol{y}(\boldsymbol{\mu}_1), \ldots, \boldsymbol{y}(\boldsymbol{\mu}_\mathbf{N})] \epsilon \mathbb{R}^{m \times N}$ the corresponding matrix of solution snapshots, where $\boldsymbol{y}_\mathbf{i} \epsilon \mathbb{R}^m$ with i = 1,..N are the field solutions computed for each one of the m points of the grid.

The POD approach consists in representing the original parameter set Ξ completed by the solution snapshots $\boldsymbol{Y}$ as a diagonal matrix $\boldsymbol{\Theta}^T \boldsymbol{\Theta} \boldsymbol{\phi}_i = \lambda_i \boldsymbol{\phi}_i$ of rank M_r, where λ_i with $i = 1, \ldots, M_r$ are the eigenvalues and $\boldsymbol{\phi}_i$ are the corresponding eigenvectors or principal modes, which can be normalized as $\boldsymbol{\varphi}_i = \frac{1}{\sqrt{\lambda_i}} \boldsymbol{\Theta} \boldsymbol{\phi}_i$ for $i = 1, \ldots, M_r$

Thus, the reduced field solution $\boldsymbol{y}^*$ in function of a generic set of parameters $\boldsymbol{\mu}$ can be expressed as:

$$\boldsymbol{y}^*(\boldsymbol{\mu}) = \sum_{i=1}^{M_r} \alpha_i(\boldsymbol{\mu}) \boldsymbol{\varphi}_i \text{ for } i = 1, \ldots, M_r \tag{13.6}$$

with $\alpha_i(\boldsymbol{\mu})$ a weight coefficient that can be determined by interpolation in function of the design database. Several methods are available in literature to determine the weight coefficients, including POD-Galerkin, zonal-POD or POD with residual minimization [7].

In a multi-fidelity strategy [8], it is possible to combine the results from vector fields of different quality.

First, the LF solution can be approximated by the POD model, obtaining from Eq. 13.6 the following expression:

$$\boldsymbol{y}^*_{LF}(\boldsymbol{\mu}) = \sum_{i=1}^{M_r} \alpha_i(\boldsymbol{\mu}) \boldsymbol{\varphi}_i \tag{13.7}$$

The POD approximation of the error between LF and HF solutions, with $\boldsymbol{\zeta}_i$ the principal modes of the error function and with $\varepsilon_\mathrm{i}(\boldsymbol{\mu})$ the weight coefficients resulting from interpolation, therefore gives:

$$\boldsymbol{e}^*(\boldsymbol{\mu}) = \sum_{i=1}^{M_k} \varepsilon_i(\boldsymbol{\mu}) \boldsymbol{\zeta}_i \tag{13.8}$$

The reduced solution for a given parameter $\overline{\boldsymbol{\mu}}$, as represented in Fig. 13.5, can therefore be given by the sum of the 2 contributions defined in Eqs. 13.7 and 13.8:

$$\boldsymbol{y}^*(\overline{\boldsymbol{\mu}}) = \boldsymbol{y}^*_{LF}(\overline{\boldsymbol{\mu}}) + \boldsymbol{e}^*(\overline{\boldsymbol{\mu}}) \tag{13.9}$$

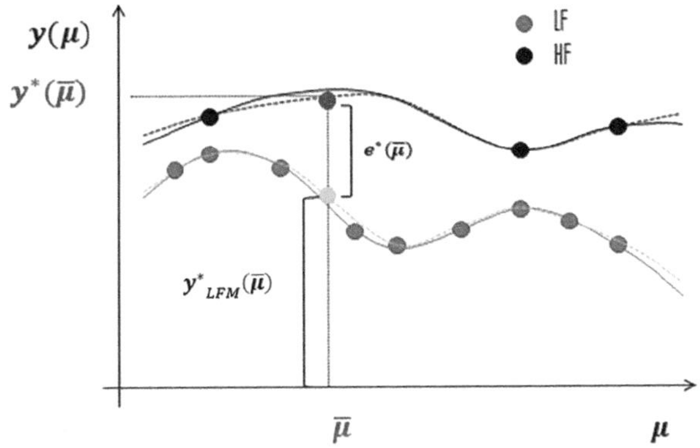

Fig. 13.5 Conceptual implementation of a multi-fidelity ROM model

13.3.1 Multi-fidelity ROM Applied to Vehicle Aerodynamics

The Aerodynamic model that has been used to apply the multi-fidelity ROM methodology is DrivAer + [9], a car model prepared by the University of Munich in collaboration with AUDI and BMW (Fig. 13.6). The model includes mirrors, underbody, and rotating wheels. The CFD analysis is performed at a reference speed of 140km/h (full turbulent flow).

The mesh model has been parameterized using Mimic mesh morphing software from Optimad (https://www.optimad.it/mimic/) for a total of 7 parameters: 2 parameters modify the shape of the pillars, 2 others the extrusion of the rear spoiler (Fig. 13.6), and other 3 the vertical and horizontal displacements of the rear bumper shape (Fig. 13.6).

Two numerical models have been defined in OpenFoam: the HF one is a RANS simulation on a grid of 15 million cells (Fig. 13.7), while the LF model is a RANS simulation on a coarser grid of 2 million cells obtained by a sensitivity analysis made on the main mesh parameters (Fig. 13.7). The simulation of the LF model is 10 times faster than the HF one, and the loss in terms of accuracy is in the order of 2%.

The multi-fidelity ROM methodology has been completely set up in the process automation workflow of modeFRONTIER (Fig. 13.8). Every component of the problem is a node connected to the others: the nodes including the input parameters values that are updated for each design are linked to the script node that automatically executes the mesh morphing with Mimic and performs on the modified mesh the CFD analysis with OpenFoam. Output responses and in particular drag and lift coefficients are finally obtained from the analysis.

Once both the LF and HF samples have been computed, a POD analysis has been performed to evaluate the principal modes for each set of solutions. In Fig. 13.9 we report the first principal modes, or snapshots, that have been identified for the LF series (top) and for the HF-LF error (bottom).

For the LF model, 200 evaluations have been performed: in this case, a basis of 89 modes shows a good convergence of the energy (Fig. 13.10), i.e. it can express with a good approximation the total energy content (the λ parameter indicates the

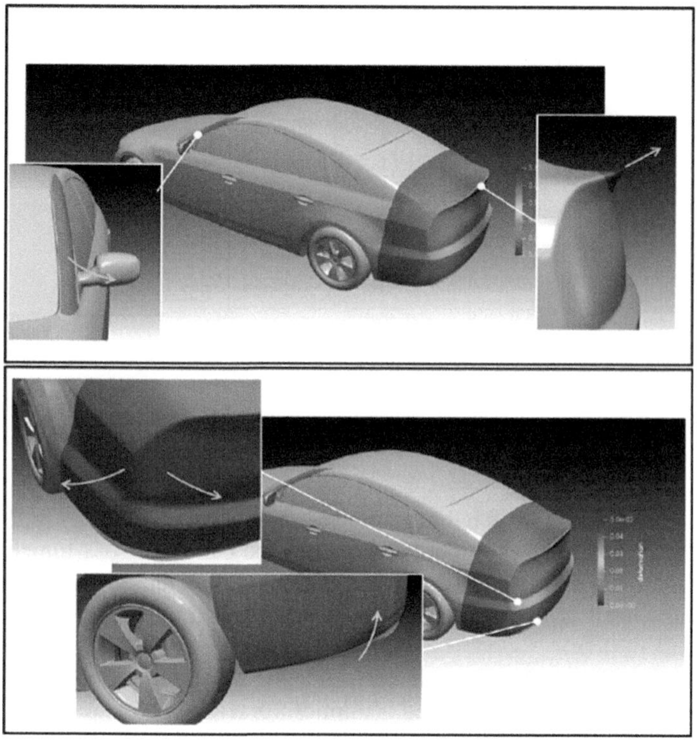

Fig. 13.6 DrivAer + model: parameterization (Mimic) of rear bumper, pillars and rear spoiler

distribution of the eigenvalues: U corresponds to Energy, p to pressure and v to velocity). For the HF-LF error model, since the number of HF evaluations is reduced, we can extract a basis of 18 modes only to expresses with a good approximation the energy content.

Once the ROM model is available after the POD analysis, it can be used directly inside VOLTA, which is the ESTECO web-based collaboration platform that allows any authorized users to evaluate directly the ROM models in function of any combination of parameters, or defining specific goals to execute an optimization. In the end, the user can visualize the input/output parameters of the selected configurations, and can also plot the CFD field computed directly by the ROM model.

To estimate the level of accuracy of the ROM versus the original CFD models, we have considered the Cd coefficient as criteria of comparison (Table 13.2): the ROM model reaches a level of accuracy even slightly higher than the LF CFD model (2% of error w.r.t HF instead of 2.1%), but the big advantage of the ROM model is that it can be actually used as a quasi real-time CFD model, since it takes just 4 min of evaluation time, versus the 13 CPU hours of the LF model and 160 CPU hours for the HF.

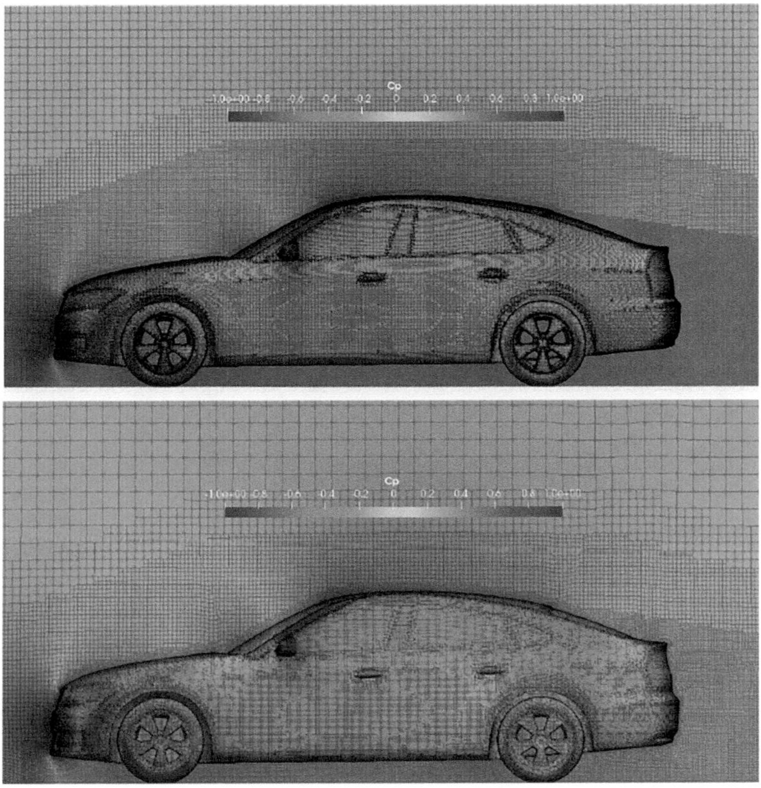

Fig. 13.7 OpenFoam analysis on refined and coarser mesh (15E6 vs 2E6 cells)

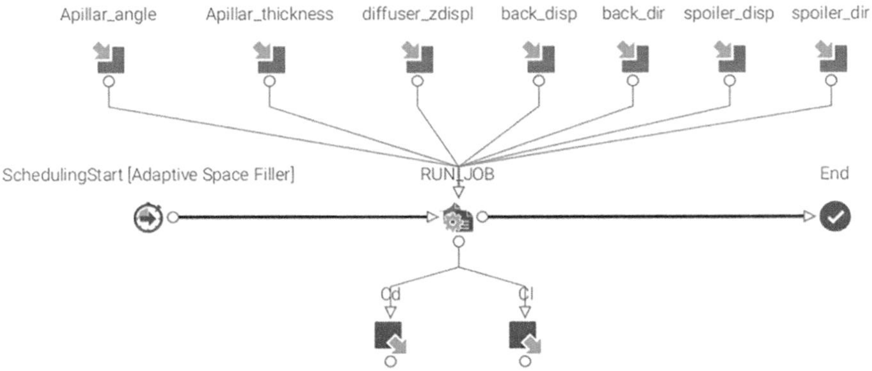

Fig. 13.8 ModeFRONTIER process automation workflow for ROM database

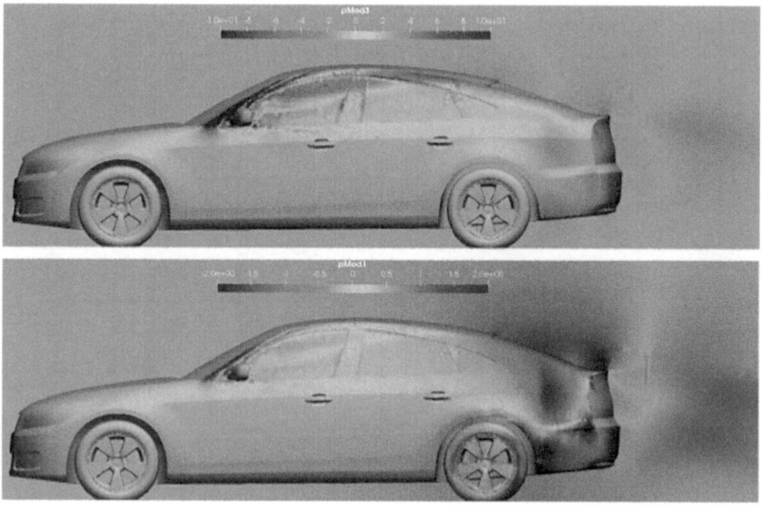

Fig. 13.9 Snapshots of first modes for LF (top) and error HF-LF (bottom) datasets

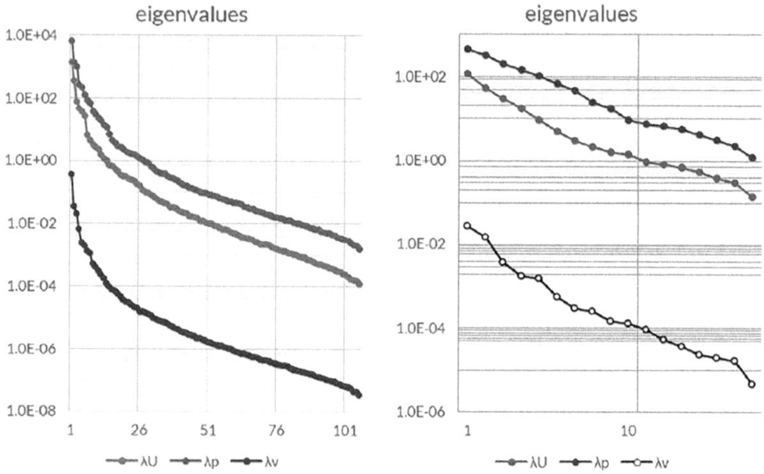

Fig. 13.10 Eigenvalues distribution (energy, pressure, and velocity) vs modes number for LF (left) and error (right) datasets

Table 13.2 Performance of ROM model and CPU training time versus HF and LF

Performance (random point)	Cd error (%)	CPU h
HFM	0	160
LFM	2.1	13
ROM	2.0	0.067 [~4 min]]

13.4 Conclusions

Two different methodologies for multi-fidelity Response Surfaces have been proposed in this paper.

Cokriging can be used to obtain an efficient RSM combining LF and HF sampling points. The Dataset Reducer DOE, based on the selection of the subset of a LF dataset that minimizes the errors for RSM training, can be used to identify the optimal set of HF sampling points for RSM training. The efficiency of the methodology has been proven by applying the methods to an aerodynamic test case.

Finally, multi-fidelity ROM (Reduced Order Model) has been introduced to build by the POD (Proper Orthogonal Decomposition) technique an equivalent CFD model in function of a reduced number of parameters. The method has been applied to an automotive aerodynamic test case, that proved the efficiency of the methodology in terms of simulation time and accuracy of the results.

References

1. Kennedy MC, O'Hagan A (2000) Predicting the output from a complex computer code when fast approximations are available. Biometrika 87:1
2. Forrester AIJ, Sóbester A, Keane AJ (2007) Multi-fidelity optimization via surrogate modelling. In proceedings of the royal society, mathematical, physical and engineering sciences
3. Le Gratièt L (2013) Multi-fidelity Gaussian process regression for computer experiments. Dissemination at Université Paris-Diderot-Paris VII
4. www.esteco.com
5. Ladson CL, Rill S (1987) High reynolds number transonic tests of an NACA 0012 airfoil in the langley 0.3-meter transonic cryogenic tunnel. NASA Technical Memorandum 100527, p 89
6. Genesio Rf, Milanese M (1976) A note on the derivation and use of reduced-order models. IEEE Trans Autom Control 21.1:118–122
7. Ullmann S, Rotkvic M, Lang J (2016) POD-Galerkin reduced-order modeling with adaptive finite element snapshots. J Comput Phys 325:244–258. ISSN 0021-9991
8. Perron C, Rajaram D, Mavris DN (2021) Multi-fidelity non-intrusive reduced-order modelling based on manifold alignment. https://doi.org/10.1098/rspa.2021.0495
9. Collin C, Mack S, Indinger T, Mueller J (2016) A numerical and experimental evaluation of open jet wind tunnel interferences using the DrivAer Reference model. SAE Int J Passeng Cars-Mech Syst 9(2016–01–1597):657–679

Chapter 14
Krueger High-Lift System Design Optimization

Emiliano Iuliano, Domenico Quagliarella, and Jochen Wild

Abstract This work describes the cooperative/competitive design process that led to the definition of the Krueger flap to be used in the numerical and experimental tests of the European project UHURA. The project requirements are particularly challenging because it is necessary to develop a device with good aerodynamic high-lift characteristics, but it is necessary to consider many constraints of structural and kinematic nature. Indeed, the kinematics for its deployment is quite complex and imposes hard constraints on the Krueger shape, and the structural characteristics must allow it to withstand considerable structural stresses in the deployment phase which is studied in the wind tunnel.

Keywords High-lift aerodynamics · High-lift system design · Design optimization

14.1 Introduction

Within the EU UHURA project [1], the need emerged to design a folding bull nose Krueger device as the target configuration for studying unsteady aerodynamic effects during deployment and retraction of such a system. Consequently, the design objec-

This work was performed when Dr. Iuliano was working at CIRA.

E. Iuliano · D. Quagliarella (✉)
Italian Centre for Aerospace Research (CIRA), Via Maiorise, Capua, Italy
e-mail: d.quagliarella@cira.it

E. Iuliano
e-mail: emiliano.iuliano@lilium.com

E. Iuliano
Lilium eAircraft GmbH, Claude-Dornier Str. 1, Geb. 335, 82234 Wessling, Germany

J. Wild
German Aerospace Center (DLR), Institute of Aerodynamics and Flow Technology, Braunschweig, Germany
e-mail: jochen.wild@dlr.de

T. Tuovinen et al. (eds.), *Advanced Computational Methods and Design for Greener Aviation*, Computational Methods in Applied Sciences 59,
https://doi.org/10.1007/978-3-031-61109-4_14

tive of generating as much as possible lift force was subject to the heavy and demanding constraint of being adequately integrated into existing wind tunnel models. Hence, the design specifications require a high-performance Krueger device that doesn't exhibit artificial flow effects besides offering high-performance. This constraint is more important than getting the ultimate performance in terms of maximum lift. As a starting point, the design experience obtained in previous projects, DeSiReH [2] and AFLoNext [3], has been used to generate an initial Krueger shape. Hence, the characteristic design parameters of the Krueger device obtained in DeSiReH have been mapped to the DLR-F15-LLE geometry [4]. The design of the Krueger flap has been obtained by a cooperative concurrent engineering approach between two UHURA partners, namely CIRA and DLR, in an iterative process. In a first loop, independent optimizations were performed based on the partner's best practice methods. Afterwards, the designs were merged by selecting beneficial aspects of both optimization results. Finally, the design was adapted to respect refined kinematics constraints. The synthesized design achieves all requirements from kinematics and achieves a high level of maximum lift coefficient. For the final design, aerodynamic forces have been derived for the Krueger panel and the bull-nose to support kinematics sizing.

14.2 Kruger Device Shape Optimization

The design of the Krueger flap has been obtained by a cooperative concurrent engineering approach with two UHURA partners, CIRA and DLR, involved. Based on a common design case definition and requirements, in an iterative process a common design has been created. In a first loop independent optimizations were performed based on partner's best practice methods. Afterwards the designs were merged by selecting beneficial aspects of both optimization results. Finally the design was adapted to respect refined constraints, especially from kinematics feasibility requirements.

14.2.1 Initial Krueger Device Shape

The initial Krueger flap design, reported in Fig. 14.1 is derived from the AFLoNext design parameters mapped on DLR-F15-LLE airfoil. Initial side constraints taken from the AFLoNext project were the front spar clearance and the shielding requirement against 7° impact at 6° angle of attack.

The design variables from AFLoNext result in a Krueger flap with a deflection angle close to the upper limit of $\delta_K = 150°$. The shallow setting originates from a relatively thin leading edge and the strong shielding requirement. The initial Krueger flap shape has been provided via Coordination Memorandum UH-11-CM03-DLR [4].

14.2.2 *Design Case Definition*

The design case specification was made to explore the broadest possible design space. Hence, the degrees of freedom for the design of the Krueger flap incorporate all shape variables for the bull nose and the inner panel shape. The Krueger flap is selected as a folding bull nose device as this incorporated a higher degree of freedom for the Krueger leading edge shape and avoids the necessity of additional devices to seal the Krueger cavity in retracted position.

14.2.2.1 Design Targets and Requirements

The target of the design is to obtain a reasonable Krueger flap shape that incorporates a highly loaded Krueger flap for high maximum lift coefficient based on a feasible gooseneck kinematics. The Krueger angle should not be too shallow in order to prevent premature separation on the upper side of the Krueger. As the Krueger flap is intended for a laminar wing airfoil, the Krueger position must achieve a position that prevents insect contamination during take-off and landing. The initial Krueger shielding was first taken from the AFLoNext project to prevent an impact of an insect with 7° relative to the surface tangent at an angle of attack of 7°. Figure 14.2 depicts the geometric interpretation of this design target.

14.2.2.2 Design Constraints

The size of the Krueger flap is limited by a request on a front spar clearance of 0.5%c. Additionally due to the type of kinematics, there is an implicit fixation of the panel

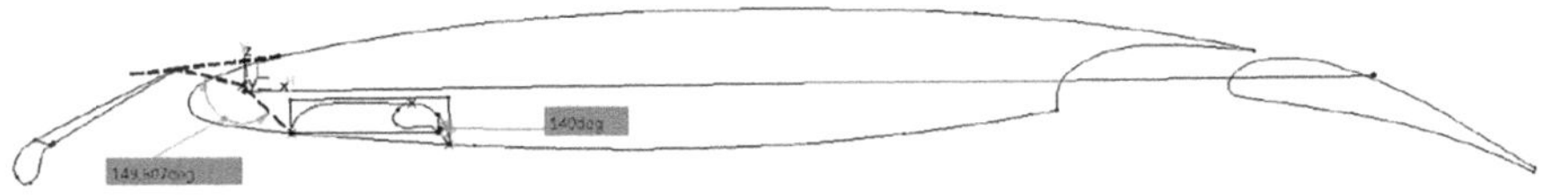

Fig. 14.1 Initial Krueger flap design

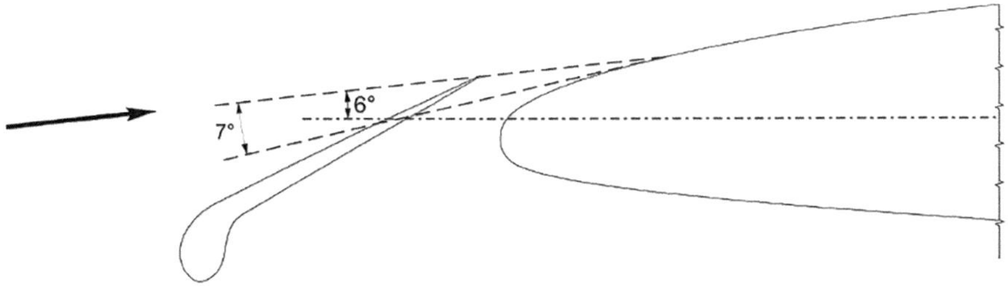

Fig. 14.2 Geometric representation of the insect shielding requirement

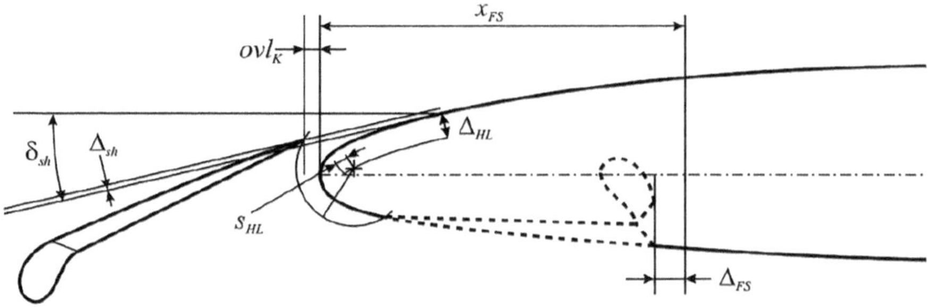

Fig. 14.3 Limitation of Krueger panel size implicitly from inematics type and target setting

trailing edge resulting from the designated setting in deflection position and the hinge line position. The hinge line is hereby restricted to be not closer than 0.556%c to the upper shell shape to allow for bearings to mount the kinematics. Figure 14.3 shows the relations of these limitations.

For the bull nose shape, a side constraint is given on the size of the cavity. In order to maintain structural integrity, the cavity needed to store the Krueger shall not extend 40% of the local airfoil thickness at any position.

14.2.3 DLR Design

DLR used its validated numerical optimization environment CHAeOPS in conjunction with the RANS CFD solver FLOWer [5]. Computational multi-block structured grids are generated by a parametric procedure that achieves self-similar mesh qualities throughout the design space. The mesh consists of 10 blocks with a total of 151,162 grid points. The mesh resolution is adopted for viscous flows to achieve a non-dimensional wall distance of the first cell in the field in the order of $y^+ = 1 \div 2$.

The flow solver FLOWer is a standard finite volume method to solve the Reynolds-averaged Navier-Stokes (RANS) equations. It applies a 2nd order central flux discretization with artificial dissipation. Steady state solutions are obtained by an explicit 5-stage Runge-Kutta time stepping scheme accelerated by implicit residual smoothing, local time stepping and a 3-level multigrid. For turbulent viscous flow the closure of the RANS equations is obtained by application of the one-equation Spalart-Almaras turbulence model modified by Edwards and Chandra. The turbulence equation is solved by a fully implicit scheme. In order to speed-up the optimization, the RANS calculations during the optimization have been computed on a coarser mesh obtained as the 2nd multigrid level.

The computational grid is a multi-block grid consisting of 10 structured blocks. The mesh consists of about 150,000 grid points. The boundary layer is resolved by boundary conforming meshes with a high density normal to the wall with at least 32 layers to resolve the boundary layer. The first wall spacing is adopted to the designated

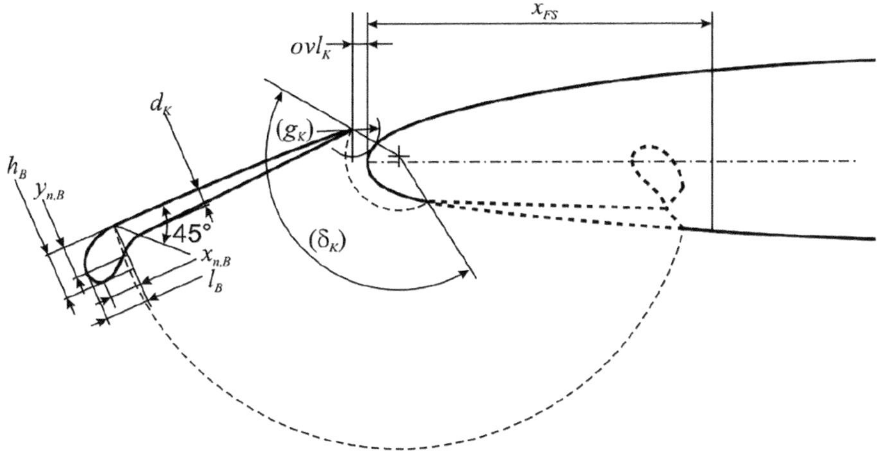

Fig. 14.4 Geometric parameterization of the Krueger flap applied in the DLR design optimization

Reynolds number to achieve a non-dimensional wall distance of $y^+ = 1 \div 5$. The farfield boundary is located at 10 airfoil chord lengths as the solver features a farfield vorticity correction based on the circulation produced by the airfoil.

Figure 14.4 depicts the geometric relation of the design parameters used in the optimization process. Together with the parameters for the hinge line coordinate and the front spar clearance shown in Fig. 14.3, 12 parameters in total are used to fully describe the geometry. As stated above, the length of the Krueger panel and the deflection angle are no direct parameters but result from the feasibility constraints of the kinematics and the designated trailing edge location in deflected position. For the panel the free shape parameter is the thickness at the junction to the bull nose. The bull nose shape is defined by four parameters defining the size and the position of the extrema in local coordinates. The extrema are GC1-continuously connected by cubic splines to the Krueger panel shape.

The optimization has been aimed at two objectives relating to first order aircraft performance indicators in high-lift flight condition. The first objective has been to minimize the stall speed V_S by maximizing the maximum section lift coefficient $c_{L,\max}$.

$$f_1 = -c_{L,\max} \ @ \ M_\infty = 0.19 \tag{14.1}$$

The second objective has been a performance increase to maximize the climb speed inidicated by the climb ratio based on section data

$$f_2 = \frac{c_D^2}{c_L^3} \ @ \ M_\infty = 0.2337 \ \left(\hat{=}1.23 \times V_S\right) \tag{14.2}$$

The optimization method applied was a mono-objective simplex method with sub-space division. The single objective has been constructed by an addition of both

Table 14.1 Summary of DLR optimization results

ID airfoil	Spalart-Allmaras		
	F_{obj} [1]	f_1	f_2
Baseline	−0.0572	−2.7331	2.6759×10^{-5}
Intermediate-optimum	−2.2568	−3.4162	1.1594×10^{-5}
Final-optimum	−3.4851	−4.3271	0.8420×10^{-5}

[1] In the table, the values of the global objective are based on the final weighting

objectives with weights attempting an approximately equal influence. A first optimization was run starting with the baseline configuration and a global objective of

$$F_{\text{obj}} = f_1 + 1000 \cdot f_2 \tag{14.3}$$

The maximum lift iterations were started at an angle of attack of 15.5°. After 160 RANS evaluations it turned out that in the optimized configurations the maximum lift was not properly detected due to a too high starting angle of attack. The starting angle of attack was reduced to 11°. Additionally, due to the improvements obtained that far, the objective function was misbalanced to favour maximum lift only as the high drag values of the initial configuration resulting from flow separations disappeared. Therefore the global objective has been refined to

$$F_{\text{obj}} = f_1 + 10000 \cdot f_2 \tag{14.4}$$

The final optimization then needed 6 optimization cycles performing 330 RANS evaluations of different configurations at a wall time of 444 h on a single workstation. Table 14.1 lists the changes in the objective functions. Figure 14.5 depicts the changes in the geometry. It is seen that the optimized Krueger is much more steeply deflected. Additionally, the bull nose significantly increased in size and exhibited a smooth curvature. The Krueger panel length is approximately similar, which is better observed when inspecting the resulting cavity in the wing.

For analysis of performance and loads, the full polar curve has been simulated on the fine grid. Figure 14.6 shows a comparison of the lift coefficient versus the angle of attack of the initial and the optimized configuration. Due to the shallow Krueger deflection of the initial configuration, the lift curve exhibits an early separation due to flow separation over the Krueger flap, which is successfully prevented with the optimized design parameters. The optimized configuration exhibits a maximum section lift coefficient of $c_{L,\max} = 4.527$ at an angle of attack of $\alpha = 26°$. In the linear range, there is only a slight drop in lift coefficient compared to the initial configuration. Such high performance is promising as it is very similar to those of a classical slat device, especially with regard to the shallow trailing edge flap deflection.

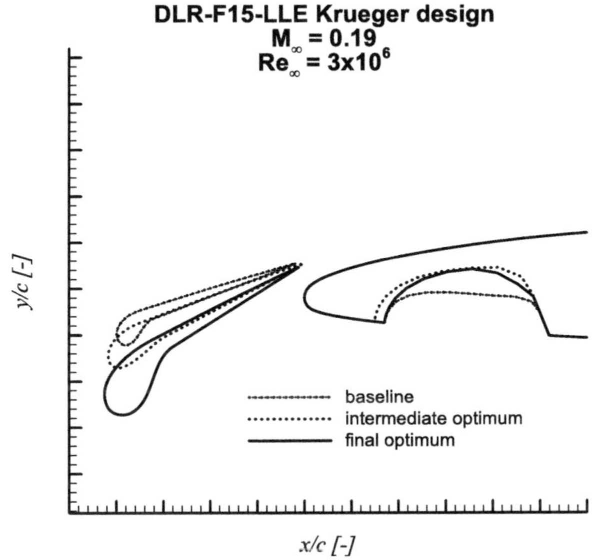

Fig. 14.5 Comparison of the shapes of the Krueger flap for the initial, an intermediate and the finally optimized configuration of the DLR design optimization

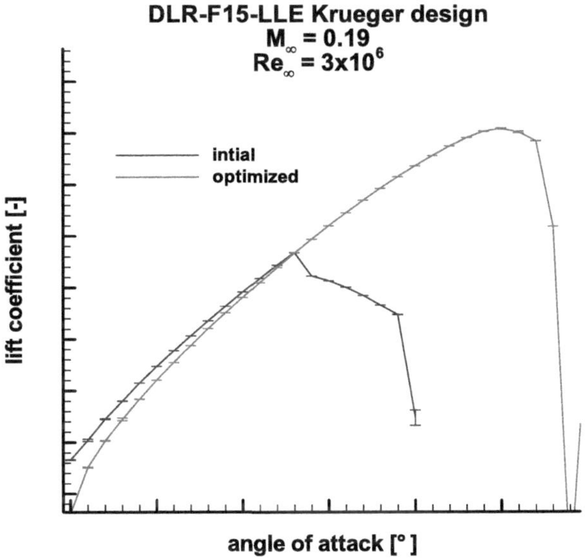

Fig. 14.6 Comparison of lift coefficient versus angle of attack for the initial and optimized configuration of the DLR design optimization

14.2.4 CIRA Design

CIRA was in charge of defining the shape of the Krueger element side by side with DLR. As the focus was on designing the Krueger shape compatibly with stowing and kinematic constraints, a dedicated CAD-free geometric parameterization was implemented to allow for feasible panel length and deployment handling. Then, a numerical optimization task was performed to improve the high-lift performance of the baseline shape. In the following sections, details are provided about both activities.

14.2.4.1 Geometry Parameterization and Constraints Handling

The shape of the Krueger element was strongly constrained as defined in 14.2.2.2 to meet kinematics and deployment considerations. Consequently, CIRA geometry parameterization has been tailored to this specific problem at hand. With reference to Figs. 14.7, 14.8 and 14.10, a parametric CAD-free procedure has been setup consisting of the following steps:

Hinge point stay-in line and location—the constrained hinge line is parameterized as a curve, and the hinge point location is taken as the corresponding curvilinear abscissa (s_h, ranging from 0 to 1, where 0 and 1 identify the lower and upper limit, respectively, and 0.5 identifies the cuspid at leading edge, see Fig. 14.7);

Krueger panel forward limit—in order to avoid the clash of the Krueger trailing edge with the pointy wing leading edge during the deployment, a limiting circle is

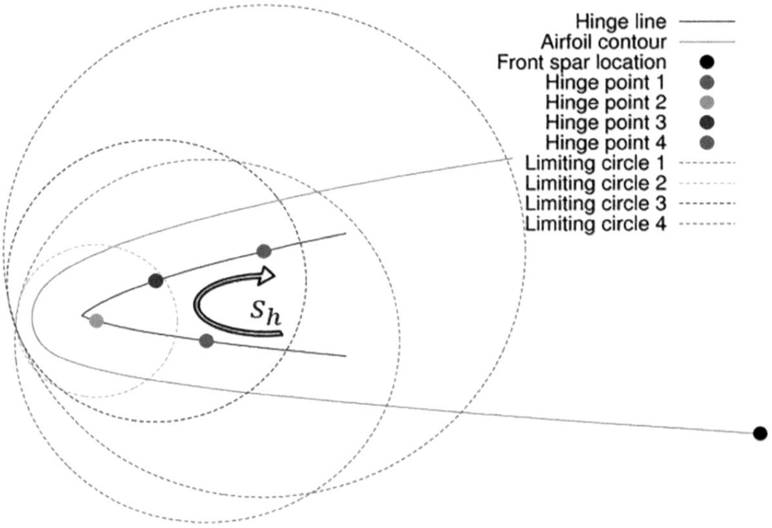

Fig. 14.7 Offset hinge line, hinge point locations and corresponding limiting circles

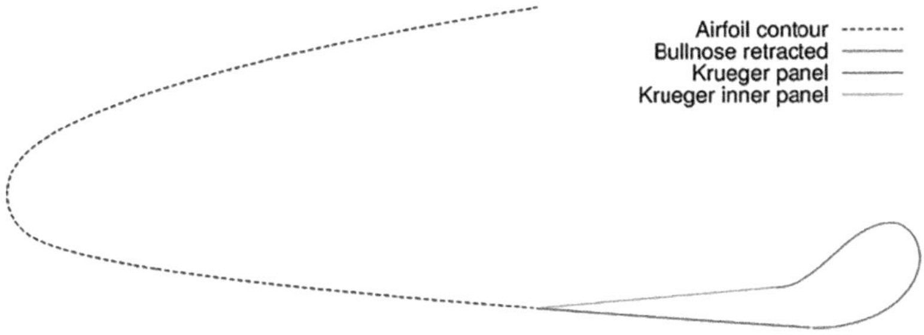

Fig. 14.8 Definition of the Krueger geometry in rectracted position. The outer panel (purple line) is defined by kinematic constraints, the reference bullnose shape (red line) is connected to it and the inner panel is defined as a line segment (green line)

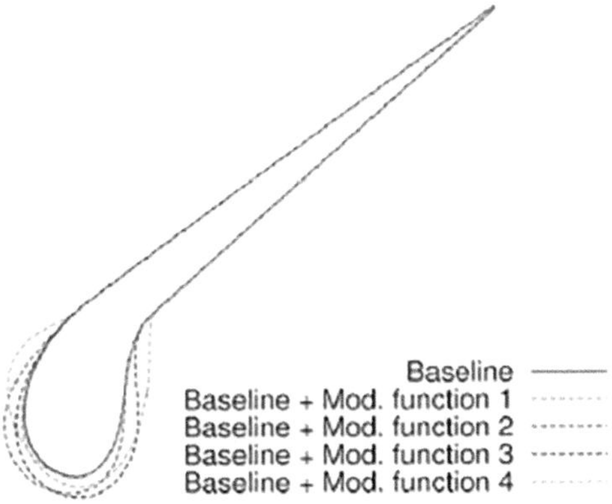

Fig. 14.9 Reference bullnose shape modified according to the selected Hicks-Henne functions

drawn, representing the trajectory of the Krueger element trailing edge during the deployment phase. The circle is centered in the hinge point and tangent to the leading edge. The radius is increased of an additional $0.005\%c$ in order to assure a minimal clearance during the deployment. Figure 14.7 shows various choices for hinge point and corresponding limiting circles;

Krueger panel definition—the Krueger outer panel (in retracted position) is automatically defined by the parameter s_h. Indeed, the panel aft limit is represented by the front spar position, the forward limit by the intersection of the limiting circle with the airfoil contour;

Baseline shape in retracted position—the (reference) bullnose shape is attached to the Krueger panel in retracted position and the Kruger inner panel is defined as a line segment connecting the bullnose to the trailing edge (see Fig. 14.8);

Bullnose shape modification—the bullnose shape is locally modified as $y_m(x) = y_b(x) + \sum_{i=1}^{4} w_i f_i(x)$, where $y_m(x)$ is the modified shape, $y_b(x)$ is the baseline reference shape, w_i are the shape control parameters and $f_i(x)$ are sinusoidal Hicks-Henne bump functions: $f_1(x) = 0.1\sin^3\left(\pi x^{0.431}\right)$, $f_2(x) = 0.1\sin^3\left(\pi x^{0.757}\right)$, $f_3(x) = 0.1\sin^3\left(\pi x^{1.357}\right)$, $f_4(x) = 0.1\sin^3\left(\pi x^{3.106}\right)$. Figure 14.9 shows an example of the geometry modifications introduced by applying the bump functions one by one. The range of variation of the weights are chosen in such a way as to keep the bullnose height within the limit imposed by the stowage space in retracted position (around 40 % of the main airfoil thickness). As it will be detailed in the following sections, this aspect represents an important trade-off between the achievable aerodynamic performance and the allocation of the Krueger device.

Modified shape in deflected position—the entire Krueger shape, modified as aforementioned, is then rotated about the hinge point by an angle δ which varies according to the hinge point location so as to satisfy the shielding requirement. With reference to Fig. 14.10, δ is the angle identified by the intersections of the limiting circle with the lower airfoil contour (Krueger panel trailing edge) and the shielding line (purple line) respectively. By construction, there exists a range of values of s_h which do not allow to satisfy the shielding constraint (no intersection between the limiting circle and the shielding line) or to reach it with a feasible rotation ($\delta < 150°$). The allowed range is found to be $s_h \in [0.55, 1.0]$, i.e. the hinge point should stay on the upper branch of the hinge line and far enough from the leading edge cuspid. This consideration is directly exploited to set the range of variation of the hinge point in the optimization loop.

The total number of design parameters is 5: the s_h parameter controls the panel length, the deflected Krueger setting and gap/overlap (in accordance with the shielding constraint), while the 4 parameters w_i controls the bullnose shape.

14.2.4.2 Shape Optimization Problem Definition

The shape optimization problem is aimed at improving the performance of the three-element configuration in high-lift conditions at single design point $\left(M_\infty = 0.15, \Re_\infty = 7 \times 10^6\right)$. The reference length for airfoil coefficients calcula-

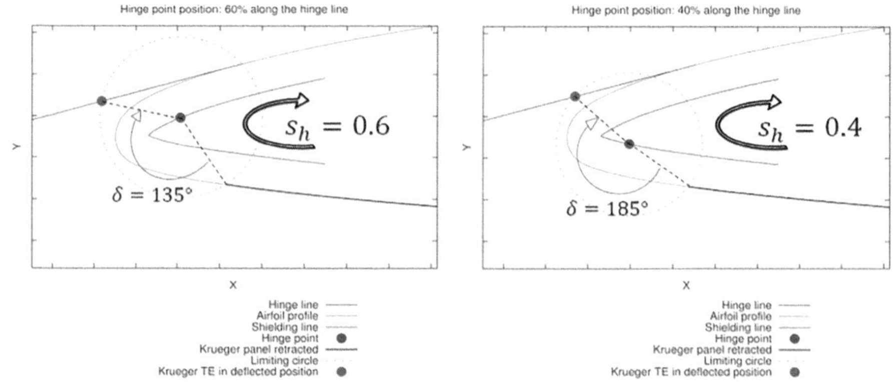

Fig. 14.10 How the hinge point location and shielding angle constraints determine the Krueger trailing edge deflected position

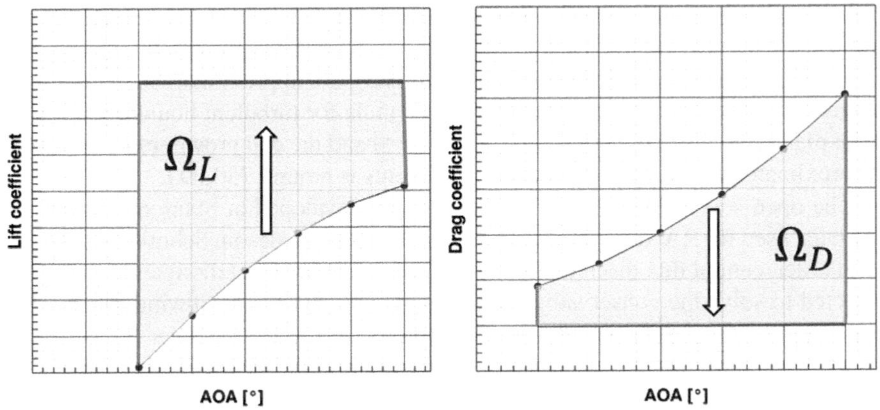

Fig. 14.11 Measure of lift and drag curve domains for objective function computation

tion is based on the projected length onto the X-axis in deflected position ($= 1.37m$ for unit chord airfoil). Instead of considering only the maximum lift coefficient as objective, a quite different approach has been adopted to improve the aerodynamic performance along a selected branch of the lift curve. With reference to Fig. 14.11, two "utopia" points (red lines) have been identified for lift and drag coefficients, namely $c_L^* = 3.2$ and $c_D^* = 0.0300$, as well as a window of observation for the angle of attack defined by $\alpha_l = 14°$ and $\alpha_u = 24°$ (lower and upper bounds, respectively). The objective function is evaluated as:

$$f_{\text{obj}} = \int_{\alpha_l}^{\alpha_u} \left[\left(c_L^* - c_L \right) + 10 \left(c_D - c_D^* \right) \right] \mathrm{d}\alpha = \mu \left(\Omega_L \right) + 10 \mu \left(\Omega_D \right) \qquad (14.5)$$

where μ is the area of the green domains in Fig. 14.11 and the weight 10 on the drag contribution has been set to give it the proper importance. In other words, such a choice of the objective function rewards both an increase in $c_{L,\max}$ and a better lift characteristic while approaching the stall condition. On the other hand, taking into account the drag performance allows penalizing the onset of separated regions. The angle of attack range $[\alpha_l, \alpha_u]$ has been discretized in steps of $\Delta\alpha = 2°$ in order to reduce the total number of CFD simulations and save computational time.

14.2.4.3 Shape Optimization Process and Results

An automatic meshing procedure has been prepared in order to obtain fixed-quality meshes around varying geometries. The procedure employs the GMSH open-source tool [6] to generate mixed quad-tri meshes. The minimum triangle element size is 0.00075 (with respect to unit chord airfoil), while the farfield is a square box located at 40 airfoil chords from the main element leading edge and its element size is set to 6.

A distance-based refinement is employed near the airfoil surface, enabling a smooth transition from near-wall to farfield elements. A boundary layer region (quad layers) is attached around the solid surfaces by specifying the approximate boundary layer height (estimated by means of Schlichting formula for turbulent boundary layer on a flat plate), the first cell height (to have $y^+ = 1$) and the cell growth ratio ($= 1.08$). Approximately, the resulting number of elements is around 750,000.

The open-source SU2 flow solver v6.2.0 [7], developed at Stanford University, has been used for RANS analyses. The JST scheme by Jameson, Schmidt and Turkel (2nd order, central flux discretization with addition of artificial dissipation) has been selected to solve the conservative variables, while a 2nd order upwind scheme has been used to solve the Spalart-Allmaras turbulent variable. An Euler implicit time scheme has been used to advance the solution in pseudo-time towards steady-state with an adaptive CFL number ranging from 2 to 50. Convergence has been assumed when either a 6-order residual drop has been reached for all equations or all residuals are lowered below 10^{-8}.

The optimization task has been split into two steps: the first was aimed at quickly exploring the design space with a global algorithm to find a sub-optimal solution, and the second targeted a refinement of the solution. In the initial phase, the Efficient Global Optimization (EGO) algorithm [8, 9] was launched to perform a balanced search of the design space by means of a Kriging surrogate model assisted by the Expected Improvement Maximization criterion. This choice has been driven by the limited dimension of the design space, which makes such an algorithm very effective. The design space was initially sampled with 36 Latin Hypercube Sampling points and then 86 EGO iterations were performed. Starting from the EGO best solution, a Covariance matrix adaptation evolution strategy (CMA-ES) [10, 11] evolutionary search has been launched with a population size of 12 candidates. This phase has required further 268 CFD evaluations of design candidates.

Optimization results are summarized in Table 14.2 in terms of objective function values and measures of lift/drag curve contribution. For the sake of completeness, even if during the optimization iterations the CFD simulations have been carried out with the Spalart-Allmaras turbulence model, the candidates under analysis (baseline, EGO and CMA-ES optima) have been further analyzed with the $k - \omega$ SST model. Results show that the optimization trend is well captured even by the SST model: both EGO and CMA-ES optima exhibit superior aerodynamic performances with respect to the reference shape. The $k - \omega$ SST turbulence model is in line with the optimization results and is more conservative in predicting the aerodynamic performance of the three-element airfoil in high-lift conditions. The chosen optimization strategy has proven to be successful as the global algorithm recovered the most significant aerodynamic issues of the reference shape while the CMA-ES algorithm worked to push the lift curve towards the "utopia" point while securing the drag levels.

Figure 14.12 depicts the optimized shapes as compared to the reference one. As a result of the design process, the optimized Krueger elements present a narrower gap and a steeper setting. The bullnose contour is thicker and more rounded, which strongly contributes to increasing the maximum lift and the stall angle. The EGO optimum features the shortest panel length and the steepest setting: this design feature

Table 14.2 Summary of CIRA optimization results

ID airfoil	Spalart-Allmaras			$k-\omega$ SST		
	f_{obj}	$\mu(\Omega_L)$	$\mu(\Omega_D)$	f_{obj}	$\mu(\Omega_L)$	$\mu(\Omega_D)$
Baseline	18.893	8.657	1.23	28.016	11.670	1.635
EGO optimum	7.266	4.830	0.244	8.039	5.412	0.263
CMA-ES optimum	6.391	3.993	0.239	7.050	4.283	0.277

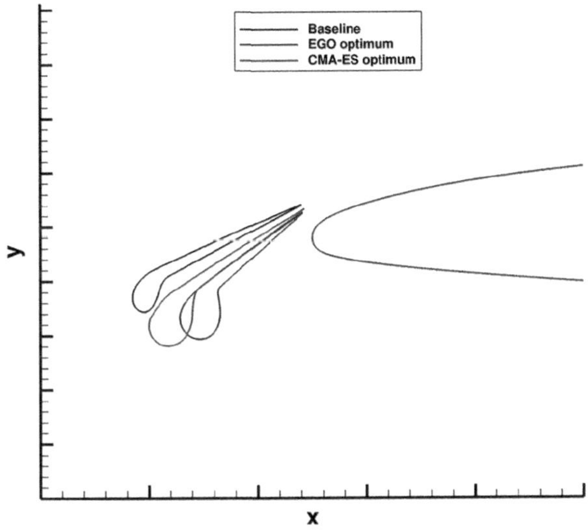

Fig. 14.12 Comparison of reference baseline and optimized shapes

seems to be in partially opposite trend with respect to the final solution, i.e. the CMA-ES optimum, and may suggest a multi-modal behavior, which makes the optimal solution not unique.

Figure 14.13 shows the lift coefficient curve versus the angle of attack for reference and optimized Krueger shapes as computed by using both Spalart-Allmaras and $k-\omega$ SST turbulence models. In order to be consistent with the data shown in the next section about Krueger shape synthesis, the aerodynamic coefficients have been scaled to unit reference length. In the linear range, a slight lift loss is observed, while at higher angles of attack the flow separation on the suction side of the Krueger element is recovered and the lift levels are significantly enhanced. The baseline $c_{L,max}$ is 3.43 at 14° angle of attack ($c_{L,max} = 3.71$ at 18° angle of attack with the Spalart-Allmaras model). As regards the CMA-ES optimum, the $c_{L,max}$ predicted with the SST model is 4.06 at 23° angle of attack ($c_{L,max} = 4.17$ at 25° angle of attack with Spalart-Allmaras model).

In Fig. 14.14 the aerodynamic flow field (Mach number contour map) is shown at 18° angle of attack for reference and CMA-ES optimum shapes with both turbulence

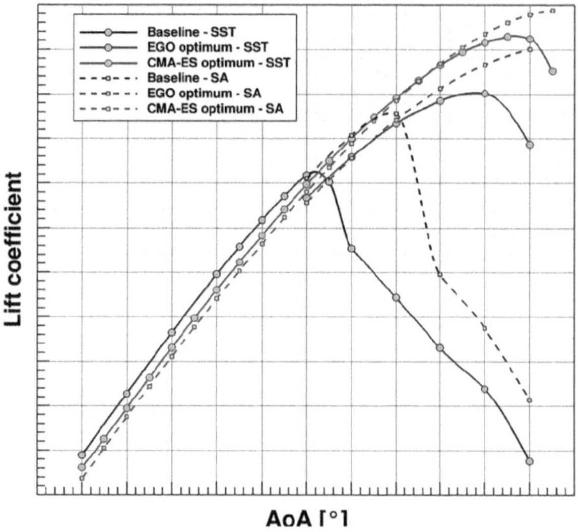

Fig. 14.13 Lift coefficient curve versus angle of attack for baseline and optimized shapes, Spalart-Allmaras (SA) and $k - \omega$ SST (SST) turbulence models

models employed. This angle of attack has been selected because it illustrates quite well two main outcomes of the reported task. On one hand, the suppression of the flow separation on the Krueger slat suction side is clearly highlighted as main design feature (e.g., looking at the figure from top to bottom); on the other hand, it shows some discrepancy in the flow prediction when changing the turbulence model (e.g., looking et the figure from left to right), as the SST model predicts a slight early stall.

14.2.5 Comparison of Designs

In this section, a cross-check analysis between DLR and CIRA designs is proposed. Before entering into the detailed analysis, a brief synthesis of the main differences between the optimization approaches adopted by the two partners is given here:

CFD method: DLR employed the in-house FLOWer multi-block structured solver, CIRA used the open-source SU2 unstructured solver;

Geometry parameterization: DLR used a CAD-based approach with 12 parameters, CIRA used a CAD-free approach with 5 parameters. In both cases, the panel length and deflection angle are not free parameters and are driven by kinematic feasibility. The main difference is related to the bullnose parameterization: DLR allowed free variation of the design parameters, CIRA explored a more limited set of allowable shapes in order to avoid issues with the Krueger stowage in retracted position. Of course, the attained performance is impacted as well, as detailed later on;

Geometry detail: DLR optimization took into account the aerodynamic effect of the Krueger open cavity shape, CIRA modeled the main airfoil with closed cavity;

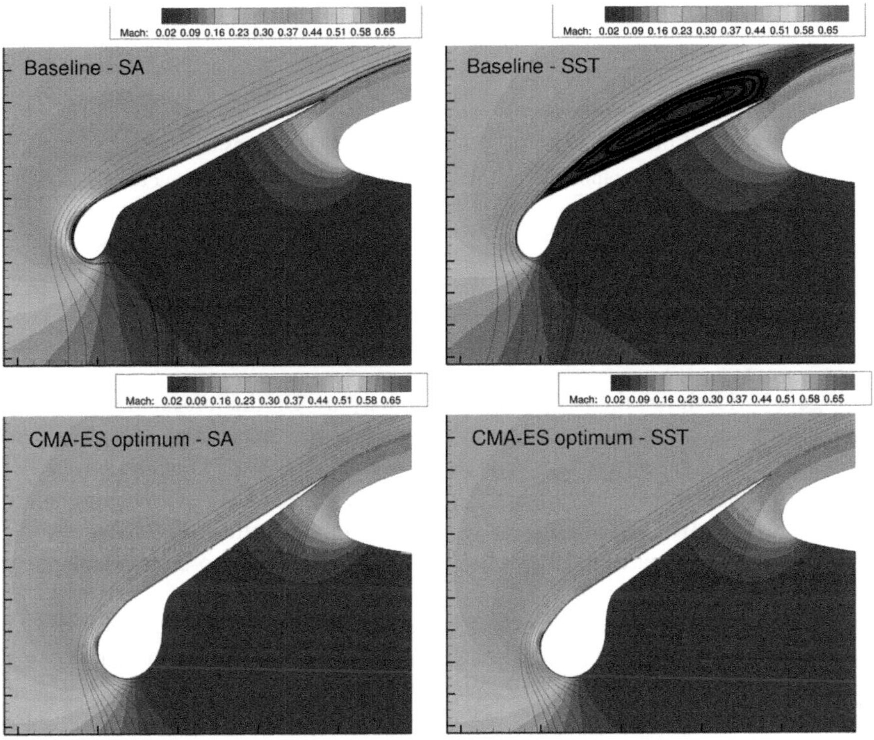

Fig. 14.14 Mach number contour map and flow streamtraces for reference and optimized shape at 18° angle of attack, Spalart-Allamaras (SA) and $k - \omega$ SST turbulence models

Optimization problem: DLR optimized for a linear combination of maximum lift and climb speed, CIRA defined a maximization problem along the higher branch of the polar curve while also taking into account the drag characteristics;

Optimization algorithm: DLR used the simplex method, CIRA performed a surrogate-based global optimization first, followed by an evolutionary search.

The aim of the crosscheck analysis is to quantify the impact of the CFD code and of the mesh type on the optimal lift curve prediction. Moreover, as DLR performed the design by including the Krueger open cavity shape within the CFD model while CIRA did not consider it, it is deemed interesting to understand if such difference may have an impact on the design. Table 14.3 provides information about selected crosscheck cases. The following comparative analyses are envisaged:

1. Code to code crosscheck: cases 2, 3, 5 and 6 are compared to investigate the effect of the CFD approach (CFD code and mesh type) on the aerodynamic performance of optimal candidates;
2. Krueger cavity check: cases 1, 2, 4 and 5 are selected to investigate the effect of the Krueger cavity.

Table 14.3 Selected cross-check cases

Case ID	Geometry	Cavity included (Y/N)	Mesh	Solver
1	CIRA optimum	N	Quad-tri unstruct	SU2 (SST)
2	CIRA optimum	Y	Quad-tri unstruct	SU2 (SST)
3	CIRA optimum	Y	Multi-block struct	FLOWer (SA)
4	DLR optimum	N	Quad-tri unstruct	SU2 (SST)
5	DLR optimum	Y	Quad-tri unstruct	SU2 (SST)
6	DLR optimum	Y	Multi-block struct	FLOWer (SA)

Figure 14.15 shows the two comparative analyses. On the left-hand side, the lift curves for cases 2, 3, 5 and 6 are drawn. The general agreement is good both in terms of quantitative comparison and trend capturing. The DLR CFD approach predicts higher lift slopes and slightly higher $C_{l,max}$ with respect to CIRA CFD approach for both DLR and CIRA optima. Likewise, the stall angle of attack for CIRA airfoil is predicted at 22° compared to 23°. At low angles of attack, code-to-code discrepancies seem to be mildly amplified.

On the right-hand side of Fig. 14.15, the lift curves for cases 1, 2, 4 and 5 are reported. The effect of the Krueger cavity on the aerodynamic performance can be considered negligible for both CIRA and DLR shapes. Some additional details are provided in Figs. 14.16 and 14.17, where the pressure coefficient is compared at low ($= 8°$) and high ($= 24°$) angle of attack. DLR optimized Krueger shape is larger than CIRA's; hence the contribution of the Krueger element to the lift generation is more important. CIRA shape, indeed, partially recovers this feature thanks to higher local curvature and increased loading. Figure 14.17 confirms that the aerodynamic disturbance related to the open cavity is purely local and therefore negligible at both low and high angle of attacks.

14.3 Kruger Device Shape Synthesis

After the initial concurrent design phase, both obtained optimum solutions have been compared in detail. In the analysis, the following aspects were identified.

- The Krueger panel of the DLR design is too long as it violates the front spar clearance constraint and even penetrates the front spar itself. This glitch has been introduced already in the initial configuration while porting the AFLoNext design to the DLR-F15-LLE shape. The AFLoNext configuration had a front spar located more downstream and in the setup of the DLR-F15 this has not been corrected to the corresponding front spar location
- DLR used the older and stronger impact shielding requirement of 7° impact at a geometric 6° angle of attack. CIRA implied in their design the relaxed 9° impact

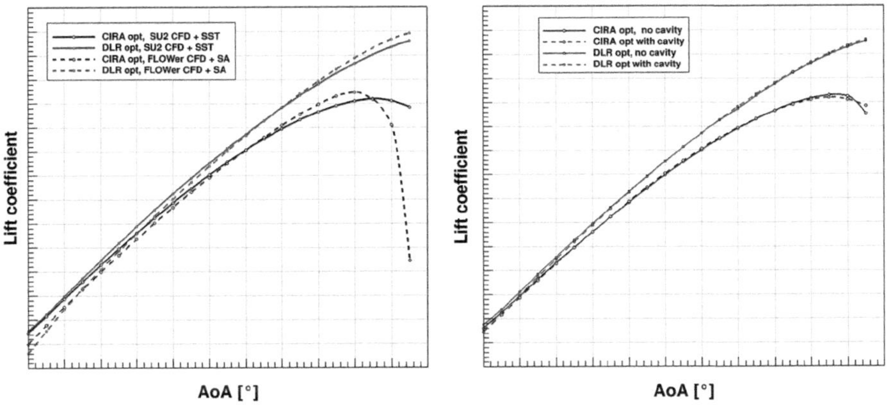

Fig. 14.15 Lift coefficient vs angle of attack for CIRA and DLR optima, impact of CFD approach (left) and Krueger cavity (right)

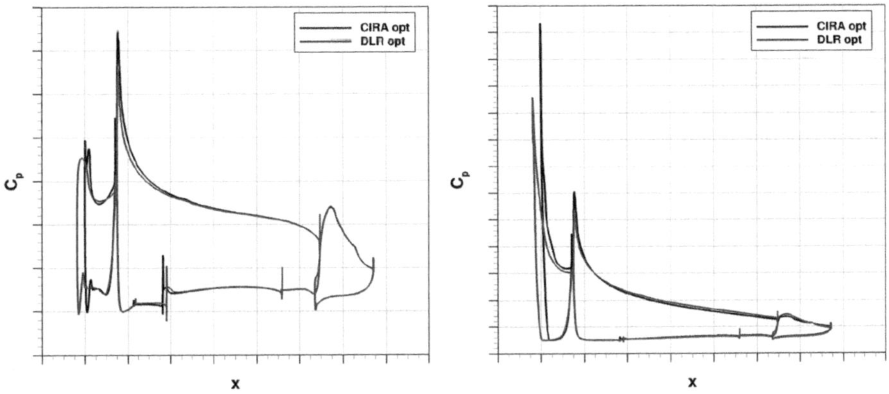

Fig. 14.16 Pressure coefficient comparison between CIRA and DLR optima, AoA= 8° (left) and 24° (right)

angle. Further on, CIRA respected that the geometric angle of attack for a wing section of a swept wing is increased by $\tan \alpha_{2D} = \tan \alpha_{3D} / \cos \theta$. For an assumed wing sweep of $\theta = 30°$ the reference angle of attack is for the 2D airfoil section $\alpha_{2D} = 6.92°$. The relaxed shielding requirement allows for a lower position and a steeper deflection closer to 140°, which is assumed to be beneficial.

- The CIRA design exhibits an unintended waviness on the upper side of the Krueger bull nose, which is generated by the use of Hicks-Henne functions and their connection to the rigid Panel shape on the upper side.

In a first synthesis, it was therefore decided to orient at the CIRA design but to reshape the bull nose to a smoother shape. To achieve this, The CIRA design was reparametrized using DLR's design parameters. Figure 14.18 shows a comparison of the original design shapes and the synthesized design, with a close-up to the bull

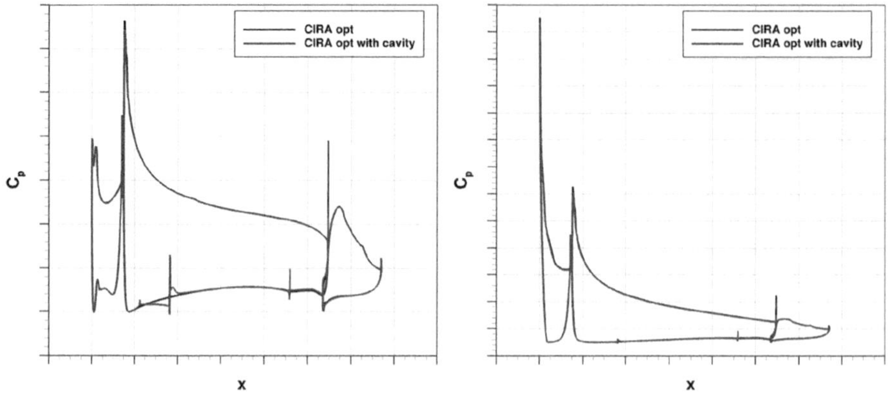

Fig. 14.17 Pressure coefficient comparison with and without cavity, AoA= 8° (left) and 24° (right)

Fig. 14.18 Comparison of the Krueger flap shapes obtained by DLR and CIRA design optimization together with the reconstructed CIRA design using DLR's parameterization approach

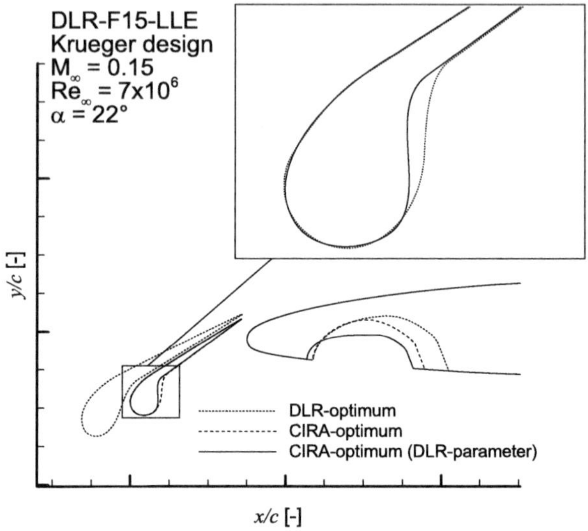

nose shape. It is seen that the DLR parameterization nicely rebuilds the CIRA design with a smooth bull nose shape.

Figure 14.19 shows a comparison of the lift coefficient of these designed configurations. The lower performance of the CIRA design can be mainly attributed to the shorter panel and the resulting smaller increase in effective airfoil chord with deflected Krueger. The smoother bull nose shape obtained by the re-parameterization is seen in the smoother stall behaviour at a similar maximum section lift coefficient. The rebuilt shape therefore retains the positive performance established by the original CIRA design. The shortfall in maximum section lift coefficient compared to the original DLR design is acceptable. On the one hand, the DLR designed Krueger is far too large and penetrates the front spar. On the other hand, it is not the purpose of

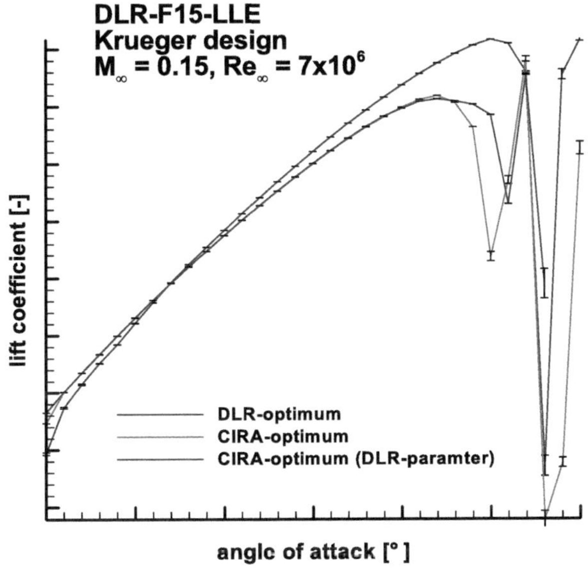

Fig. 14.19 Comparison of the lift coefficient versus angle of attack for DLR and CIRA designs and the reconstructed CIRA design using DLR's parameterization approach

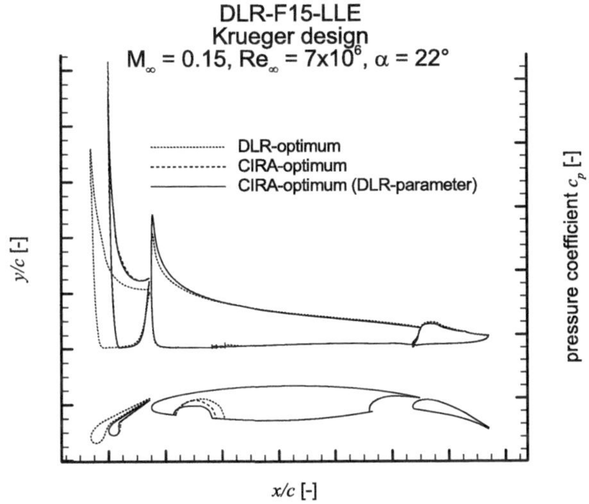

Fig. 14.20 Comparison of the pressure distributions for DLR and CIRA designs and the reconstructed CIRA design using DLR's parameterization approach

this work to design the ultimate Kreuger design but to provide a reasonable Krueger device with a high performance. A maximum section lift coefficient of above 4 is still a high-performing high-lift airfoil design.

Figure 14.20 shows a comparison of the pressure distributions for all three designs at an angle of attack close to stall of the CIRA design. Again, the reproduction of the CIRA design by the DLR parameters is good in general. At the Krueger a slightly lower suction peak is visible showing the effect of the smoother bull nose shape.

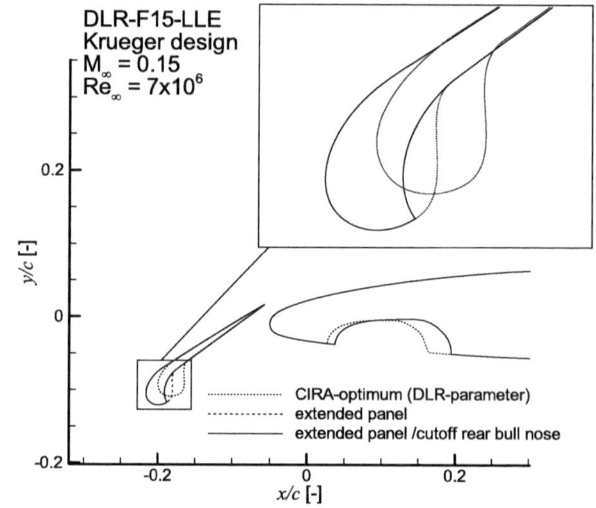

Fig. 14.21 Final changes at Krueger design to incorporate refined requirements from kinematics integration

14.4 Kruger Device Shape Alignment To Kinematics Requirements

The initial Krueger device and the synthesized design have been analysed in UHURA by ASCO regarding the feasibility of the corresponding kinematics [12]. While all the initial requirements have been met by the synthesized design, two new issues showed up, which would not prevent integration but a slight change would provide high benefits for the integration. At first, a longer Kruger panel in terms of the downstream edge would avoid an otherwise necessary cut-out in the lower skin allowing passage of the drive lever arm. At second, a cut-off of the rear bull-nose shape would allow for a higher retraction angle of the bull nose, which in turn would reduce the space allocation and provide a strengthening of leading edge ribs to support the aerodynamic loads on the leading edge. The design was changed accordingly by reducing the offset to the downstream Krueger edge to the front spar to a minimum of 0.5%c. The rear shape of the bullnose was changed. The GC2-continuous spline was replaced by a sharped edge connection. Figure 14.21 shows the adopted geometry with a close up to the changes Krueger leading edge. Within this iteration, the hinge point has additionally been refined to even more closely correspond to the CIRA-design based position of the Krueger device in deflected position.

The change in aerodynamics is shown in Fig. 14.22. In general, the extension of the Krueger panel leads to a sharper wing stall at a slightly increased maximum lift coefficient, which probably may be prevented by a less shallow deflection. The cut-off of the rear bull nose reduces stall incidence by 1°C with a slight reduction of maximum lift coefficient. The maximum lift coefficient of the final design is similar to the optimal CIRA design value.

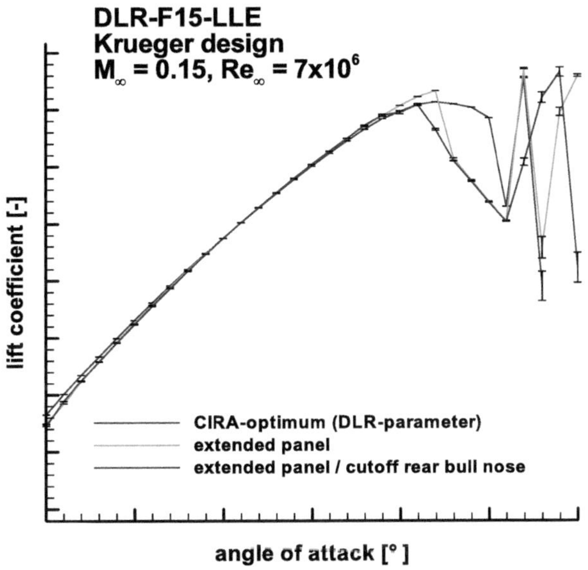

Fig. 14.22 Lift versus angle of attack for the finally refined Krueger device with respect to kinematics requirements

Figure 14.23 compares the pressure distributions at a high angle of attack close to the maximum lift coefficient. The extension of the panel is accompanied by a slight reduction of the suction peak but at a similar pressure gradient. At the main wing the pressure rise past the suction peak is visibly steeper while the suction peak level is unchanged. The cut-off of the rear bull-nose has no significant effect on the pressure distribution.

As the aim of the task was to provide a reasonable Krueger device at a high aerodynamic performance, it is judged by all partners that the obtained result is sufficient for the aim of the project. The Krueger device shows a high maximum lift coefficient at a high corresponding stall angle. Compared to the initial configuration, a premature stall is prevented. Nevertheless, some potential for a smoother stall onset still exists.

14.5 Conclusions

The design optimization work described here can be considered a success both from a methodological point of view and from the results obtained. From the point of view of the optimization method, we opted for a hybrid competitive-cooperative approach in which two UHURA partners, namely DLR and CIRA, first proceeded to an optimization phase in which, independently of each other and interpreting the requirements and constraints of the problem in a slightly different way, they produced two improved solutions compared to the baseline. In the second phase, instead of choosing and judging which of the two solutions was better, they decided to

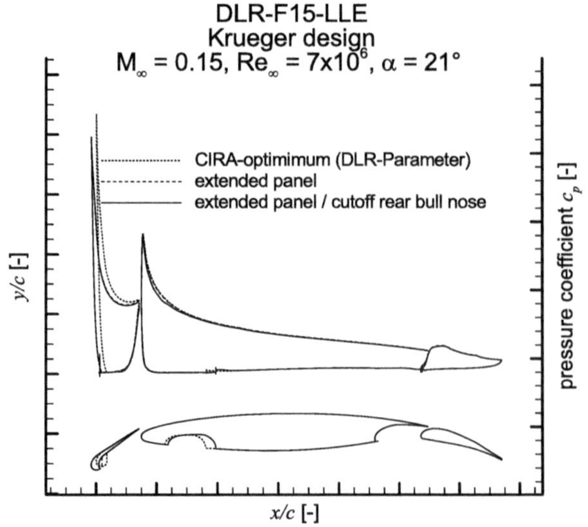

Fig. 14.23 Comparison of pressure distributions the finally refined Krueger device with respect to kinematics requirements

propose a third one that combined the favorable characteristics of both. The advantage of this approach is evident when more complex and computationally demanding problems are to be solved. In these cases, the initial design choices can profoundly influence the final result. The hybridization of two solutions obtained following non-overlapping approaches can improve the design process performance and efficiency. Finally, from the point of view of the results obtained, the design of the Krueger flap has proved to satisfy all the design requirements very well and, as evidenced by the volume of numerical and experimental works produced within UHURA, has allowed a considerable advancement of the technical knowledge on aerodynamic and structural problems inherent in the deployment of high-lift devices.

Acknowledgements The project leading to this presentation has received funding from the European Union's Horizon 2020 Research and Innovation Framework Programme under the research and innovation programme under grant agreement No 769088.

References

1. UHURA project. https://uhura-project.eu/
2. Strüber H, Wild J (2014) Aerodynamic design of a high-lift system compatible with a natural laminar flow wing within the DeSiReH project. In: 29th Congress of the international council of the aeronautical sciences. St, Petersburg, Russia, ICAS 3-932182-80-4
3. Franke DM, Wild J (2014) Aerodynamic design of a folded krüger device for a HLFC wing. In: 19. DGLR-Fachsymposium der STAB, vol 2014 of Mitteilung zum STAB Jahresbericht, pp 28–29
4. Wild J (2018) Initial geometry of DLR-F15-LLE with Krueger flap for WP 2, tech report UHURA Coordination Memorandum UH-11-CM03-DLR, DLR

5. Wild J (2002) Validation of numerical optimization of high-lift multi-element airfoils based on navier-stokes-equations. In: 20th AIAA applied aerodynamics conference
6. Geuzaine C, Remacle J-F (2009) Gmsh: A 3-D finite element mesh generator with built-in pre- and post-processing facilities. Int J Numer Meth Eng 79(11):1309–1331
7. Economon TD, Palacios F, Copeland SR, Lukaczyk TW, Alonso JJ (2016) SU2: an open-source suite for multiphysics simulation and design. AIAA J 54(3):828–846
8. Jones DR, Schonlau M, Welch WJ (1998) Efficient global optimization of expensive black-box functions. J Global Optim 13:455–492
9. Adams BM, Bohnhoff WJ, Dalbey KR, Ebeida MS, Eddy JP, Eldred MS, Hooper RW, Hough PD, Hu KT, Jakeman JD, Khalil M, Maupin KA, Monschke JA, Ridgway EM, Rushdi AA, Seidl DT, Stephens JA, Swiler LP, Winokur JG (2021) Dakota, a multilevel parallel object-oriented framework for design optimization, parameter estimation, uncertainty quantification, and sensitivity analysis: version 6.15 user's manual. Sandia Technical Report SAND2020-12495
10. Hansen N (2006) The CMA evolution strategy: a comparing review. In: Lozano JA, Larrañaga P, Inza I, Bengoetxea E (eds) Towards a new evolutionary computation: advances in the estimation of distribution algorithms. Springer, Berlin, Heidelberg, pp 75–102
11. Hansen N, Ostermeier A (2001) Completely derandomized self-adaptation in evolution strategies. Evol Comput 9(2):159–195
12. Vervliet A (2019) Kinematic constraints linked to Krueger panel, actuation and deployment mechanisms. Tech. Rep. UHURA Deliverable D12-1, ASCO
13. Wild J (2018) Loads on the initial DLR-F15-LLE Krueger flap, techreport UHURA Coordination Memorandum UH-11-CM05-DLR, DLR
14. Ciobaca V, Dandois J (2017) High Reynolds number high-lift airfoil testing with flow control. In: 35th AIAA applied aerodynamics conference
15. Wild J (2022) High-lift aerodynamics, 1st edn. CRC Press

Chapter 15
Toward Integrated Aircraft Design with Carbon Fiber Reinforced Thermoplastic (CFRTP)

Yoshiaki Abe, Keiichi Shirasu, Shugo Date, Tomonaga Okabe, and Shigeru Obayashi

Abstract This article summarizes our most recent progress in developing a simulation-based tool for aircraft design using carbon-fiber reinforced plastics (CFRP) (Date et al. in Aerosp Sci Technol 124(107565), 2022) and virtual testing of open-hole tension/compression (OHT/OHC) strength of carbon-fiber thermoplastics (CFRTP) (Shirasu et al. in J Compos Mater 56:2211–2225, 2022). The effects of fiber properties on the static aeroelastic design of CFRP wings were investigated in the first topic, which used a multi-scale framework and two-way coupled aeroelastic analysis with structural sizing. T700S, T800S, and T1100G fibers with varying stiffnesses and strengths were studied. The stress-strain responses and damage initiation/propagation mechanisms of the CFRTP with an open-hole were investigated experimentally and numerically in the second topic. The stress-strain response of OHT and OHC obtained experimentally was well reproduced by considering the material nonlinearity of thermoplastic resin using extended finite element model (XFEM) tools for the evaluation of stress strain response as well as damage and failure mechanisms in OHT and OHC tests.

Keywords Aircraft design · Virtual testing · CFRP · CFRTP

Y. Abe (✉) · S. Date · S. Obayashi
Institute of Fluid Science, Tohoku University, Sendai, Miyagi 9808577, Japan
e-mail: yoshiaki.abe@tohoku.ac.jp

K. Shirasu
Department of Finemechanics, Tohoku University, Sendai, Miyagi 9808579, Japan

S. Date · T. Okabe
Department of Aerospace Engineering, Tohoku University, Sendai, Miyagi 9808579, Japan

T. Okabe
Department of Materials Science and Engineering, University of Washington, Seattle, WA 98195, USA

T. Tuovinen et al. (eds.), *Advanced Computational Methods and Design for Greener Aviation*, Computational Methods in Applied Sciences 59,
https://doi.org/10.1007/978-3-031-61109-4_15

15.1 Introduction

One of the key technologies for next-generation aircraft design is carbon fiber reinforced plastics (CFRPs). CFRP has already been used in primary structures such as wings and fuselages: for example, the composite materials account for nearly half of the structural weight of the Boeing 787 and Airbus A350XWB.

Despite its promising properties, further applications of CFRP are hampered by their complex fracture mechanism and a lack of fundamental understanding of the material's properties. Failure through fiber breaking and damage propagation beginning with transverse cracks, in particular, are highly dependent on the properties and combinations of fibers and matrix. Those fracture mechanisms are observed and initiated at the microscale, typically at the submilimeter scale, which is generally separated from aircraft design by macroscale analyses [1, 9, 10, 12, 13, 17, 19]. As a result, there is a large gap between progress in the development of prediction models of CFRP strength and their application to the numerical study of aircraft design. This disparity motivates us to develop a multi-scale framework and investigate the impact of material property changes on the structural weight and aerodynamic performance of aircraft wings.

Furthermore, in aerostructural design, the use of thermoplastics in composite materials, such as carbon fiber reinforced thermoplastics (CFRTP), is attempted [24]. The CFRTP is advantageous in terms of toughness, reusability, and rapid mass production with a shorter manufacturing cycle time, resulting in lighter and more efficient composite aircraft structures. However, detailed experimental or numerical studies of the mechanical properties of CFRTPs, such as open-hole tensile/compressive (OHT and OHC), interlaminar fracture toughness, impact resistance and compression after impact strength, and fracture behavior, are still lacking, and virtual testing techniques for the mechanical properties of CFRTPs must be developed.

This article summarizes two articles [4, 18] motivated by the background above. The first article [4] investigates the effects of fiber selection on the aerodynamic performance and structural weight of CFRP wings when static aeroelasticity is taken into account. As an example, consider a transonic aircraft wing comparable to the Boeing 777 class. Three different carbon fibers (T700S, T800S, and T1100G) were used to investigate the effects of fiber selection. The aerodynamic performance and structural weight of the various fibers were assessed using a multiscale design approach that takes static aeroelasticity into account. The second article aims to investigate the strength and damage initiation/propagation mechanisms of the CFRTP using OHT and OHC tests as well as quasi-3D extended finite element model (XFEM) analysis.

The rest of this article is organized as follows. Section 15.2 summarizes the development of the multiscale design framework for CFRP wing, as well as the effect of fiber selection on wing performance. Section 15.2.1 introduces a schematic of the multiscale design framework developed in this study, then Sect. 15.2.2 represents the problem specification of the wing design. Section 15.2.3 shows the results of the structural design of the wing using different carbon fibers. Section 15.3 introduces both experimental and numerical studies of OHT and OHC tests of CFRTP laminates.

Section 15.3.1 represents the problem specifications of the OHT and OHC tests using the CFRTP; Sect. 15.3.2 shows the experimental and numerical results. Section 15.4 discusses the both results and perspectives in the future study. Finally, Sect. 15.5 concludes this article.

15.2 Multiscale Framework for CFRP Aircraft Design

This section summarizes [3, 4], which developed the multiscale framework for CFRP wing design and investigated the effects of fiber selection on the aerodynamic and structural weight.

15.2.1 Multiscale Design Approach for Composite Wing

A schematic of the multiscale design approach is illusttated in Fig. 15.1. As an input parameter in the analysis, the approach used material properties obtained from lower-level analyses. To estimate the stiffness and strength of the unidirectional (UD) lamina, a series of microscale analyses were performed. First, the periodic unit cell (PUC) analysis was used to determine the stiffness properties [21–23]. The longitudinal tensile and compressive strength of the lamina were then estimated using the spring element model [14] and the buckling model proposed by Sun and Jun [20], respectively. Finally, the PUC analysis was used to calculate the transverse tensile and compressive strengths, as well as the inplane shear strength [15].

The results of the microscale analysis are then passed to the macroscale analysis in Fig. 15.2a. The first step is to create the computational meshes required for the aerodynamic and structural analyses. In the aerodynamic assessment, the inviscid

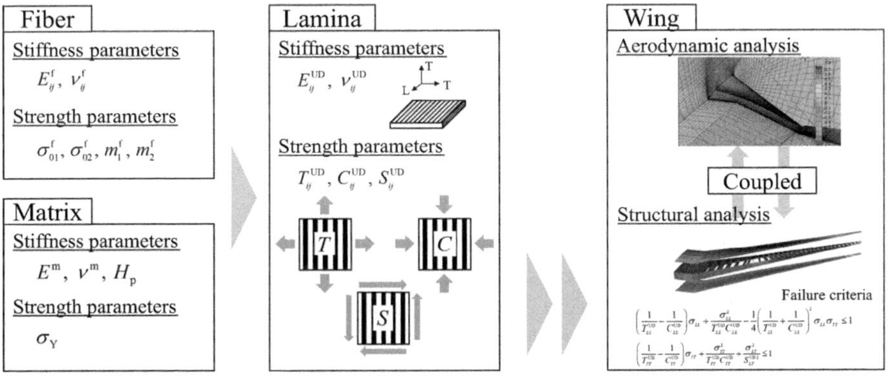

Fig. 15.1 Schematic of the multiscale framework for CFRP aircraft design [4]

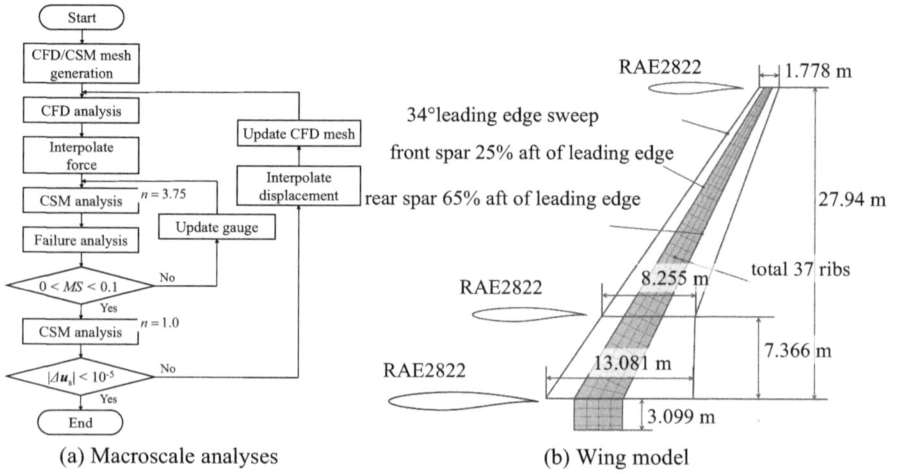

Fig. 15.2 Flowchart of the macroscale analyses and geometrical specification of the main wing [4]

compressible flow was simulated using the finite-volume method, and the static linear finite element (FE) analysis was used to obtain stress and displacement in the wing-box structure described in the following subsection. To simplify the problem, laminates were assumed to be quasi-isotropic with a stacking sequence of $[0/45/-45/90]_{sym.}$. The shell elements had the properties of a unidirectional lamina rotated in each ply, whereas the beam elements had the homogenized properties of quasi-isotropic laminates from classical laminate theory. The radial basis function (RBF) method [16] was used to transfer load and displacement between aerodynamic and structural models. To obtain a realistic gauge for the FE properties, the structural analysis and sizing processes were repeated iteratively. The fully stressed design (FSD) algorithm was used to modify the gauges, which were updated based on the margin of safety (*MS*). For both the strength and local buckling criteria, *MS* was calculated. The Christensen's stress criterion [2] was used to predict the strength of each lamina in composite shell elements. Finally, the Aitken's relaxation [8] was used to accelerate the aeroelastic system's convergence.

15.2.2 Problem Specification

As shown Fig. 15.1b, a transonic aircraft wing was considered. External geometry is defined by RAE2822 airfoil sections with no twist angle and a dihedral angle of 4°. The MITC4 shell elements were used to model the skins, ribs, and spar webs, while beam elements were used to model the spar flanges. Three different fiber materials, T700S, T800S, and T1100G, were used to estimate the effects of fiber properties on aerodynamic performance and structural weight. The resin's properties were matched to those of #3900-2B, which was used in conjunction with T800S as a carbon/epoxy composite system in the Boeing 787. T1100G and T700S have higher and lower stiffness/strength, respectively, than T800S. The cruise condition was set to 0.84 Mach number. The angle of attack (AoA) was changed to keep the lift coefficient at 0.50. The maximum load factor was set to 3.75 (limit load factor of 2.5 times a safety factor of 1.5). The load in the cruise condition was multiplied by the load factor in this study.

15.2.3 Results

The drag polars of various fibers are shown in Fig. 15.3a. All of the fibers had approximately the same drag polar, however the symbol locations were different for each fiber. The 1100G-based composite wings have a larger AoA than the T700S and T800S-based wings. This is due to the fact that a wing with stiffer fibers yields larger lift at smaller AoA. Furthermore, it should be noted that the difference in fibers results solely in a difference in AoAs.

The active constraint was then compared for different fibers with the same lift coefficient of 0.5 but different angle of attack. Figure 15.3b represents the active constraint with a minimum *MS* in all constraints; thus, the failure mode in Fig. 15.3b drives the sizing process. Except in the tip region, the matrix dominant strength is mainly active on the lower surface. The tip region is designed to prevent buckling failure. The buckling constraint is active on almost every panel on the upper surface, with the exception of the T1100G-based composite near the root. Near the root of the T1100G-based composite wing, fiber-dominant strength is active. Because the upper and lower surfaces are loaded in compression and tension, these trends are to be expected. Figure 15.4a, b show the equivalent thickness of skins for each fiber on the lower and upper surfaces. These figures show that the lower and upper surfaces of the T1100G-based composite wing have a low thickness distribution. Because of its

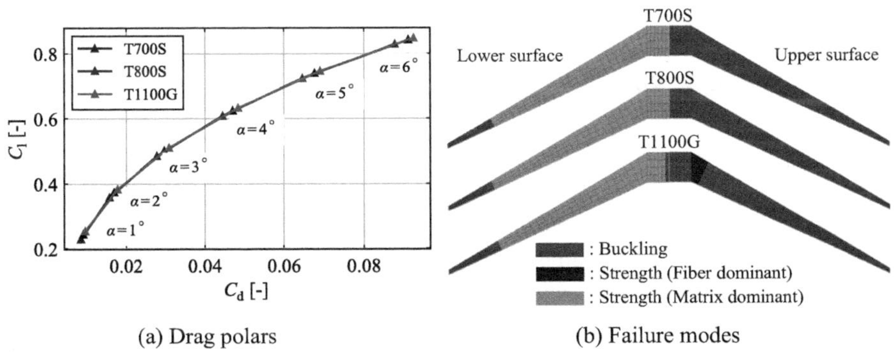

Fig. 15.3 Drag polars and failure modes with different fibers [4]

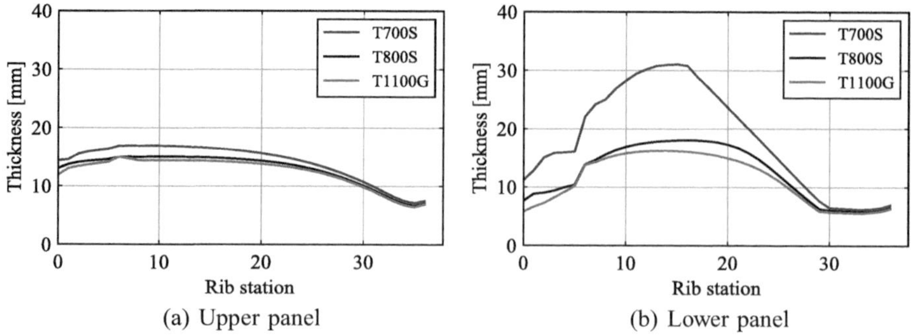

Fig. 15.4 Thickness distributions on upper and lower panels with different fibers at the same lift coefficient of 0.5 [4]

higher tensile strength on the lower surface, where matrix-dominant strength is most active, a stiffer fiber is advantageous for weight reduction. Due to its higher buckling strength, the stiffer fiber has an advantage for weight reduction on the upper surface, where the buckling constraint becomes primarily active. Subsequently, the structural weight of the T700S-based wings was 8069 kg, 7268 kg for the T800S-based wings, and 6937 kg for the T1100G-based wings. The wing based on the T1100G is 14% lighter than the wing based on the T700S and 9.9% lighter than the wing based on the T800S.

15.3 Virtual Testing for Evaluation of Mechanical Properties of CFRTPs

This section summarizes [18], which studied experimentally and numerically the stress-strain responses and damage initiation/propagation mechanisms of the CFRTP with center hole.

15.3.1 Problem Specification

Unidirectional T700G carbon-fiber-reinforced LM-PAEK prepregs were used to make CFRTP panels (TC1225; Toray Advanced Composites, United States). The T700G/LM-PAEK prepregs were consolidated using both a hot-press and an autoclave (AC), and the fiber volume fraction of each consolidated laminate was 51% and 50%, respectively. In line with the SACMA standard, open-hole samples with a stacking sequence of $[45/0/{-45}/90]_{2s}$ were tested in an Instron 5985 tensile testing machine with a 250 kN loading cell (SRM 3R-94 and 5R-94). To model the OHT and OHC strength of CFRTPs, the quasi-3D XFEM was employed. The modeling

Table 15.1 Material properties used in XFEM [18]

Laminate properties			Experimental procedure
Longitudinal Young's modulus (tension)	E_{11+}	127.5 GPa	Non-hole tension, $[0]_8$
Longitudinal Young's modulus (compression)	E_{11-}	102.1 GPa	Non-hole compression, $[0]_8$
Transversal Young's modulus (tension)	$E_{22+}(=E_{33+})$	8.4 GPa	Non-hole tension, $[90]_{16}$
Transversal Young's modulus (compression)	$E_{22-}(=E_{33-})$	7.6 GPa	Non-hole compression, $[90]_{16}$
Longitudinal shear modulus	$G_{12}(=G_{13})$	4.5 GPa	In-plane shear, $[\pm 45]_{2s}$
Transversal shear modulus	G_{23}	3.0 GPa	$=E_{22}/2(1+\nu_{23})$
Longitudinal Poisson's ratio	ν_{12}, ν_{13}	0.32	Non-hole tension, $[0]_8$
Transversal Poisson's ratio	ν_{23}	0.4	Assumed value
Coefficient of longitudinal thermal expansion	α_1	-0.9×10^{-6} K^{-1}	Thermomechanical analysis, $[0]_8$
Coefficient of transversal thermal expansion	α_2	3.8×10^{-5} K^{-1}	Thermomechanical analysis, $[90]_8$
Longitudinal tensile strength	T_{11}	3106.7 MPa	Non-hole tension, $[0]_8$
Longitudinal compressive strength	C_{11}	928.2 MPa	Non-hole compression, $[0]_8$
Transverse tensile strength	T_{22}	77.8 MPa	Non-hole tension, $[90]_{16}$
Transverse compressive strength	C_{22}	175.5 MPa	Non-hole compression, $[90]_{16}$
Longitudinal shear strength	S_{12}	202.5 MPa	In-plane shear, $[\pm 45]_{2s}$
Longitudinal shear stress at 5% strain	$S_{12}^{5\%}$	67.0 MPa	In-plane shear, $[\pm 45]_{2s}$
Interlaminar properties			**Experimental procedure**
Mode I maximum traction	t_{Ic}	77.8 MPa	$= T_{22}$
Mode II (and III) maximum traction	t_{IIc}	100.4 MPa	Short beam shear, $[0]_{28}$
Mode I fracture toughness (initiation)	G_{Ic}	1.50 N/mm	Double cantilever beam tests and FEA
Mode I fracture toughness (propagation)	G_{IR}	2.35 N/mm	Double cantilever beam tests and FEA
Mode II fracture toughness (initiation)	G_{IIc}	2.35 N/mm	End-notched flexure tests and FEA
Mode II fracture toughness (propagation)	G_{IIR}	3.0 N/mm	End-notched flexure tests and FEA
Mode interaction factor	α	1.0	

scheme employed in this investigation was based on a recent study on OHT and OHC simulations of CFRP with CF/epoxy, which implemented the elastoplastic constitutive law [6, 7]. Table 15.1 summarizes the material parameters of the CFRTP used in XFEM for the OHT and OHC simulations, as determined by experiments and FEA [18].

15.3.2 Results

The experimental and computational stress-strain curves for OHT and OHC tests are shown in Fig. 15.5. The strain was determined from the displacement of the nodes at a distance of 25.4 mm from the center of the hole to compare the stress-strain diagrams obtained from the analysis and the experiment. The experimental OHT data (black lines) show an almost linear elastic regime up to a fracture strength of 472–479 MPa at 0.89–0.95 strain. The Weibull statistical model [5] describes fiber breakage as the primary failure mechanism in OHT, and the Weibull modulus (m) is set to 41. This failure mode is discussed in more detail below. The computational model (red broken line in Fig. 15.5a) accurately predicts the stress-strain response, i.e., plastic deformation occurs at 0.6% strain and the failure strength coincides well with the experimental value. The experimental stress-strain response grew linearly up to a strain of 0.4% in the OHC tests. The stress increased nearly linearly until failure at a strain of 0.4%, though the slope of the stress-strain curve reduced slightly. When the critical energy release rate for kink-band propagation (G_{kink}) was set to 36 N/mm, a comparison of the stress-strain curves obtained through experiments and simulations (red broken line in Fig. 15.5b) revealed that the stress-strain response was accurately captured using the computational model (the LaRC03 fracture criterion [11] was used for kinking initiation in OHC). The in-situ damage observation results of the OHT specimen before fracture are shown in Fig. 15.6a, where the initial damage detected by the high-speed digital microscope occurred in the 45° ply right before the peak load. The simulated internal damage distribution, given in Fig. 15.6b, suggests that the quantity of fracture propagation is minor, and the Weibull criterion was satisfied without any delamination; this accords reasonably well with the experimental results. The in-situ damage observation data from damage start to propagation are summarized in Fig. 15.7. A high-speed digital microscope detected the initial damage in 0° ply at 88% of the peak load (Fig. 15.7a), corresponding to 0°-ply kinking. Following that, similar damage was seen at the other 0° plys with fracture noise (Fig. 15.7b), and the damage was seen to develop in the width direction right before fracture (Fig. 15.7c, d). The numerical damage distributions are depicted in Fig. 15.7e–h. Damage onset was observed in the 0° plies and adjacent plys at 87% of peak stress, which is close to the experimental values, and the direction of damage propagation at final failure was consistent with the experimental results.

15.4 Discussion

Based on Sect. 15.2, the multiscale framework can be used to estimate the structural design of the CFRP wing from the fiber and matrix properties. The results using fibers with various stiffness well demonstrated the effects of the fiber on the structural design, wherein the total weight with the T1100G/3900-2B attained the lightest structure. While all of the wing design results are based on CFRPs, the present

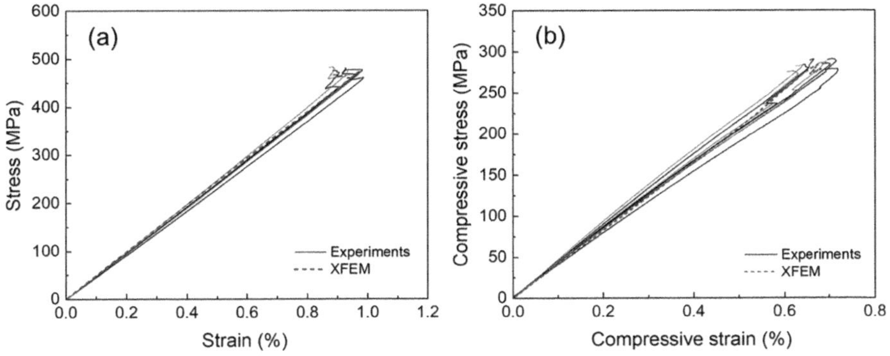

Fig. 15.5 Experimental and numerical stress-strain curves for **a** OHT and **b** OHC tests [18]

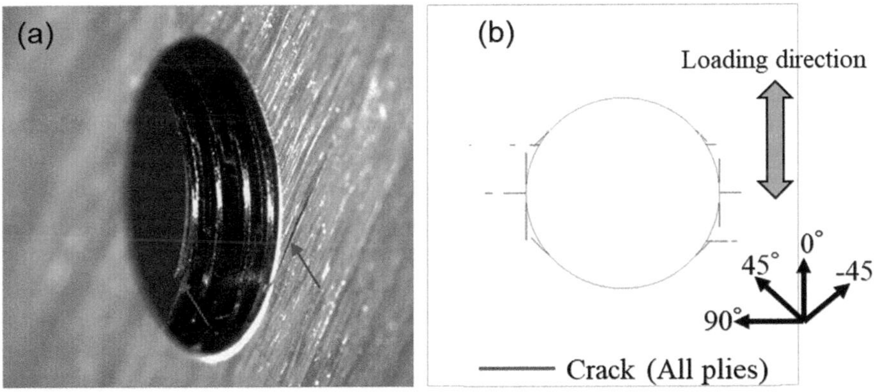

Fig. 15.6 **a** In-situ microscope observation results at 100% of peak load and **b** predicted damage distributions after fracture [18]

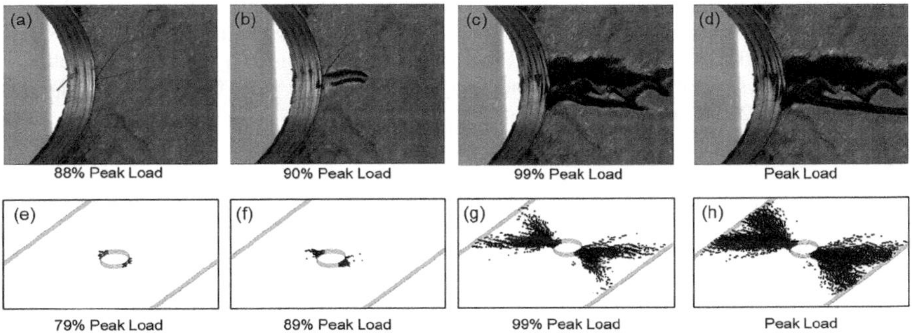

Fig. 15.7 **a–d** In-situ high-speed digital microscope observation results and **e–h** predicted damage distributions [18]

framework is also applicable to CFRTPs, and the experimental results of the OHC and OHT experiments can be directly incorporated into the macroscale analysis. Using the CFRTP (T700G/LM-PAEK), Sect. 15.3 provides experimental and computational results for the OHT and OHC tests. The developed FEA tools, including XFEM, sufficiently predicted the experimental results, motivating us to incorporate these tools into the wing design framework with CFRTPs in the future. Notably again, the experimental results can be incorporated directly into the existing wing design tool devised in Sect. 15.2.

15.5 Conclusions

This article summarized our progress in the multiscale framework for CFRP aircraft design [4] and virtual testing of OHT and OHC strength of CFRTP [18], which will be integrated as a aircraft design tool using CFRP and CFRTP in the future. In multiscale framework of aircraft design, the effects of fiber properties on the aerodynamic performance and structural weight of composite aircraft wings were investigated for T700S, T800S, and T1100G fiber materials. The results showed that the T1100G-based composite wing could reduce structural weight by 14% and 9.9%, respectively, when compared to the T700S- and T800S-based composite wings. This weight savings was achieved primarily by reducing the thickness of both the upper and lower panels. In virtual testing, the predicted stress-strain response, fracture strength, and damage scenario and distributions for OHT and OHC are reasonably consistent with the experiment. As a result, we conclude that the analytical model employed in this study is capable of simulating the stress-strain response and damage behavior of CFRTP in response to OHT and OHC.

Acknowledgements This article is based on the results obtained from the project, JPNP15006, commissioned by the New Energy and Industrial Technology Development Organization (NEDO). This study was also supported by the Council for Science, Technology and Innovation (CSTI), Cross-ministerial Strategic Innovation Promotion Program (SIP), "Materials Integration" for revolutionary design system of structural materials (Funding agency: JST), and JST FOREST Program (Grant Number JPMJFR2124, Japan).

References

1. Brooks TR, Martins JR, Kennedy GJ (2019) High-fidelity aerostructural optimization of tow-steered composite wings. J Fluids Struct 88:122–147
2. Christensen RM (1997) Stress based yield/failure criteria for fiber composites. Int J Solids Struct 34(5):529–543
3. Date S, Abe Y, Yamamoto T, Okabe T (2021) Fluid-structural design analysis for composite aircraft wings with various fiber properties. J Fluid Sci Technol 16(1):JFST0009
4. Date S, Abe Y, Okabe T (2022) Effects of fiber properties on aerodynamic performance and structural sizing of composite aircraft wings. Aerosp Sci Technol 124(107565)

5. Hallett S, Green B, Jiang W, Wisnom M (2009) An experimental and numerical investigation into the damage mechanisms in notched composites. Compos Part A: Appl Sci Manuf 40:613–624
6. Higuchi R, Okabe T, Nagashima T (2017) Numerical simulation of progressive damage and failure in composite laminates using XFEM/CZM coupled approach. Compos Part A: Appl Sci Manuf 95:197–207
7. Higuchi R, Warabi S, Yoshimura A, Nagashima T, Yokozeki T, Okabe T (2021) Experimental and numerical study on progressive damage and failure in composite laminates during open-hole compression tests author links open overlay panel. Compos Part A: Appl Sci Manuf 145:106300
8. Irons BM, Tuck RC (1969) A version of the Aitken accelerator for computer iteration. Int J Numer Methods Eng 1(3):275–277
9. Kennedy G, Martins J (2012) A comparison of metallic and composite aircraft wings using aerostructural design optimization. In: 12th AIAA aviation technology, integration, and operations (ATIO) conference and 14th AIAA/ISSMO multidisciplinary analysis and optimization conference, no September
10. Liu B, Haftka RT, Akgün MA (2000) Two-level composite wing structural optimization using response surfaces. Struct Multidiscipl Optim 20(2):87–96
11. Maimí P, Camanho P, Mayugo J, Dávila C (2007) A continuum damage model for composite laminates: Part I—Constitutive model. Mech Mater 39:897–908
12. Martins J, Kennedy G, Kenway GK (2014) High aspect ratio wing design: optimal aerostructural tradeoffs for the next generation of materials. In: 52nd aerospace sciences meeting, pp 1–24, no January
13. Naik NK, Kumar RS (1999) Compressive strength of unidirectional composites: evaluation and comparison of prediction models. Compos Struct 46(3):299–308
14. Okabe T, Sekine H, Ishii K, Nishikawa M, Takeda N (2005) Numerical method for failure simulation of unidirectional fiber-reinforced composites with spring element model. Compos Sci Technol 65(6):921–933
15. Okabe T, Kumagai Y, Higuchi R, Nishikawa M (2015) Onset of matrix cracking in fiber reinforced polymer composites: a historical review and a comparison between periodic unit cell analysis and analytic failure criteria. Adv Struct Mater 64:587–601
16. Rendall TCS, Allen CB (2008) Unified fluid-structure interpolation and mesh motion using radial basis functions. Int J Numer Methods Eng 74(10):1519–1559
17. Schultheisz CR, Waas AM (1996) Compressive failure of composites, Part I: Testing and micromechanical theories. Prog Aerosp Sci 32(1):1–42
18. Shirasu K, Tsuyuki J, Higuchi R, Onodera S, Okabe T (2022) Experimental and numerical study on open-hole tension/compression properties of carbon-fiber-reinforced thermoplastic laminates. J Compos Mater 56:2211–2225
19. Starnes JH Jr, Haftka RT (1979) Preliminary design of composite wings for buckling, strength, and displacement constraints. J Aircr 16(8):564–570
20. Sun CT, Jun AW (1994) Compressive strength of unidirectional fiber composites with matrix non-linearity. Compos Sci Technol 52(4):577–587
21. Suquet PM (1985) Local and global aspects in the mathematical theory of plasticity. Appl Sci Publ (Elsevier)
22. Suquet PM (1987) Elements of homogenization theory for inelastic solid mechanics. ci.nii.ac.jp
23. Terada K, Suzuki K, Ohtsubo H (1997) FFT-based homogenization algorithm using digital images. Mater Sci Res Int 3(4):231–236
24. Veldman S, Kortbeek P, Wőlcken PRH, Kos J, Fernandez Villegas I (2020) Development of a multifunctional fuselage demonstrator. In: Proceedings of aerospace Europe conference (AEC 2020-481)

Chapter 16
Challenges and Current Solutions of Refrigerated Transportation

Pekka Neittaanmäki, Kirill Akimov, Veronika Akimova, and Ronja Tuovinen

Abstract This paper addresses issues arising from cold chain logistics with primary focus on its environmental impact. In particular, the sources of emissions, including carbon dioxide (CO_2) and other serious pollutants contributing to global warming, are considered. We overview methods for reducing these emissions, specifically the methods that can be integrated without drastically altering existing temperature-controlled transport systems. We discuss strategies for both completely and partially mitigating emissions associated with on-road refrigeration. These strategies include the use of passively refrigerated transport packaging to eliminate the need for active refrigeration and the enhancement of current refrigeration methods with phase change materials. Moreover, we present an approach to minimise the emissions that may arise from inadequate logistics management delving into the principle behind the CO_2-based optimisation of distribution routes. Additionally, we overview mathematical optimisation techniques used in identifying environmentally preferable solutions in managing temperature-sensitive logistics. We also note a potential area where artificial intelligence could be applied to enhance the optimisation of cold chain logistics.

Keywords Refrigerated transportation · CO_2 emissions · Cold chain

P. Neittaanmäki
Lappeenranta University of Technology, Yliopistonkatu 34, 53850 Lappeenranta, Finland
e-mail: pekka.neittaanmaki@jyu.fi

K. Akimov · V. Akimova
University of Helsinki, Yliopistonkatu 3, 00014 Helsinki, Finland
e-mail: kirill.akimov@helsinki.fi

V. Akimova
e-mail: veronika.akimova@helsinki.fi

R. Tuovinen (✉)
University of Lappeenranta, Yliopistonkatu 34, 53850 Lappeenranta, Finland
e-mail: ronja.tuovinen@gmail.com

T. Tuovinen et al. (eds.), *Advanced Computational Methods and Design for Greener Aviation*, Computational Methods in Applied Sciences 59,
https://doi.org/10.1007/978-3-031-61109-4_16

16.1 Introduction

As the world becomes more interconnected so does the transportation of food, medicine, and everyday goods. As a result, one can receive a vaccine that was produced in Germany, buy bananas that were grown in Philippines and cook fish that was imported from Norway. None of this would be possible if refrigerated transportation did not keep these items cool all throughout the journey. This so-called cold chain accounts for all the actions taken to maintain a certain stable temperature no matter what weather conditions are faced outside in order to keep the products safe for consumption.

This article examines the current state of cold chain logistics with an emphasis on environmental concerns. It also provides insights into the practical implementation of temperature-controlled distribution in Finland. It identifies the origins of the emissions associated with cold chain transport. The discussion includes an analysis of methods aimed at reducing emissions, highlighting a novel approach to ensure transportation temperature that does not require an additional energy supply and underscores the importance of monitoring temperature compliance.

Furthermore, the article presents various strategies to minimize emissions in cold chain logistics. One such strategy involves optimizing the sequence of delivery stops to reduce emissions effectively. The review extends to scholarly investigations into this concept within the domain of temperature-controlled transportation. The focus is to determine which optimization methods have been extensively researched and which may possess untapped potential for enhancing the sustainability of the cold chain.

The article is organised as follows: Sect. 16.2 presents an overview of a cold chain logistics. Section 16.3 specifies the emission sources in cold chain transportation. Section 16.4 demonstrates the ways of minimisation of refrigeration-related emissions. In Sect. 16.6, the minimisation of mismanagement-related emissions is discussed. Finally, the conclusion and a short discussion of this article are provided in Sect. 16.7.

16.2 Cold Chain

Cold chain is the actions taken to maintain a product at a certain temperature through the journey from producer to the consumer. The cold chain consists of five parts, like shown in Fig. 16.1.

In this article we will focus mainly on the problems caused by transportation. However since transportation and storage face some similar obstacles, we will discuss the solutions for both parts, when relevant.

We reached out to a distributor[1], who distributes temperature-controlled products. He has experience of distributing from the storage facility to shops and restaurants in

[1] Artturi Neittaanmäki.

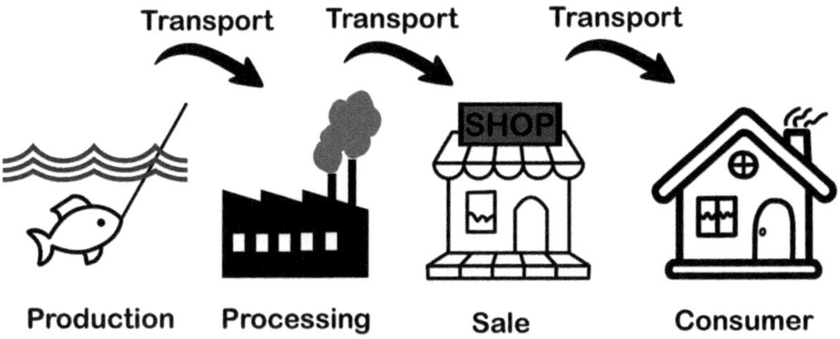

Fig. 16.1 Different parts of cold chain. The example is about the cold chain of fish

cities as well as in smaller villages. The car can handle a few different temperatures at once by locating the products on different places. The refrigeration is on from the moment the driver fetches the vehicle until the moment when the last product has been distributed. The driver is responsible for the temperature and is required to inform the customer in case the temperature has risen above a certain limit, meaning that the produce is not safe anymore. The products are often distributed in boxes, platforms or carriages with wheels. They often need to be transported back once the products have been loaded off them, which takes room in the truck.

It is important to have an unbroken cold chain since products that have gotten warmer or colder than the wanted temperature might not be safe for consumption any longer. This is especially important when transporting medicine or easily perishable products like meat or fish. This cold chain can be created in many ways that vary from passive cooling (e.g. insulating cold boxes) to active cooling (e.g. VCR, or vapor compression refrigeration).

16.2.1 Environmental Impact

Cold chain has a huge impact on the environment and economy. This is why regulations for different parts of it are getting stricter. It is estimated that Vapor Compression Refrigeration systems cause approximately 10% of greenhouse gas emissions and consumes 15% of the electrical energy. It holds about 80% of the market share for refrigeration (both commercial and industrial) and air conditioning systems [1]. According to a report made by Food and Agriculture Organization of the United Nations in 2017, approximately one third of the food is wasted during the production, processing, transportation and consumption. Majority of it is wasted during production and post-harvest, especially in the developing countries or by the consumer, especially in Europe and North America and East Asia. On average 4% is lost during distribution [2]. However the number lost due to transportation (any time the food is moved from place to place) can be larger, since the products need to be transported from producer to processing facilities, from there to storage facilities and only from there to distribution.

On top of the emissions, refrigerated transportation is a big source of noise pollution. This is a problem especially in proximity to cities and villages, since noise pollution has been linked to increased stress levels, disturbed sleep and cardiovascular diseases [3]. Some of the cold transportation happens during night when there is the least amount of traffic. This however leads to the engine running loudly during offload and that can disturb the sleep of people living nearby.

Because of climate change many organizations want to reduce the emissions they or their members produce. For cold transportation this means either reducing the number of products that are transported across the globe or reducing emissions caused by this transportation. Since the first option is highly unlikely if we look at trends in globalization, the reduced emissions of transportation methods are the way to go. For example, EU set a regulation that by 2030 new trucks need to reduce emissions by 45% compared to 2019 [4].

16.2.2 Economical Impact

Cold chains are not ideal for economics either. Loss of products is not good for the producer, because they need to work extra to cover for the percentage that gets lost. The cold chain is also expensive to maintain. Equipment for the cold chain is more expensive than their non-temperature-controlled counterparts and often requires installation right into the van, truck or storage room. This brings up costs to the final product that is sold to the consumer. The transportation is also inefficient, since in most systems one cannot transport products that require a vastly different temperature. This means that multiple trucks can be driving nearly the same route half-empty because they are transporting things that cannot be stored together in the same temperature. Most of the systems release gas as a side product of the refrigeration systems and thus are not suitable for air transportation in large quantities [5].

16.2.3 Vapor Compression Refrigeration

Vapor compression refrigeration is the most commonly used method in refrigeration, cooling or air conditioning. It accounts for approximately 80% of the market share [1]. This is due to the the variety of different power supply methods, e. g. taking the power straight from different parts of the engine or from a separate generator. The VCR cycle looks something like this [6]:

1. Evaporator: evaporates refrigerant by absorbing thermal energy from the environment. The evaporator is located in the are which is cooled by the cooling effect of the absorption.

2. Compressor: compresses the refrigerant to the maximum pressure of the cycle. This happens by absorbing energy from an external source (e.g. the engine)
3. Condenser: Allows the refrigerant to condense from gas to liquid by releasing thermal energy in the environment. The condenser is usually located outside the area which is being cooled (e.g. outside the truck).
4. Expansion device: Drops the pressure of the refrigerant, which cools it down. This Low-pressure and low-temperature refrigerant is directed back to the evaporator and the cycle starts over again.

There are a few things we should consider regarding VRC systems. First issue is the refrigerant. Study estimated that around 5–25% of the annual charge per year of refrigerant in vapor compression transportation refrigeration units get leaked [7]. Refrigerants are often hydrofluorocarbon refrigerants with high Global Warming Potentials [7]. EU has set legislation to phase out from using these refrigerants, which leads to growing market for alternatives [8]. The second thing to consider is efficiency. This is especially visible when the systems is closely linked to the engine. One of the weakest points for VCR systems is on and off loading of the truck. During on and off loading the doors to the cooled area will be open for longer periods of time and thus the cooled air mixes with the warmer air from the outside. The truck needs to stay running to counter-power the mixing and keep products from spoiling [9]. Similar efficiency problem arises when the load is not full. The system still needs to keep the entire truck cold, even though it is only half full. The efficiency of the engine drops to 1–10% in these two cases [9].

16.3 Emission Sources in Cold Chain Transportation

In cold chain transportation, most CO_2 emissions arise from fuel combustion inside internal combustion engines. This type of engines is widely used in refrigerated transport as main engines to propel trucks and as auxiliary engines to ensure the refrigeration. The purpose of the internal combustion engine is to convert the chemical energy of the fuel into more convenient mechanical energy [10], which can directly be used by:

- the powertrain to rotate the wheels
- a generator to produce electric current for an electric refrigeration unit
- the compressor of the refrigeration unit.

The energy conversion is carried out as ignition (reduction-oxidation reaction), in which most carbon of the fuel compounds forms CO_2 while some carbon forms compounds that will eventually transform to CO_2 [10].

The amount of CO_2 emissions released from the main or an auxiliary engine depends on the fuel composition and the amount of combusted fuel [10]. This amount depends on the energy content of the fuel [10], efficiency of the engine [11], and desired power output [12].

- In case of the main engine, the desired power output is mainly determined by needs of the powertrain, which consumes energy to accelerate or compensate speed losses caused by air resistance, tire rolling resistance, and gravity (when travelling uphills) [12].
- In case of the auxiliary engine, the desired power output is mainly determined by needs of the refrigeration unit, which consumes energy to lower the temperature of the freight compartment and compensate temperature rises caused by heat transfer from the environment to cargo area (due to inefficient thermal insulation) and warm air flows inside the freight compartment during unloading [13].

In addition to the fuel combustion, the refrigeration is a source of emissions in the cold chain transportation. Most refrigerated systems are based on heat absorption (endothermicity) of evaporation and heat emission (exothermicity) of condensation [14]. In order for the refrigerant to condense at an outside temperature that is usually much higher than its dew point [15], the dew point is raised by increasing the pressure [14]. However, high pressure causes leakages of the refrigerant that is most commonly R-404A or R-452A in the refrigerated road transport [16]. These substances are mixtures of different refrigerants that are chemically pure hydrofluorocarbons. These compounds are considered greenhouse gases [17]. For a more detailed description of the formulations of R-404A and R-452A refer to Fig. 16.2a and c.

The amount of leaked refrigerant may be converted to the corresponding amount of CO_2 emissions based on the climatic impact of the refrigerant when released into the atmosphere (see Fig. 16.2). The conversion factor may be defined as a ratio of the respective effects on the amount of energy trapped in the atmosphere. This rate is called global warming potential (GWP) and is commonly used in policies [18], particularly in the context of refrigerant. However, greenhouse gases do not influence global warming instantly after their emission and the GWP value varies with respect to the interval under consideration [18]. Commonly, the accumulated global warming effect over the following time horizons is considered: 20 years (GWP-20), 100 years (GWP-100) and 500 years (GWP-500) [19].

In addition to the primary emissions that are directly emitted during cold chain transportation, there are secondary emissions. These arise from the carbon footprint of the products that are lost due to inadequate temperature-controlled transportation. Namely, flawed cold chain logistics force production to produce more products and emit more CO_2 in order for production to not only meet consumer demand but also a certain percentage of transportation losses. The losses may occur due to a complete absence of refrigeration during transport, which is more relevant to developing countries, or due to malfunctions in refrigeration units, breakdowns of refrigerated trucks, and traffic accidents, which is more relevant to developed countries [21].

(a) Mass share of the constituents of R-404A

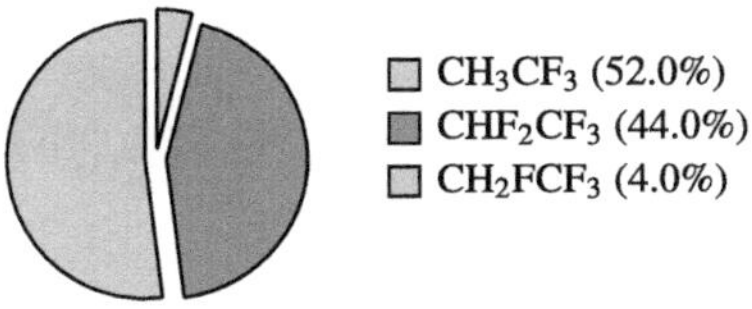

(b) GWP values of R-404A

Constituent	GWP-20	GWP-100	GWP-500
CH_3CF_3	7840	5810	1940
CHF_2CF_3	6740	3740	1110
CH_2FCF_3	4140	1530	436
R-404A mixture:	**≈ 7210**	**≈ 4730**	**≈ 1510**

(c) Mass share of the constituents of R-452A

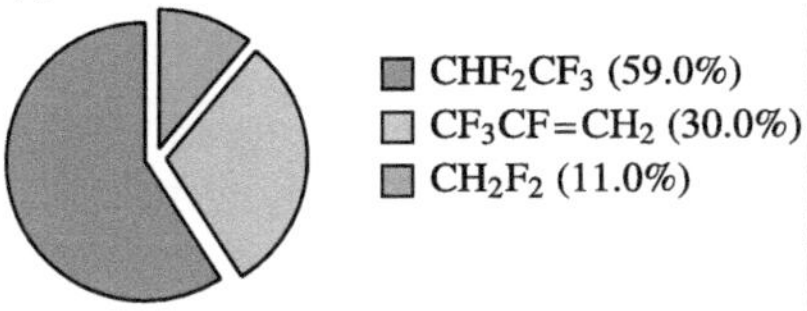

(d) GWP values of R-452A

Constituent	GWP-20	GWP-100	GWP-500
CHF_2CF_3	6740	3740	1110
$CF_3CF{=}CH_2$	1.81	0.501	0.143
CH_2F_2	2690	771	220
R-452A mixture:	**≈ 4270**	**≈ 2290**	**≈ 680**

Fig. 16.2 Global warming potentials over 20 years (GWP-20), 100 years (GWP-100) and 500 years (GWP-500) for R-404A and R-452A, expressed in CO_2 equivalents. Calculations are based on the composition of each refrigerant [15] and the GWP-20 values of their respective compounds [20]. Constituents are color-coded for convenience

16.4 Minimisation of Refrigeration-Related Emissions

Due to the high demand on more efficient and environmentally friendly solutions related to refrigerated transportation and cold chains there has been an increase in new innovation and research on the topic. In this section we will discuss some of them.

16.4.1 Passive Cooling

Well-insulated box with a cold gel mat inside it (see Fig. 16.3), presents a considerable solution for temperature controlled transportation [5]. The box ensures that the cooling provided by the cold gel mat wont mix with the air outside box, giving a perfect solution for on and off loading. The insulation would also allow the transportation of multiple products that require different temperatures from each other. This would drastically reduce the used energy and thus lower the emissions. The box would also include a sensor that can collect and share data and keep track of the temperature. The downside of the box is that it needs to be transported back instead of simply recycling like cardboard boxes it at the final destination. Some solutions include a box that can be disassembled for the journey back, when it would take less space.

Fig. 16.3 Example of a cold box according to the new patent [5]

16.4.2 *Phase Change Materials*

Phase change materials absorb thermal energy by changing the phase from solid to liquid or liquid to gas and thus keep the products at a wanted temperature. There are a few different PCMs types to choose from:

- Organic: Paraffin or non paraffin
- Inorganic: metals or salt hydrates
- Eutectic solutions: combinations of multiple materials [6].

Phase change materials can often be used alongside VCR systems. This means that the cooling effect is given by both, PCM and VCR. A study showed that when the car was moving at a low speed or stationary the energy consumption was reduced by 17% due to the PCM [22]. The cold packs or tubes that have PCM within are generally easy to use and don't require a lot of installation.

The PCM materials do have some downsides to them as well. The containers are often not optimized to be used in refrigerated transport and thus it can be hard to place the PCM into the box evenly [6]. The biggest issue however comes from having only one possible temperature, the melting or vaporizing point of the material, and only one duration, the time it takes for the material to melt. This limits the range of temperatures as well as the range of materials that can be used.

16.5 Minimisation of Transportation-Related Emissions

The International Energy Agency (IEA) [23] expects that replacement of internal combustion engine trucks with those powered with electric batteries and hydrogen fuel cells will significantly reduce emissions by 2050. The IEA notes that the battery power is more promising for medium-duty trucks taking shorter and regular routes while the hydrogen power is more promising for heavy-duty trucks taking long-haul routes. To achieve reasonable results in emission reduction, the IEA estimates that trucks should be electrified faster. In 2022 only 1.2% of trucks sold worldwide were electric. However, the sale share of electric trucks should increase to 33% by 2030, according to the IEA.

The passive cooling concept described in Sect. 16.4.1 holds the potential to lower emissions from cold chain logistics by integrating intermodal transportation. The IEA [24] asserts that transferring freight from road to rail not only improves energy efficiency of transportation but also lowers its emissions. The IEA underscored by the fact that rail transport is uniquely characterised by its extensive electrification, with up to 50% of rail freight transport being electrically powered. Furthermore, the IEA regards trains as the only viable alternative to trucks for land-based transportation. However, refrigerated freight cars are no longer being manufactured in Europe [16], which may significantly hinder the adoption of rail transport in cold chain logistics. The use of cold boxes, as described in Sect. 16.4.1, may enable the transportation of perishable products at ambient temperature in conventional cars, including alongside other types of freight.

16.6 Minimisation of Mismanagement-Related Emissions

Mismanagement of cold chain logistics may lead to inefficient fuel consumption and consequently to higher CO_2 emissions. A rough example of poor management may be distributing products one by one. Obviously, it is more efficient to deliver the whole order at once. A similar conclusion may be drawn at the order level: delivering several orders at a time is more efficient than delivering them separately. However, making further decisions might not be that straightforward. For instance, it could be challenging to determine which locations to visit at a time and, more crucially, to decide the sequence in which to tour them. These questions are dealt with in the vehicle routing problem (VRP).

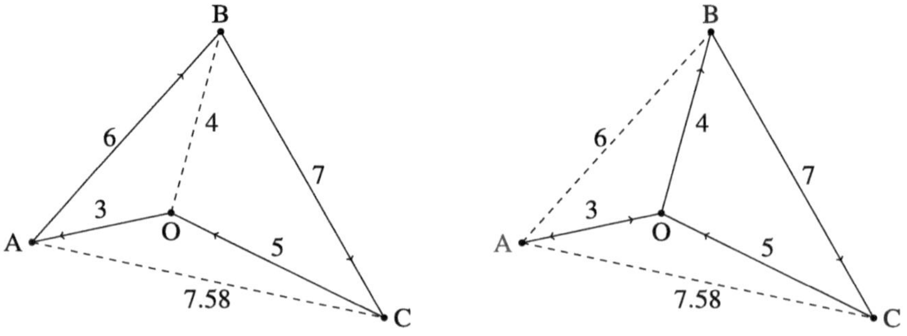

(a) When all customers can be served at a time, the route $O \to A \to B \to C \to O$ is the shortest one with its length 21. Other routes visiting each location once have lengths 21.58 and 22.58

(b) When two customers at most can be served at a time, the route $O \to A \to O \to B \to C \to O$ is the shortest one. Its length is 22. Other routes visiting each customer once have lengths 23, 23.58, and 24.

Fig. 16.4 Visual illustrations of the TSP (left) and VRP (right). O represents a distribution centre. A, B, and C represent customers

16.6.1 Modelling the Vehicle Routing Problem for Emission Minimisation

The VRP is thought to be first presented by G. B. Dantzig and J. H. Ramser in 1959 [25], and it is originally formulated as the problem of lessening the total length of distribution routes in a petrol supply network [26]. Namely, the authors suggest an approach to allocate refuelling stations to tankers in accordance with the ability of the tankers to hold demands of the stations. Subsequently, they apply the travelling salesman problem (TSP) to each tanker—in other words, rearrange the stations in the visiting order that minimises the length of the tour.

Both TSP and VRP may be represented visually as problems of joining up dots on the paper (see Fig. 16.4). In the TSP the challenge is to loop the dots together such that the length of the loop is a minimum. In contrast, the VRP is a more challenging problem of minimising the total length of loops constructed as follows: First, the points representing customers are identified and labeled with numbers in order to represent their demands. Then, the labeled points are painted with different colours so that the sum of the labeled numbers does not exceed a limit (truck capacity) within any group of dots sharing the same color. Finally, the dot representing the distribution center is looped with each group of points.

The VRP may be reformulated into the problem of minimising CO_2 emissions in a temperature-controlled distribution network when the amount of CO_2 emitted on a route is considered instead of the length of the route [12]. However, this approach complicates the problem due to the following fact: while the shortest path between any two locations is a constant that can be found beforehand, and moreover, is assumed to be known in the original VRP, the minimum amount of CO_2 emitted during a conveyance depends on several factors and cannot be determined with certainty in advance. Some of the challenges are listed below.

- *Load-dependence*: The heavier the freight is the more fuel is consumed for its transportation and consequently more CO_2 emitted [11, 12]. Consequently, the amount of CO_2 being emitted on the path to the customer depends on the order of the customer in the distribution sequence.
- *Vehicle-dependence*: Heterogeneous fleets are originally considered in the context of different capacities of trucks [27]. However, the fuel efficiencies of the trucks may differ as well [11], which may result in different amounts of fuel consumed and CO_2 emitted by different trucks under equal conditions [12].
- *Road-dependence*: The steepness of the road may result in higher fuel consumption and higher emissions [12].
- *Speed-dependence*: Every vehicle has an optimum driving mode that minimises fuel consumption. Inevitable deviations from this during the delivery process increase the amount of emissions per unit of travel. The impact of speed limits on emissions is discussed in [12].
- *Driver-dependence*: Speed fluctuation, especially rapid accelerations and decelerations, may result in higher fuel consumption (emissions) compared to moderate driving [11].
- *Time-dependence*: Travelling during rush hours can significantly affect the speed of the truck resulting in inefficient fuel consumption and higher emissions compared to deliveries during off-peak hours [28].

Many realities further increase the level of complexity of the routing optimisation. For example, limitation of truck drivers' working hours may impose *time windows*—in other words, constraints on delivery times [29]. That is why the most CO_2-efficient routing strategy may prove inappropriate for a practical application and either increasing vehicle speeds or involving more vehicles may be needed.

16.6.2 Solving the CO_2-Oriented Vehicle Routing Problem for Cold Chain Logistics

The CO_2-oriented VRP in temperature-controlled logistics is most commonly solved by means of mathematical optimisation, where the optimisation problem is represented in the form of one or more objective functions and constraints. Consequently, the objective of the problem becomes to either maximise the value of the function (in case of profit) or minimise it (in case of environmental impact), subject to the given constraints. For that purpose, appropriate values are sought for the variables, commonly referred to as decision variables.

The decision variables denote decisions affecting the final characteristics of the model, such as economic, environmental and social impact. In the context of optimizing temperature-controlled transportation with respect to CO_2 emissions, it is common to simultaneously search for optimal routing, location-allocation, and inventory.

Table 16.1 A classification of solution approaches used to solve the VRP with respect to CO_2 emissions in temperature-controlled transportation

Applied method	Method type			References
	Exact	Metaheuristic	Hybrid	
ε-constraint method	✓			[30, 32]
Revised multi-choice goal programming method	✓			[31]
Cycle evolutionary genetic algorithm		✓		[35, 36]
Particle swarm optimization (standard and modified)		✓		[34]
Many-objective gradient evolution algorithm		✓		[40]
The ribonucleic acid ant colony optimization algorithm		✓		[38]
An improved ant colony algorithm with a multi-objective heuristic function		✓		[39]
An adaptive genetic algorithm		✓		[37]
Non-dominated sorting genetic algorithm II		✓		[33]
Hybrid genetic algorithm (with heuristic rules)			✓	[43]
A matheuristic algorithm based on the iterated local search algorithm and a mixed integer programming			✓	[41]
A variable neighborhood search combined with the non-dominated sorting genetic algorithm II and techniques for order preference by similarity to an ideal solution			✓	[44]
Hybrid solution technique (based on possibilistic linear programming and fuzzy weighted goal programming approach)			✓	[42]

These three elements are interconnected so to optimize the model as a whole it is important to look for values of decision variables within a single problem rather then subsequently solve the vehicle routing problem, location-allocation problem, and inventory problem. For example, routing decisions may impact inventory levels, and the placement of facilities (location-allocation) can affect both routing and inventory decisions.

In the field of cold chain logistics, the desired values of the decision variables are most commonly sought using various optimisation techniques, including exact [30–32], metaheuristic [33–40], and a combination of both, referred to as hybrid methods [41–44]. For more details see Table 16.1. Recent research advancements show a trend towards solving VRPs by means of artificial intelligence, particularly using deep reinforcement learning [45, 46]. However, to the best of our knowledge, this trend has not yet been extensively applied in optimising logistics for temperature-sensitive goods.

16.7 Conclusion

In this article we showed how the cold chain is a crucial part of today's world and what are it's weak points. However, refrigerated transportation has a considerable impact on the environment and that the current systems need to improve in order to meet the growing needs to reduce emissions while the needs for transportation are growing.

The majority of the systems used currently are Vapour Compression Refrigeration (or VCR) systems. Other common systems is the use of Phase Change Materials, especially when combined with a VCR system to provide extra help. There are also some newer solutions, one of them being well-insulated cold box with a cold-gel mat and a sensor inside. These new innovations and solutions could provide the much needed help to reduce the emissions of cold transportation.

We discussed the optimization of cold transportation by using the Vehicle Routing Problem. The biggest issue with this optimization was that CO_2 emissions are not a constant, but depend on the traffic situation and can be drastically changed with an external influence. One of the answers to that problem was to use real-time data about traffic to solve and update the route.

We have shown that the refrigerated transportation is a complex system. It can provide us with a lot of valuable data, that can be used to analyze and optimize the cold chain and refrigerated transport. In the future, it will be vital to use this data to train AI methods, that could be used to further reduce emissions. This will be the next step in our research regarding this topic.

Acknowledgements Thanks to Lappeenranta University for their support. Thanks to Artturi Neittaanmäki for his report on local cold transport in Finland.

References

1. She X et al (2018) Energy-efficient and -economic technologies for air conditioning with vapor compression refrigeration: a comprehensive review. Appl Energy 232:157–186. ISSN: 0306-2619. https://doi.org/10.1016/j.apenergy.2018.09.067. https://www.sciencedirect.com/science/article/pii/S0306261918313734
2. Food and Agriculture Organization of the United Nations (2017) The future of food and agriculture: trends and challenges. Italy, Rome. ISBN: 978-92-5-109551-5. https://fao.org/agrifood-economics/publications/detail/en/c/471795/
3. World Health Organization (2011) Burden of disease from environmental noise Quantification of healthy life years lost in Europe. https://www.who.int/publications/i/item/9789289002295
4. European Commission (2022) Reducing CO_2 emissions from heavy-duty vehicles. https://climate.ec.europa.eu/eu-action/transport/road-transport-reducing-co2-emissions-vehicles/reducing-co2-emissions-heavy-duty-vehicles_en. Accessed 04 Oct 2023
5. Proskin J (2020) Method for temperature controlled transport, EP 2 494 293 B1. European Patent Office

6. Maiorino A, Petruzziello F, Aprea C (2021) Refrigerated transport: state of the art, technical issues, innovations and challenges for sustainability. Energies 14(21). https://doi.org/10.3390/en14217237
7. Rai A, Tassou SA (2017) Environmental impacts of vapour compression and cryogenic transport refrigeration technologies for temperature controlled food distribution. Energy Convers Manage 150:914–923. ISSN: 0196- 8904. https://doi.org/10.1016/j.enconman.2017.05.024
8. European Union (2014) Regulation (EU) No 517/2014 of the European parliament and of the council of 16 april 2014 on fluorinated greenhouse gases and repealing Regulation (EC) No 842/2006 Text with EEA relevance. Official Journal of the European Union. Retrieved from EUR-Lex. https://eur-lex.europa.eu/legal-content/EN/TXT/?uri=CELEX:32014R0517
9. Fard SM, Khajepour A (2016) An optimal power management system for a regenerative auxiliary power system for delivery refrigerator trucks. Appl Energy 169:748–756. ISSN: 0306-2619. https://doi.org/10.1016/j.apenergy.2016.02.078
10. Garg A, Kainou K, Pulles T (2006) Chapter 1: introduction. In: Eggleston HS et al (ed) 2006 IPCC guidelines for national greenhouse gas inventories, Energy, vol 2. IGES, Japan
11. National Research Council (2010) Technologies and approaches to reducing the fuel con sumption of medium and heavy-duty vehicles. The National Academies Press, Washington, DC. https://doi.org/10.17226/12845. http://nap.nationalacademies.org/12845
12. Bektasş T, Laporte G (2011) The pollution-routing problem. Transp Res Part B Methodol 45(8):1232–1250. ISSN: 0191-2615. https://doi.org/10.1016/j.trb.2011.02.004
13. Poks A et al (2020) Wholistic simulation of an all-electric refrigerated delivery vehicle:1–6. https://doi.org/10.23919/SICEISCS48470.2020.9083498
14. Dinçer I (2017) Refrigeration systems and applications, 3rd edn. John Wiley & Sons, Incorporated. ISBN: 9781119230755
15. Designation and Safety Classification of Refrigerants (2022) ANSI/ASHRAE 34. American National Standards Institute/American Society of Heating, Refrigerating and Air-Conditioning Engineers. ISSN: 1041-2336
16. Air Conditioning Refrigeration and Heat Pumps Technical Options Committee (RTOC) (2022) Transport refrigeration. In: 2022 Report of the refrigeration, air conditioning and heat pumps technical options committee: 2022 assessment. United Nations Environment Programme, pp 125–136. ISBN: 978-9914-733-93-8. https://ozoneunep.org/system/files/documents/RTOC-assessment
17. Matthews JBR et al (2021) Annex VII: glossary. In: Masson-Delmotte V et al (eds) Climate change 2021: the physical science basis. Contribution of Working Group I to the sixth assessment report of the intergovernmental panel on climate change. Cambridge University Press, Cambridge, United Kingdom and New York, NY, USA, pp 2215–2256. https://doi.org/10.1017/9781009157896.022
18. Intergovernmental Panel on Climate Change (IPCC) (2013) Anthropogenic and Natural Radiative Forcing. In: Climate change 2013—the physical science basis : working group I contribution to the fifth assessment report of the intergovernmental panel on climate change. Cambridge University Press, pp 659–740. https://doi.org/10.1017/CBO9781107415324.018
19. Forster P et al (2021) The Earth's energy budget, climate feedbacks, and climate sensi- tivity. In: Masson-Delmotte V et al (ed) Climate change 2021: the physical science basis. contribution of working group I to the sixth assessment report of the intergovernmental panel on climate change. Cambridge University Press, Cambridge, United Kingdom and New York, NY, USA, pp 923–1054. https://doi.org/10.1017/9781009157896.009
20. Smith C et al (2021) The Earth's energy budget, climate feedbacks, and climate sensitivity supplementary material. In: Masson-Delmotte V et al (ed) Climate change 2021: the physical science basis. Contribution of working group I to the sixth assessment report of the intergovernmental panel on climate change. Intergovernmental Panel on Climate Change. https://www.ipcc.ch/
21. High Level Panel of Experts on Food Security and Nutrition (2014) Food losses and waste in the context of sustainable food systems. Report Committee on World Food Security, Rome

22. Shafiei SE, Alleyne A (2015) Model predictive control of hybrid thermal energy systems in transport refrigeration. Appl Therm Eng 82:264–280. ISSN: 1359-4311. https://doi.org/10.1016/j.applthermaleng.2015.02.053
23. International Energy Agency (IEA) (2023) Net zero roadmap: a global pathway to keep the 1.5 C goal in reach. https://www.iea.org/reports/net-zero-roadmap-a-global-pathway-to-keep-the-15-0c-goal-in-reach. License: CC BY 4.0. Paris
24. International Energy Agency (2019) The future of rail, p 175. https://www.oecd-ilibrary.org/content/publication/9789264312821-en
25. Laporte G (2009) Fifty years of vehicle routing. Transp Sci 43(4):408–416. https://doi.org/10.1287/trsc.1090.0301
26. Dantzig GB, Ramser JH (1959) The truck dispatching problem. Manage Sci 6(1):80–91. https://doi.org/10.1287/mnsc.6.1.80
27. Golden B et al (1984) The fleet size and mix vehicle routing problem. Comput Oper Res 11(1):49–66. https://doi.org/10.1016/0305-0548(84)90007-8
28. Franceschetti A et al (2013) The time-dependent pollution-routing problem. Transp Res Part B Methodol 56:265–293. ISSN: 0191-2615. https://doi.org/10.1016/j.trb.2013.08.008
29. Kok AL et al (2010) A dynamic programming heuristic for the vehicle routing problem with time windows and European community social legislation. Transp Sci 44(4):442–454. https://doi.org/10.1287/trsc.1100.0331
30. Golestani Maryam et al (2021) A multi-objective green hub location problem with multi item-multi temperature joint distribution for perishable products in cold supply chain. Sustain Prod Consum 27:1183–1194. https://doi.org/10.1016/j.spc.2021.02.026
31. Jouzdani J, Govindan K (2021) On the sustainable perishable food supply chain network design: A dairy products case to achieve sustainable development goals. J Clean Prod (2021). https://doi.org/10.1016/J.JCLEPRO.2020.123060
32. Soysal M, Bloemhof-Ruwaard JM, van der Vorst JGAJ (2014) Modelling food logistics networks with emission considerations: the case of an international beef supply chain. Int J Prod Econ. https://doi.org/10.1016/J.IJPE.2013.12.012
33. Gharehyakheh A et al (2020) A multi-objective model for sustainable perishable food distribution considering the impact of temperature on vehicle emissions and product shelf life. Sustainability (Switzerland) 12(16). https://doi.org/10.3390/su12166668
34. Li Y, Lim M, Tseng M (2019) A green vehicle routing model based on modified particle swarm optimization for cold chain logistics. Ind Manag Data Syst. https://doi.org/10.1108/IMDS-07-2018-0314
35. Qin G, Tao F, Li L (2019) A vehicle routing optimization problem for cold chain logistics considering customer satisfaction and carbon emissions. Int J Environ Res Public Health. https://doi.org/10.3390/IJERPH16040576
36. Wang S et al (2017) Optimization of vehicle routing problem with time windows for cold chain logistics based on carbon tax. https://doi.org/10.3390/SU9050694
37. Wang Z, Wen P (2020) Optimization of a low-carbon two-echelon heterogeneous- fleet vehicle routing for cold chain logistics under mixed time window. Sustainability. https://doi.org/10.3390/SU12051967
38. Zhang LY et al (2019) Low-carbon cold chain logistics using ribonucleic acid-ant colony optimization algorithm. J Clean Prod. https://doi.org/10.1016/J.JCLEPRO.2019.05.306
39. Zhao B et al (2020) Cold chain logistics path optimization via improved multi-objective ant colony algorithm. IEEE Access. https://doi.org/10.1109/ACCESS.2020.3013951
40. Zulvia FE, Kuo RJ, Nugroho DY (2020) A many-objective gradient evolution algorithm for solving a green vehicle routing problem with time windows and time dependency for perishable products. J Clean Prod. https://doi.org/10.1016/J.JCLEPRO.2019.118428
41. Babagolzadeh M et al (2020) Sustainable cold supply chain management under demand uncertainty and carbon tax regulation. Transp Res Part D Transp Environ. https://doi.org/10.1016/J.TRD.2020.102245
42. Tirkolaee EB, Aydin NS (2022) Integrated design of sustainable supply chain and transportation network using a fuzzy bi-level decision support system for perishable products. https://doi.org/10.1016/J.ESWA.2022.116628

43. Wang S, Tao F, Shi Y (2018) Optimization of location-routing problem for cold chain logistics considering carbon footprint. Int J Environ Res Public Health. https://doi.org/10.3390/IJERPH15010086
44. Wu D et al (2022) Research on the time-dependent split delivery green vehicle routing problem for fresh agricultural products with multiple time windows. https://doi.org/10.3390/AGRICULTURE12060793
45. Mustakhov T, Akhmetbek Y, Bogyrbayeva A (2023) Deep reinforcement learning for stochastic dynamic vehicle routing problem. https://doi.org/10.1109/ICECCO58239.2023.10147154
46. James JQ, Yu W, Gu J (2019) Online vehicle routing with neural combinatorial optimization and deep reinforcement learning. IEEE Trans Intell Transp Syst 20(10):3806–3817. https://doi.org/10.1109/TITS.2019.2909109

Chapter 17
Effect of Computational Generative Product Design Optimization on Part Mass, Manufacturing Time and Costs: Case of Laser-Based Powder Bed Fusion

Patricia Nyamekye, Rohit Lakshmanan, and Heidi Piili

Abstract Light weight and high-performance metals are usually desired for transport applications for economic and environmental benefits, especially in automotive and aerospace. Conventional manufacturing (CM) methods used in manufacturing motor vehicle and aircraft components often lead to compromises, for example by adding non-function excess materials for ease of manufacturing or incur extra cost with the removal of enormous start up preforms. The manufacturability limitations of CM and supporting digital software previously constrained product design and manufacturing flexibilites to allowable geometrical features, tools and fixtures. Additive manufacturing (AM) is a manufacturing method allowing unprecedented design and manufacturing freedom and flexibilities. The evolvement AM from being a mere consumer goods, prototyping, and tooling manufacturing method to include functional end-use parts help optimize performance and economic value. AM allows to manufacture lightweight and intricate geometrical designs which are otherwise impossible via CM methods. AM uses a layerwise approach to manufacture components using digitally defined data. AM is often criticized, especially the metal based, for the need of supports, post-processing, high investment cost, high energy consumption, low build rate, and poor surface quality. Hollows and overhangs designs require sacrificial supports that are removed prior to or during the post-processing phase.

P. Nyamekye (✉)
School of Energy Systems, Lappeenranta-Lahti University of Technology LUT, Yliopistonkatu 34, FI-53851 Lappeenranta, Finland
e-mail: patricia.nyamekye@lut.fi

R. Lakshmanan
LUT Business School, Lappeenranta-Lahti University of Technology, Yliopistonkatu 34, FI-53851, Lappeenranta, Finland
e-mail: rohit.lakshmanan@lut.fi

H. Piili
Department of Mechanical and Materials Engineering, University of Turku, Joukahaisenkatu 3-5 B, FI-20520 Turku, Finland
e-mail: heidi.piili@utu.fi

T. Tuovinen et al. (eds.), *Advanced Computational Methods and Design for Greener Aviation*, Computational Methods in Applied Sciences 59,
https://doi.org/10.1007/978-3-031-61109-4_17

Technological advances in AM such as generative design and process simulations have helped users reduce or omit some of the limitations to adoption. Digital tools allow a swift optimizing of the process parameters and manufacturing including part placement, build orientation and support structures prior physical build. Simulation-assisted product design for AM parts allow generative optimized design iterations in consideration of predefined AM parameters and rules and end-application requirements. This study provides an overview and exemplary application of digital tools in product design (simulation and optimization). A case study demonstrating product designs highlights the potential benefits of comparable optimized plate designs to mass, time and cost savings using laser assisted powder bed fusion (L-PBF), one of the subcategory of AM. The aim of this study was to highlight a strategic adoption of L-PBF via offered potentials of intricacy, lightweight and hierarchical gradient design via simulation assisted product design for AM parts. This study proposes a strategic adoption plan which potentially maximizes resource efficiency via digital tools in AM. The results provide insight into the potential benefits of L-PBF and demonstrate the potential of such approach to enhance the confidence in adopting PBF for metals.

Keywords Additive manufacturing · Powder bed fusion · Design for additive manufacturing · Design optimization · Digital tools

17.1 Introduction

Manufacturing methods can either be additive (e.g., powder bed fusion), subtractive (e.g., machining) or formative (e.g., welding, molding) depending on the mode by which materials are removed or formed. Subtractive methods produce parts by removing material from workpieces. Additive manufacturing, AM (aka 3D printing) produces parts by adding material layer-by-layer in contrast to subtractive methods [1]. Some aspects of conventional manufacturing (CM) systems have not kept up with advancements of current industry 4.0 technologies and fail to meet modern engineering and operational requirements such as weight reductions, speedy lead times and minimum order quantity omission. Advanced manufacturing technologies, such as additive manufacturing [2] and advanced machining [3, 4] are exemplary methods to simultaneously produce uniquely discrete metal parts while satisfying such operational requirements. AM potentially combines optimal shape, microstructure, and process conditions to influence functionalities including improved mechanical (strength), aesthetics (geometry adherence), sustainability (environmental footprints) among others with precise material distribution and resource optimization to achieve such requirements [5].

Industrial engineering activities are both business and engineering centered in practice [6]. In carrying out a component design, one challenge is whether design features can be fabricated so that they will satisfy intended functional requirement and offer economic benefits. This require often multidisciplinary optimization of product design, particularly in AM where there is flexibility to integrate functions and parts without adding to costs, time or manufacturing steps [7, 8]. An optimal design is often required to satisfy different functionalities such as improved mechanical properties (e.g., ultimate tensile stress), customization, resource and cost efficiencies, or a combination of these, depending on the use case [2, 6]. Product design in AM considers the selection of materials, processing parameters, postprocessing, quality assurance, software, and hardware in an integrated digital workflow. The advent of design for additive manufacturing (DfAM) along AM is fostering the uptake of the unique capabilities of AM [9–11] by defining the design opportunities, feedstock requirement, process constraints and cost savings [9, 12]. The flexibility and swiftness to create optimized designs via AM are propelled by the capabilities of digital tools (computational design and simulation). Computational design and modelling software allow a seamless workflow in AM as well other modern manufacturing systems. Simulation assisted product design allow the exploration of different design configuration such as allowable geometrical features, size, material, process parameters, and build plans. Digital tools help achieve optimal control and optimum designs [6]. Optimal control systems offer a means to increase throughput with automated system controls. Optimum designs via simulation assisted product design for AM parts offer means to create high value metal parts to fulfil design goals and maximize throughput [6]. Some attainable benefits with metal AM parts are functionally-graded designs, design consolidation, design complexity, better-optimized products, reduced time to market, localized manufacturing and resource effectiveness [5, 13, 14]. The success of the adoption of metal AM may not advance solely on the offered benefits. There is the need to gain a prior understanding of the distinct capabilities and limitations of the process for effective and broader adoption [9]. Many adopters of AM do so with a set of conventional product designs without prior knowledge of AM design aspects [15]. The efficient and effective use of computational software and the readiness to implement supportive strategies at the management level is equally essential to ensure that metal AM offers the anticipated advantages.

The adoption of metal AM faced some hindrances despite the numerous benefits, for example, limitations, availability, and high costs of software and AM machines, quality control, and lack of sufficient knowledge [2, 7, 9, 16–18]. Metal based laser assisted PBF (L-PBF), a subcategory of AM, is particularly limited due to reasons such as post-processing need, limited materials, numerous affecting processing parameters, product standards and quality management [7]. The high cost of well-sophisticated digital tools that give the possibility to vast design, the limited number of materials and the lack of suitable standards continue to hinder the wide spread of AM [15, 19]. The continuous development of suitable low-cost software, machine systems, new materials and standards such as ISO/ASTM 52,911–1 and ASME Y14.46–2022 are promising to democratize and streamline metal AM for a wider adoption [15, 20, 21]. Metal AM refers to a layer-by-layer manufacturing of three-dimensional physical components from a digital model using metal material.

Computational tools for product design for AM allow to create functionally-driven designs with no or negligible manufacturing difficulties [8]. The right usage of software solutions along the right usage of AM machines enables first-time-right (FTR) parts with desired performance. The concept of FTR in advanced manufacturing systems is key to reducing the amount of downtime, faulty products, and extra costs while maintaining functions. Ensuring that the component can offer new value enables businesses to strategize for unexplored markets through new product designs and industry architectures, requiring well-informed decision-making regarding which comparable methods and digital tools (e.g., AM or computer numeric control machining (CNC-machining); SolidWorks or nTopology) give the needed advantage to thrive in ongoing competitive environments.

AM allows on-demand manufacturing to meet market needs, thereby replacing the conventional operational models of make-to-order (MTO) and make-to-stock (MTS). AM enable decentralized working [22] along digital thread for components creation, inventory and data management [5] capable of swift response to dynamic industrial value chains. Such disruptions are facilitated through a single-seamless strand of data that stretches from the initial design to the finished part, referred to as the digital thread. Digital tools are useful in AM as the entire process chain can be controlled and managed using data from geometry preparation, design optimization, validation, and reworking, build preparation, pre-/actual- post-build simulation, and quality assurance/documentation. Simulation assisted product design for AM is one working method revolutionizing industrial AM product design. This method combines the abilities of digital tools to design optimized parts in conformity to AM-specific design considerations to ensure their printability [23].

Metal AM process need conscious optimal designing and manufacturing planning to overcome limitations along the design, manufacturing and use phases. This study focuses on product design optimization options in L-PBF. The study investigates the impact of integrated computational product design and simulation offered efficiencies in terms of mass saving, production time and costs.

17.2 Metal Additive Manufacturing

Additive manufacturing (AM) is the *process of joining materials to make parts from 3D model data, usually layer upon layer, as opposed to subtractive manufacturing and formative manufacturing methodologies*. ISO/ASTM 52,900:2021 [23] categorizes AM into seven main groups, namely, powder bed fusion (PBF), binder jetting (BJ), material jetting (MJT), directed energy deposition (DED), material extrusion (MEX), sheet lamination (SHL) and vat photopolymerisation (VPP).

(1) PBF: is *an AM process in which thermal energy selectively fuses regions of a powder bed.*
(2) BJT: *is an AM process in which a liquid bonding agent is selectively deposited to join powder materials*

(3) MJT: is *an AM process in which droplets of build material are selectively deposited*
(4) DED: is *an AM process in which focused thermal energy is used to fuse materials by melting as they are being deposited*
(5) MEX: is *an AM process in which material is selectively dispensed through a nozzle or orifice*
(6) SHL: is *an AM process in which sheets of material are bonded to form a part*
(7) VPP: is *an AM process in which liquid photopolymer in a vat is selectively cured by light-activated polymerization* [23].

The different AM categories can manufacture shapes with a vast range of single or multi materials, including polymers, metals, ceramics, composites and paper [5, 13, 24].

Metal products are possible via PBF, BJT, MJT, and SHL. Metal AM refers to a layer-by-layer manufacturing of three-dimensional physical components from a digital model using metal material. Metal based AM parts are mostly applied for high-valued sectors including aerospace, health and the automotive [7]. L-PBF, is the most matured and widely used for industrial application [14, 25] and is suitable for several metals, including aluminum alloys, stainless steel, and titanium alloys, nickel alloys, metal-ceramic composites [26]. Some of the key advantages for the use of L-PBF are high resolution, dimensional accuracy, biomimetic design flexibility for better-optimized metal products [13, 14, 27]. L-PBF also offers advantages such as the flexibility of throughput [28] by controlling the build volume [29] and build rate with multi-laser systems [30]. The more efficiently the platform space is used for combined build, the better a production lead time [5]. L-PBF, via the use of optimized product design (including use of suitable software, tools, and personnel training) can help create components that offer lifetime cost-effectiveness. Studies [2, 14] have shown that optimized product design can reduce total manufacturing cost example by 54% via the reduction of feedstock, preparation, actual-build and postprocessing time [27].

17.2.1 Digital Tools for Product Design in Metal AM

Natural patterns and shapes are poised with beauty, complexity, are lightweight and aesthetically pleasing, though, mostly non-prismatic geometries [13, 31]. Shapes can broadly be classified either as organic or geometric. Organic shapes originate from nature and thus have no defined reading for the shapes making them more flexible but less accurate. Geometric shapes are formed based on defined readings making them more accurate but less inflexible. The modern-day industrial era requires a combination of both shape characteristics. Design and engineering are constantly finding new ways to maximize time, resource efficiency, performance, and overall cost. However, the available methods do not allow full uptake of such benefits as only the permissible designs based on available tools, fixtures and materials are possible. CM

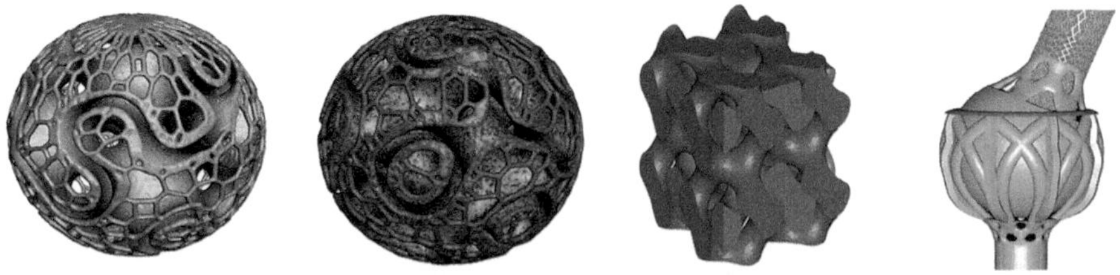

Fig. 17.1 Examples of feasible simulation assisted product design for AM parts, color variation indicates multi-materials. Reproduced from [21]

methods mandatorily considers first available tools, fixtures and machine constraints before designing a part with computer aided design (CAD) software. Designers need to redesign or use multi-manufacturing steps to make permissible organic-like geometric designs. AM conversely offers design freedom through computer generative designing to create biomimetic designs (e.g., nature-mimicked designs (see Fig. 17.1)) that are both more efficient as well as aesthetically pleasing beyond cognitive design abilities. Examples of such geometries include honeycomb, lattice and web-like structures.

As Fig. 17.1 shows AM can feasibly manufacture unique geometry with multi-transitional features (shape, hierarchical structures, multi-material) of varying complexities [21]. Computer and numerical-based software help simulate complex design problems, accurately predict performance, and can run swift iterations to optimize product geometric shape, parameters, materials, simplify workload, and reduce time. Product design in AM significantly benefits from the ability of digital tools and DfAM guidelines to create successful products. The capability of computational design, simulation software and digital-centricity of AM allow to simulate complex design problems and accurately predict performance to generate optimal designs and achieve FTR manufacturing of such design [14]. FTR manufacturing is propelled in manufacturing via digital tools for designing, validating, manufacturing, monitoring, and testing. Simulation-driven product design is promising to the creation of non-defective products; objectively to eliminate reworks, minimize waste and operational costs. Through computational tools digital counterparts serve as a model for (1) initial design, (2) finite element analysis (FEA) and/or computational fluid dynamic (CFD) study, (3) selecting build parameters, materials and (4) planning and simulate the build process prior physical commitment to maximize efficiency. System interoperability of digital software allows for the exchange of files for further studies or optimizing (e.g., SolidWorks CAD model to nTopology for lattice optimization). Pre-selection and testing of build orientations and processing parameters virtually offer means to optimize manufacturing time and reduce /omit faulty build cycles.

Some AM machine systems are equipped with monitoring systems that observe a manufacturing process via different sensors or cameras and halt the process in case of any identified output fluctuations [6, 32, 33]. Other AM machine are equipped with both monitoring and feedback controllers that offer closed-loop system to produce the desired output with adaptive processing parameters controls [33]. Examples of such systems integrated in AM machines are VELO3D intelligent fusion and Interlayer Realtime Imaging & Sensing System (IRISS®). AM machine systems equipped with such systems automatically adjust parameters to correct fluctuating condition and optimize performance. These ensure continuous manufacturing without the need to halt the build process in the event of any fluctuations [34, 35]. Creating optimized designs via generative design or topology optimization software eases the workload and offers better time efficiency for creating organic and geometrically complex shapes. These advanced simulation-driven design tools are promising to achieve features which are otherwise not possible with parametric modeling [31].

17.2.2 Cost Structure in Metal AM

Some major contributing factors to the costs of metal AM products are design costs, build time, machine costs, and feedstock costs, pre-and postprocessing costs [2, 19, 27, 30, 36, 37]. Efficient use of feedstock reduces the part cost, and this can be achieved with reduced production runs, parts weights and support structures. The build time, which depends on the part height and build volume, is closely tied to machine cost [27, 30]. In L-PBF, the build rate directly contributes to manufacturing cost. The longer it takes to build a part, the more the total cost of manufacturing. Likewise, using unsuitable feedstock can potentially increase defects and repetitive build cycles. The use of high quality feedstock and multiple lasers (high power) offer a means to increase the build rate though at a much higher cost [27, 38]. A consideration of the potential lifetime usage of more expensive feedstock and L-PBF machine amount to high-cost savings due to its ability to increase build rate [27].

Different industrial and academic studies have demonstrated how the different components (e.g., machine costs, raw material) [30], activities (designing, actual manufacturing, post-processing,) [39] and quality assurance (testing, certification) [5, 25] affect AM costs. The cost model for L-PBF is demonstrate in study [30] where build time and cost for L-PBF parts are estimated using Eqs. 17.1 and 17.2, respectively [30].

$$Build\ time = \frac{\frac{(Part\ Volume)}{(Build\ Rate*80\%)} + \frac{(Max\ Part\ Height)}{(Layer\ Height)} * (Recoat\ Time)}{3600} \tag{17.1}$$

$$Total\ AM\ Part\ Cost\ (€) = \frac{(Build\ Time)x(Machine\ Operating\ Cost)+Material\ cost}{(1-Pre-(or\ post-)Processing\ Cost\%)} \tag{17.2}$$

There are several cost discrepancies between different types metal AM methods. A comparison of common metal AM may differ by example, part resolution, part size, type and form of feedstock. The degree of flexibility, quality of the constituent, part size, need for supports and build rate can affect the cost for the different metal AM sub-categories. Metal AM machine systems can make components of varying complexities, resolutions, and sizes, which all can affect the overall manufacturing cost [40]. While BJ or PBF may be a good option for high complexity and resolution, parts size limits their application commonly to about 1 to 10 cm. DED, on the contrary, can make parts up to 1 m but often with lower resolution and may not be suitable for high complex designs. Other factors such as machine systems, software solutions, feedstock, need or no need of supports, and quality assurance also contribute to cost differences. Costs in metal AM can sometimes remain hidden until encountered in the product design. Such costs in metal L-PBF include qualification testing, overhead costs such as support staff, monitoring, consumable costs (for example, gases), build plate removal, lightweighting and cleaning [41]. To avoid operational surprises, adopters of AM need to critically understand how such factors can affect the overall cost of their product before investing in the process.

The manufacturing phase of metal AM can maintain a constant cost structure while increasing complexity and batch size [8, 38] compared with comparable manufacturing methods such as machining. A constant cost in AM implies that production cost remains relatively constant per part irrespective of change in product design (shape, intricacy, or other) for comparable build volume utilization. The cost stability implies that component designs that may be more difficult to manufacture conventionally [14] e.g. lattice [42] and scaffolds-like designs [43] may be considered for AM. Manufacturing products with intricate designs are often labor intensive and may require more tooling and lead time when using CM methods [5]. CM, however, may be a cost-effective option for making components requiring minimal intricacy and larger build volume. Business models can be used to evaluate cost performance based on different use scenarios to identify key cost drivers (e.g., product design, raw materials, energy, etc.) and analyze the impacts on profitability and cost saving opportunities to make informed decisions that are more likely to result in positive outcomes. Utilizing such analyses can guide companies to make optimal choices by comparing alternative designs or even competing manufacturing methods. A breakeven point (BEP) analysis for instance can be used to select the best points at which comparable manufacturing methods are most cost-effective. BEP diagram marks the border at which total revenues equal total expenses [44]. Value chain analysis (VCA) also can help identify aspects of the supply chain needed to optimize either for cost or differentiation advantage [45]. Strengths, weaknesses, opportunities, and threats (SWOT) models of metal AM prior to adoption could give empirical insights into potential difficulties and identify the best opportunities that may be used along the value chain to create long-term cost benefits.

17.2.3 Strategic Exploring and Exploiting of Value Changers and Linkages to Maximize Use of Simulation Assisted Product Design for AM Parts

Metal AM adoption may be enhanced through expert-led technology and knowledge transfer to support cognitive design abilities of workforce. Learning suitable design optimization methods, quality assurance, constraints of the specific AM subcategory and product application can accelerate positive outcomes. An understanding of product design in AM is prerequisite to the full exploitation of offered potentials to aid its adoption into mainstream manufacturing [9]. The acquisitioning and knowledge of digital tools are essential to optimal product designs creation and manufacturing planning to accelerate functionality, lead time reductions, manufacturing cost efficiencies as well as to offset unwanted outcomes in making informed decisions to boost AM adoption. Applying simulation driven software with AM does not necessarily translate to increased efficiency or successful FTR builds. The adoption of AM must aim to enhance FTR production by rightful utilization of example mindsets, raw materials, acquired software, hardware, and process controls skills and knowledge applicable to AM [9]. Value changers including management, operation and engineering need to ensure right actions are taken to achieve the needed product quality [7]. The study of Pradel et al. [9] highlights the role of intensive data, statistics and optimization to improve quality in AM. Management needs to monitor process efficiency (right things) and process effectiveness (results). Efficiency is the right use of the right resources, such as the right usage of available resources (machines, equipment, materials, software, workforce, time). Efficiency involves resource consumption improvement alongside a decrease in waste. Effectiveness at the managerial level ensures there are right things (e.g., software, machines, materials) and demonstrates how well a procedure satisfies a request. Managers in manufacturing industries must ensure that the skills of personnel are appropriately used with tangible and intangible resources to satisfy the objectives of the company as well the desires of customers. Different strategies such as communication, research support, training to increase workforce skills, and adequate digital software are needed to control resource efficiency and cost-efficacy and promote the well-being and inclusiveness of all personnel. Good communication among the value changers (management, engineering, production, etc.) within the work chain could guarantee successful implementation and working strategies for benchmarking the process a broader adoption [15].

17.3 Case Study

This case study briefly introduces topology and lattice design optimization influences on effective stiffness and stresses using computational design and simulations analysis of static stresses and displacement. This study demonstrates the influence of simulation assisted product design of AM parts for lightweighting, manufacturing time and costs optimization. Virtual design cases with stainless steel 316L was performed to compare time and costs according to the study of Simpson [30]. The aim of this study was to elucidate the role of simulation driven-software in L-PBF and to evaluate the potential to create optimized designs (i.e., to reduce weight and increase stiffness) without comprising the performance of metal components. The goal of the design was to reduce mass as possible (50% minimum) and increase stiffness for the final parts without compromising on the ability to withstand the used force. The design optimization was done with two different design optimization methods and computer-based engineering design software (1) topology optimization (SolidWorks software) and (2) lattice optimization (nTop software). While the former software is only limited to topology optimized designs, the latter is equipped to create both optimization options. These software were used in (1) creating geometry, (2) defining material properties (e.g., stainless steel 316L), (3) design constraints (e.g., plate was fixed from one of the short edges), (4) define structural loading (e.g., 100 N force along the edge of the plate), (5) running static studies in this case study. The basic solid plate and the two optimized designs (topology and lattice) will be referred to as Part A, Part B and Part C, respectively hereafter. A basic one-body model Part A is a simple rectangular plate with dimensions 100 mm by 50 mm by 5 mm for length, height, and width respectively.

The effect of build orientation and build platform utilization on manufacturing time was also evaluated using Materialize Magics and EOS EOSPRINT suitable for the selected material. The stepwise approach to the design optimization and virtual manufacturing is shown in Fig. 17.2.

As Fig. 17.2 shows, the workflow considered in this case study including modelling of the CAD model, simulating, optimizing and validating of the optimized design, pre-build, and actual build planning. This study uses the cost and time equations as in study [30] for L-PBF. The Eqs. 17.1 and 17.2 consider material, part volume (actual and support if any), part height, machine operating cost, pre-and post-processing cost. The material properties of SS 316L and processing parameters used are shown in Table 17.1.

As Table 17.1 shows, the virtual planning of the build considers two different build orientations for the different parts varying the maximum build height as 50 mm and 100 mm, respectively, for horizontal and vertical orientations. For equal comparison, other parameters were kept the same for this case. A practical case might, however, increase the build rate and layer height to optimize the build time. The values used also bring the estimated and software values a par. EOSPRINT and Materialise Magics were used for the actual build preparation and virtual time estimation.

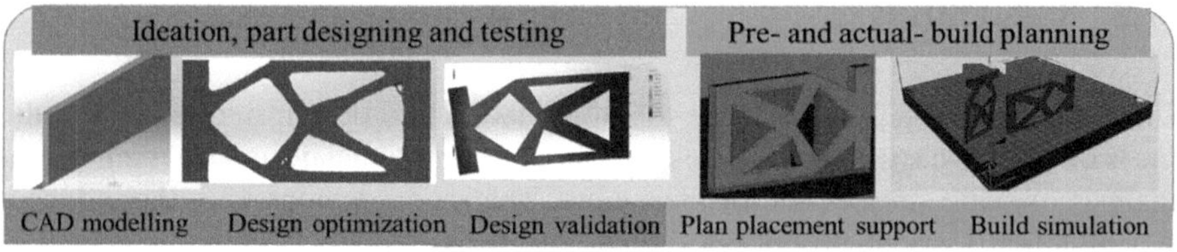

Fig. 17.2 Illustration of simulation assisted product design for AM parts workflow for case example

Table 17.1 Material properties and use input parameters used for the different scenario simulation (* Horizontal / **Vertical) orientations)

Density (kg/m^3)	7900	System manufacturer and user-defined inputs Supports structures not considered due to lack of license
Modulus of elasticity (GPa)	210	
Poisson's ratio	0.3	
Build rate (mm^3/s)	3.0	
Layer height (mm)	0.04	
Recoat time (s)	10	
Maximum part height (mm)	50* / 100**	
Part volume m/*p* (mm^3)	Part A, 25,000; Part B, 15,075; Part C, 9860	

17.3.1 Results

Optimally complex and optimized designs via the generative designs that feature intricate geometries are possible. The relative time difference based on the three designs and the two build orientations are shown in Table 17.2.

The result of the design optimizations in Part B and Part C and build time are shown in Table 17.3.

The results of comparable time for the different build scenarios are summarized in Tables 17.2 and 17.3. The Table 17.3 shows weights, stress, and displacement for the different plates based on the optimized orientation for a single build. The results highlight how topology and lattices improves resource efficiency and performance. The result of this study is validated by the software-generated build time and the calculated build time for both horizontal and vertical orientations. The orientation and part volume can be optimized to control the build time with comparable process parameters as shown in Tables 17.2 and 17.3. The reduced weight and the ability to manufacture parts of increased complexity can reduce the build volume, thus the time and costs in agreement with studies [8, 38]. The optimized components potentially also offer use-phase reduced cost and energy use as lightweight components consume less amount of energy. The differences in stresses for the optimized parts require empirical testing to validate their ability to withstand operational forces against that of the base plate.

Using virtual design optimization studies highlights potentials of digital tools for cost benefits in L-PBF without the risks of trial and error and associated laboratory experiments costs. Such approach may guide decision-making and help strategic planning to facilitate the acceptance of metal L-PBF. Figure 17.3 is a schematic of how the required efforts may be used to subdue or propagate the realization of the benefits of metal L-PBF adoption.

Figure 17.3 depict how communication flow affect costs or increase efficacy benefits. The implementation of metal L-PBF can affect productivity within the value chain. The implementation with Fig. 17.3a shows a one-way flow of information and the lack of needed DfAM efforts. This strategy may improvise traditional product

Table 17.2 Tabulation of effect of the used build orientations and build volume on estimated build time

Build scenario i & iii horizontal, ii & iv vertical	Time (h)	
Ai, Single horizontal	5,40	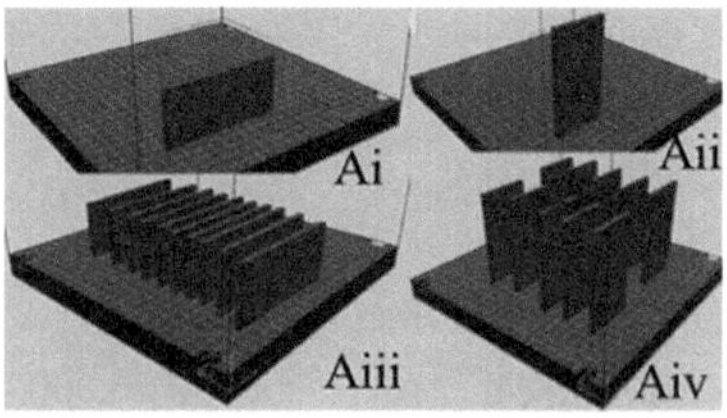
Aii, Single vertical	9,10	
Aiii, Ten pieces horizontal	25,26	
Aiv, Ten pieces vertical	29,05	
Bi, Single horizontal	4,51	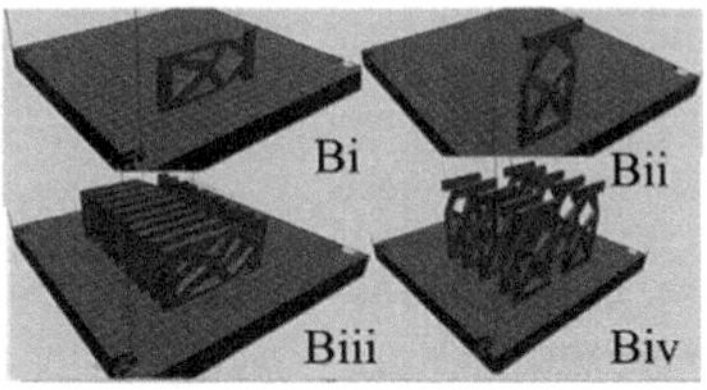
Bii, Single vertical	8,21	
Biii, Ten pieces horizontal	17,11	
Biv, Ten pieces vertical	20,55	
Ci, Single horizontal	4,28	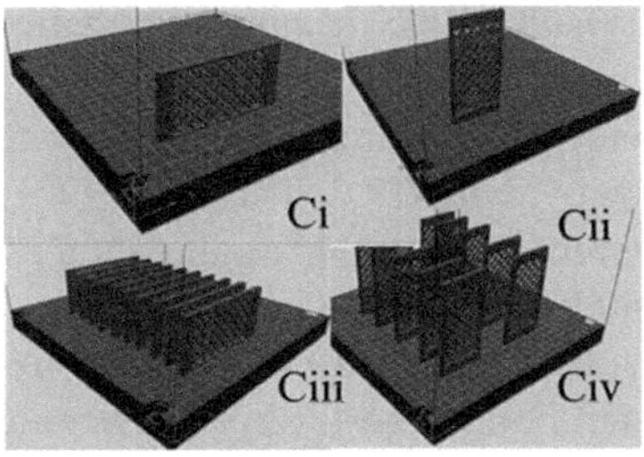
Cii, Single vertical	7,58	
Ciii, Ten pieces horizontal	13,29	
Civ, Ten pieces vertical	17,12	

design guideline for AM product design without the process-specific considerations. Top management may not understand the need for simulation assisted product design for AM parts. This lack of understanding may negatively affect the decision-making in such instances based on successful or failed build rates without considering the root causes. The adoption plan in Fig. 17.3b shows a reciprocal two-way communication between the different operational levels and adequate optimized product design efforts. The strategy involves all levels of personnel to understand specific metal PBF constraints and make decisions considering both successful and failed components. Decision-makers can integrate digital tools along the value chain to enhance value, time, energy and material efficiency. This model could potentially yield material efficiency and create components of a higher value that can offset the much-publicized high energy consumption in metal L-PBF.

Table 17.3 The main result of the case study for the base, topology, and lattice plates. in the optimal horizontal orientation

Variable	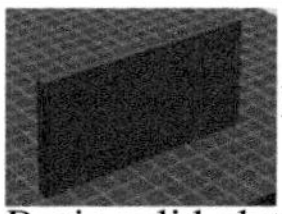 Part A Basic solid plate	 Part B Topology optimized	 Part C Lattice optimized
Mass (g)	198	119	77.9
Mass saving (%)	–	39.8	60.7
Von misses stress (MPa)	6.04	8.77	27.0
Virtual build time (h)	5.40	4.51	4.28
Estimated build time (h)	6.37	4.79	4.61
Virtual combined build time (h)	25.3	17.1	13.3
Estimated combined build time (h)	32.4	20.9	14.9
Estimated cost 1part with * (€)	276	228	214

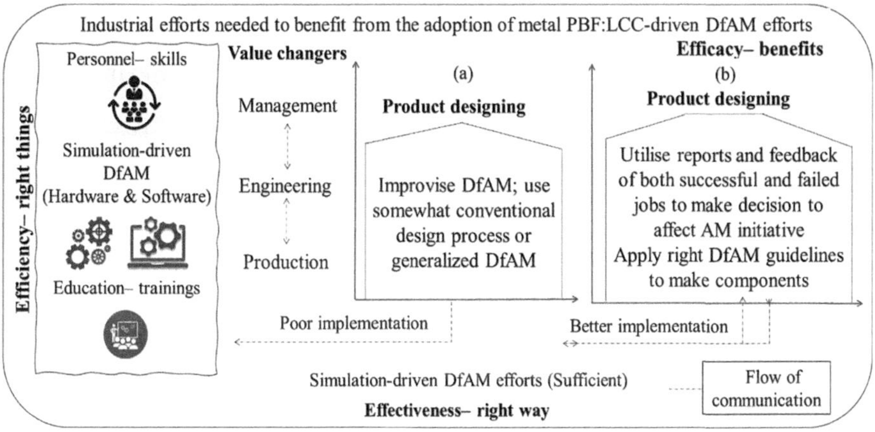

Fig. 17.3 Representation of **a** one-way and **b** two-way information flow to the adoption of metal L-PBF

17.3.2 Discussion

Metal AM continues to disrupt different sectors that benefits from flexibilities of customization, lightweighting and functional graded designs. Computer generative designs uses user defined input and computer algorithms to incorporate features such as intricate internal shapes, conformal flow channels, organic-geometric transitional shapes with process-dependent properties, and heterogeneous material properties. Such materials and parts benefit by their quality, resource efficiency, operational excellence, and cost-effectiveness. These capabilities influence high-end manufacturing industries such as medicine, aviation, and transport sectors to maximize metal product performance.

Literature survey conducted in this study shows that there is limited knowledge and expertise of optimized product design for AM parts and associated digital tools capable of leveraging the full potentials of metal AM. Re-imagining know-how of conventional CAD to generative designing is needed to influence a good exploration and exploitation metal AM benefits. The mindsets of each value changer in the value chain must adapt along the technological advancement in AM. Advancing the understanding of product design for AM parts and having the necessary computational skillsets can optimize metal AM parts for a strategic adoption. Available AM suitable software are sometimes limiting to all process steps as such there is the need for continuous development of these software. Efficient and high-fidelity simulation algorithms will be required in order to analyze and synthesize complex shapes, constraints and specifications [2, 46]. Digital tools in AM (1) eases workloads and reduces developmental time, (2) reduces manufacturing steps and costs (3) gives the freedom to generate complex gradient designs. Inefficient use of digital tools can prolong the design and manufacturing phase time and even result in unsatisfactory part designs. Right use of product design software and other supporting activites digital tools can profoundly implicate on the identification of (1) when to utilize AM (2) how to exploit offered AM opportunities and (3) how to mitigate the specific process limitations.

The adoption of AM requires an analysis of comparable iterated design alternatives to identify the best fits of application and other performance, e.g., manufacturing and use costs. There is a need to understand the cost structure and related factors that can influence economic choices where cost is of essence. Metal AM must be used to create optimized designs that will allow cost reductions via seamless digital integration of e.g., design creation, quality assurance and data management of the final components. The adoption of AM require effective communication between administration, design engineers and production team for a better implementation.

17.4 Conclusions

In this study the design optimization effects through time and costs with topology and lattices have been investigated. A significant mass, time and cost decrease were obtained in order of 61%, 21%, 32% respectively for one part of the most optimal design (Part C) in horizontal build orientation. Digital tools (computational design and simulation software) and system interoperability in AM allow a quick iteration of product design to create aesthetically pleasing, lightweight, resource-efficient, cost-effective, and structurally sound parts. These benefits safely equate to savings in mass, time and costs (see Table 17.3). Digital tools simplify the design and manufacturing phases in AM by auto generating ready to or nearly ready-to-build part and to optimize processing parameters (see Fig. 17.2). Digital tools in AM improves working and resource efficiency with the right planning e.g., part placement, supports generation, and automation of supporting reports for documentation, marketing, or communicating with among working teams. The importance of this study is the elucidation of how AM via generative design optimizes products designs to reduce resource consumption in respect to achieving aspects of goal 12 (responsible consumption and production,) of the 17 sustainable development goals.

This study is limited only to virtual experimental study without actual manufacturing of the parts which would be needed to ascertain the identified offered benefits to support industrial uptake of AM. Recommendations for further studies could include a practical study for selecting design/process alternatives based on the created optimized design with empirical data. This would validate if the virtual and calculated time and cost evaluations in this study confirm to actual practical data.

Acknowledgements The authors would like to thank the Analytics-based Management for Business and Manufacturing Industry platform at LUT University for supporting this interdisciplinary research. The authors acknowledge LUT laser material processing and additive manufacturing for proving the software used in the case study, and for assisting in the modelling and simulations (especially Vesa Tepponen and Sami Westman). Authors would also acknowledge Department of Mechanical Engineering (Faculty of Technology) in Turku University for their support and help to be able to conduct this study.

References

1. Baumers M, Tuck C, Wildman R, Ashcroft I, Hague R (2017) Shape Complexity and Process Energy Consumption in Electron Beam Melting: A Case of Something for Nothing in Additive Manufacturing? J Ind Ecol 21:S157–S167. https://doi.org/10.1111/jiec.12397
2. Wiberg A, Persson J, Ölvander J (2019) Design for additive manufacturing – a review of available design methods and software. Rapid Prototyp J 25:1080–1094. https://doi.org/10.1108/RPJ-10-2018-0262
3. Tapie L, Mawussi B, Bernard A (2012) Topological model for machining of parts with complex shapes. Comput Ind 63:528–541. https://doi.org/10.1016/j.compind.2012.02.005
4. Gupta K (2020) A review on green machining techniques. Procedia Manuf 51:1730–1736. https://doi.org/10.1016/j.promfg.2020.10.241
5. Blakey-Milner B, Gradl P, Snedden G, Brooks M, Pitot J, Lopez E et al (2021) Metal additive manufacturing in aerospace: A review. Mater Des 209:110008. https://doi.org/10.1016/j.matdes.2021.110008
6. Arora JS. The Basic Concepts. Introd. to Optim. Des. Fourth Edi, London: Elsevier Inc.; 2017, p. 3–18. https://doi.org/10.1016/b978-0-12-800806-5.00001-9.
7. Yang H, Rao P, Simpson T, Lu Y, Witherell P, Nassar AR et al (2021) Six-Sigma Quality Management of Additive Manufacturing. Proc IEEE 109:347–376. https://doi.org/10.1109/JPROC.2020.3034519
8. Fraunhofer. Effects of additive manufacturing on logistics and aftersales (Auswirkungen von generativer Fertigung auf Logistik und Aftersales) 2016. https://docplayer.org/23435311-Fraunhofer-austria-research.html (accessed May 5, 2022).

9. Pradel P, Zhu Z, Bibb R, Moultrie J (2018) A framework for mapping design for additive manufacturing knowledge for industrial and product design. J Eng Des 29:291–326. https://doi.org/10.1080/09544828.2018.1483011
10. Rosen DW. Design for additive manufacturing: A method to explore unexplored regions of the design space. 18th Solid Free. Fabr. Symp. SFF 2007, 2007, p. 402–15.
11. Leary M, Downing D, Lozanovski B, Harris J. Design principles. Fundam. Laser Powder Bed Fusion Met., Elsevier; 2021, p. 119–54. https://doi.org/10.1016/B978-0-12-824090-8.00013-5.
12. Culleton M, Mcdonnell D, Shipley H, Trimble D, Lupoi R. Dimensional Accuracy and Surface Finish in Additive Manufacturing Design for additive manufacturing (OPTIMIZED PRODUCT DESIGN): The importance of software within selective laser melting (SLM). Jt. Spec. Interes. Gr. Meet. between euspen ASPE, KU Leuven, BE: Euspen; 2017, p. 1–4.
13. Reichardt A, Shapiro AA, Otis R, Dillon RP, Borgonia JP, McEnerney BW et al (2021) Advances in additive manufacturing of metal-based functionally graded materials. Int Mater Rev 66:1–29. https://doi.org/10.1080/09506608.2019.1709354
14. Leary M. Digital design for AM. Des. Addit. Manuf., Elsevier; 2020, p. 33–90. https://doi.org/10.1016/b978-0-12-816721-2.00003-8.
15. Edwards D. ASME updates 3D printing standard to streamline production 2022:1–3.
16. Schrade P. 10 key skills-for-additive-manufacturing. EOS Gmb 2022. https://www.eos.info/en/blog-articles/skills-for-additive-manufacturing.
17. Ustundag A, Cevikcan E. Industry 4.0: Managing The Digital Transformation. Springer Nature; 2018. https://doi.org/10.1007/978-3-319-57870-5.
18. Thomas-Seale LEJ, Kirkman-Brown JC, Attallah MM, Espino DM, Shepherd DET (2018) The barriers to the progression of additive manufacture: Perspectives from UK industry. Int J Prod Econ 198:104–118. https://doi.org/10.1016/j.ijpe.2018.02.003
19. Bosio F, Aversa A, Lorusso M, Marola S, Gianoglio D, Battezzati L, et al. A time-saving and cost-effective method to process alloys by Laser Powder Bed Fusion. Mater Des 2019;181. https://doi.org/10.1016/j.matdes.2019.107949.
20. ISO/ASTM. EN ISO/ASTM 52911–1 Additive manufacturing — Design Part 1: Laser-based powder bed fusion of metals. 2019.
21. ASME. Y14–46_2022, Product Definitionfor AdditiveManufacturing 2022:52.
22. Framinan JM, Perez-Gonzalez P, Fernandez-Viagas V (2022) An overview on the use of operations research in additive manufacturing. Ann Oper Res. https://doi.org/10.1007/s10479-022-05040-4
23. ISO/ASTM. ISO/ASTM 52900:2021 Additive manufacturing — General principles — Fundamentals and vocabulary. vol. 2. 2021.
24. Fu E, Wentland L (2022) A survey of 3D printing technology applied to paper microfluidics. Lab Chip 22:9–25. https://doi.org/10.1039/d1lc00768h
25. Hajalfadul M, Baumers M. Building a quality cost model for additive manufacturing. Univ Khartoum Eng J 2021;11. https://doi.org/10.53332/kuej.v11i1.148.
26. Vafadar A, Guzzomi F, Rassau A, Hayward K (2021) Advances in metal additive manufacturing: A review of common processes, industrial applications, and current challenges. Appl Sci 11:1–33. https://doi.org/10.3390/app11031213
27. Simpson TW. The Value of Design for Additive Manufacturing (OPTIMIZED PRODUCT DESIGN) : Additive Manufacturing Magazine. Gardner Bus Media, Inc 2020. https://www.additivemanufacturing.media/blog/post/the-value-of-design-for-additive-manufacturing-optimized product design (accessed June 3, 2022).
28. Pal S, Kokol V, Gubeljak N, Hadzistevic M, Hudak R, Drstvensek I. Dimensional errors in selective laser melting products related to different orientations and processing parameters. Mater Tehnol 2019;53:551–8. https://doi.org/10.17222/mit.2018.156.
29. Salmi M, Ituarte IF, Chekurov S, Huotilainen E (2016) Effect of build orientation in 3D printing production for material extrusion, material jetting, binder jetting, sheet object lamination, vat photopolymerisation, and powder bed fusion. Int J Collab Enterp 5:218–231
30. Simpson TW. Industrializing AM: A Simple Cost Equation. Addit Manuf 2020. https://www.additivemanufacturing.media/articles/industrializing-am-a-simple-cost-equation.

31. Dassault Systèmes. Working with complex geometry and organic shapes. SOLIDWORKS 2020. https://www.solidworks.com/sites/default/files/2019-06/3DS-2019-EBOOK-Sub-D-modeling-for-SW-2020.pdf (accessed May 10, 2022).
32. Craeghs T, Clijsters S, Kruth JP, Bechmann F, Ebert MC (2012) Detection of Process Failures in Layerwise Laser Melting with Optical Process Monitoring. Phys Procedia 39:753–759. https://doi.org/10.1016/j.phpro.2012.10.097
33. Liu C, Law ACC, Roberson D, Kong Z (James). Image analysis-based closed loop quality control for additive manufacturing with fused filament fabrication. J Manuf Syst 2019;51:75–86. https://doi.org/10.1016/j.jmsy.2019.04.002.
34. Sciaky Inc. Advanced closed-loop control system for optimizing the EBAM® Metal 3D-printing process 2022. https://www.sciaky.com/additive-manufacturing/iriss-closed-loop-control.
35. VELO3D. VELO3D Intelligent Fusion™ - YouTube 2019. https://www.youtube.com/watch?v=-DHaGnAvbVw (accessed November 7, 2020).
36. Jiménez M, Romero L, Domínguez IA, Espinosa MDM, Domínguez M. Additive Manufacturing Technologies: An Overview about 3D Printing Methods and Future Prospects. Complexity 2019;2019. https://doi.org/10.1155/2019/9656938.
37. Yi L, Ehmsen S, Glatt M, Aurich JC (2021) Modeling and software implementation of manufacturing costs in additive manufacturing. CIRP J Manuf Sci Technol 33:380–388. https://doi.org/10.1016/j.cirpj.2021.04.003
38. Barroqueiro B, Andrade-Campos A, Valente RAF, Neto V (2019) Metal additive manufacturing cycle in aerospace industry: A comprehensive review. J Manuf Mater Process 3:52. https://doi.org/10.3390/jmmp3030052
39. Rickenbacher L, Spierings A, Wegener K (2013) An integrated cost-model for selective laser melting (SLM). Rapid Prototyp J 19:208–214. https://doi.org/10.1108/13552541311312201
40. Huckstepp A. Metal Binder Jetting | Digital Alloys 2019. https://www.digitalalloys.com/blog/binder-jetting/ (accessed October 14, 2020).
41. Ray JT (2006) Calculating the cost of Additive Manufacturing. Ind Paint Powder 82:19–24
42. Ramadani R, Pal S, Kegl M, Predan J, Drstvenšek I, Pehan S et al (2021) Topology optimization and additive manufacturing in producing lightweight and low vibration gear body. Int J Adv Manuf Technol 113:3389–3399. https://doi.org/10.1007/s00170-021-06841-w
43. Tofail SAM, Koumoulos EP, Bandyopadhyay A, Bose S, O'Donoghue L, Charitidis C (2018) Additive manufacturing: scientific and technological challenges, market uptake and opportunities. Mater Today 21:22–37. https://doi.org/10.1016/j.mattod.2017.07.001
44. Mitchell C. Breakeven Point (BEP). Investopedia 2022. https://www.investopedia.com/terms/b/breakevenpoint.asp.
45. Vijayan G, Kamarulzaman NH, Mukherjee A, Vaiappuri SKN. Strategic value creation in a supply chain. Handb. Res. Glob. Supply Chain Manag., IGI Global; 2016, p. 186–204. https://doi.org/10.4018/978-1-4666-9639-6.ch011.
46. Chaudhry S, Soulaïmani A. A Comparative Study of Machine Learning Methods for Computational Modeling of the Selective Laser Melting Additive Manufacturing Process. Appl Sci 2022;12. https://doi.org/10.3390/app12052324.

Chapter 18
A Review of Microscale and Mesoscale Simulation of Laser Powder Bed Fusion

Aditya Gopaluni, Heidi Piili, Ashish Ganvir, and Antti Salminen

Abstract Additive manufacturing (AM) is an advanced method of manufacturing complex parts layer by layer based on digital model. Pandemic showed that AM was the go-to technology for the production of emergency medical equipment. Recent growth in post-pandemic world around software of AM shows that AM can be a revolutionary change for several industries. Laser powder bed fusion (L-PBF) is used to produce parts with high resolution and good part properties; it is one of most widely used AM technology for metal parts e.g., for aerospace industry. L-PBF is based on laser beam and material interaction where the powder material is melted and then solidifies. This occurs in a short time frame of the order of 0.02 s and makes the whole process challenging to be studied in real time. Studies have shown the development of numerical methods and the use of simulation software can help to understand the laser beam and material interaction. This is key in understanding the material behavior under melting and mechanical properties of the part produced by L-PBF because mechanical properties of part are directly linked with the solidification of the melted powder material. A detailed study of the laser beam and material interaction is needed on a microscale and mesoscale level as it provides a better understanding and helps in the development for the given material for the L-PBF process. The main conclusion from this review is the need to develop a methodology to use simulation at microscale and mesoscale level to understand the laser beam and material interaction and improve the efficiency of L-PBF process using the relationship between the raw material parameters, melt pool characteristics and part properties.

Keywords Digital manufacturing · Additive manufacturing · AM · Laser powder bed fusion · L-PBF · Metals · Microscale · Mesoscale · Simulation · Software

A. Gopaluni (✉) · H. Piili · A. Ganvir · A. Salminen
Faculty of Technology, Department of Mechanical Engineering, University of Turku, Joukahaisenkatu 3-5 B, FI, 20520 Turku, Finland
e-mail: aditya.gopaluni@utu.fi

T. Tuovinen et al. (eds.), *Advanced Computational Methods and Design for Greener Aviation*, Computational Methods in Applied Sciences 59,
https://doi.org/10.1007/978-3-031-61109-4_18

18.1 Introduction

The SARS Cov-2 virus spread which was officially termed as a pandemic showed that AM provided a solution for the disrupted supply chain in the industry [1]. There was high demand of large amount of personal protection equipment (PPE) and other medical equipment and several industries stepped up to use AM to meet the demands [2]. Recent studies show also that growth of AM software indicates the increasing interest of the industry in AM as a process.

This is especially valid for aerospace industry because the aerospace industry requires parts with high functionality, performance and durability such as turbomachinery blades [3]. The usage of AM software at process and part scales enables the aerospace industry to reduce the material wastage and improve the efficiency of part function.

Additive manufacturing (AM) is a manufacturing technique that allows freedom in design as parts are manufactured layer by layer [4]. AM is a layer-by-layer based production approach, which gives certain freedom to design e.g., for fabrication of complex parts with complex internal features [5]. Layer-wise nature of AM enables also fabrication of light-weight parts, and this is why one of first industrial applications in AM was within aerospace industry [3].

Among the different AM processes, laser-based powder bed fusion (L-PBF) has demonstrated the highest resolution for producing complex parts using metals since the layers are very thin, around 0.04 mm [6]. L-PBF involves melting and solidification of metal powder by laser beam [4]. This process is challenging as it has many different parameters affecting it [7].

The different parameters affecting interaction between laser beam and material within the L-PBF process are: material parameters, process parameters and equipment parameters [8]. Material parameters are related to the powder material used for manufacturing and process parameters are related to the setting up the laser and powder material for the interaction phenomenon [9]. Equipment parameters are related to the L-PBF machine which is used for manufacturing [10]. Table 18.1 is a list of the different parameters concerning the L-PBF process.

Table 18.1 consists of the list of different parameters namely material, process and equipment related. Figure 18.1 is a description of the interaction between laser beam and material, and this is an important part of the L-PBF process [16].

Table 18.1 Parameters affecting L-PBF process

Parameter types	List of parameters	References
Material	Packing density, Particle size distribution, Powder layer thickness; Etc	[11, 12]
Process	Laser power, Powder spreading speed, Laser beam speed; Etc	[13]
Equipment	Gas flow, Build volume, Recoater design; Etc	[14, 15]

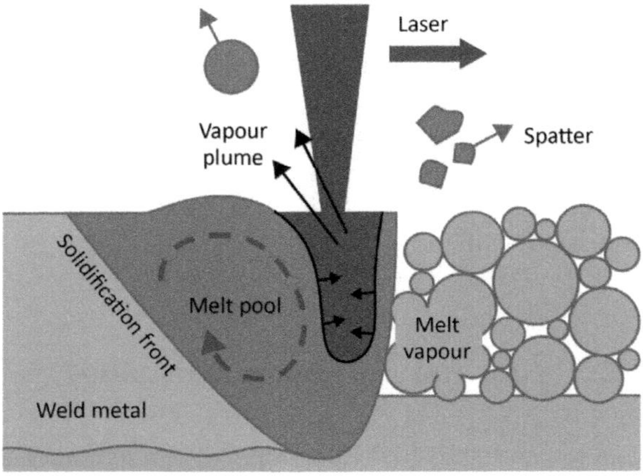

Fig. 18.1 Laser beam and material interaction [16]

As it can be observed from Fig. 18.1, the laser beam melts the metal powder which forms melt pool. As melt pool moves with laser beam, molten metal solidifies behind it to form solid metal along the path travelled by the laser beam [16].

The interaction between laser beam and metal material happens fast, as beam moves at c. 1000 m/s and this is where the formation of the melt pool happens [17]. The diameter of laser beam is only ca. 0.1 mm, and melt pool formed is slightly larger than that [18]. The stability of the process and the quality of the parts produced depends heavily on this phase of the process.

Characteristics of melt pool like size and cooling rate determine the initial microstructure and material properties for a given material, as in [19]. This has been established in Yang et al. where characteristics of melt pool like size and cooling rate determine the initial microstructure and material properties for a given material.

Figure 18.2 shows input and output parameters of the L-PBF process from the point of view of the raw material, the melt pool and AM part [20].

As it can be seen from Fig. 18.2, the first in L-PBF process is the raw material, which is metal powder. The input parameters for this stage are the powder layer thickness, layering speed and packing density, which become output parameters for the next building block [20]. This is followed by the melt pool which forms upon

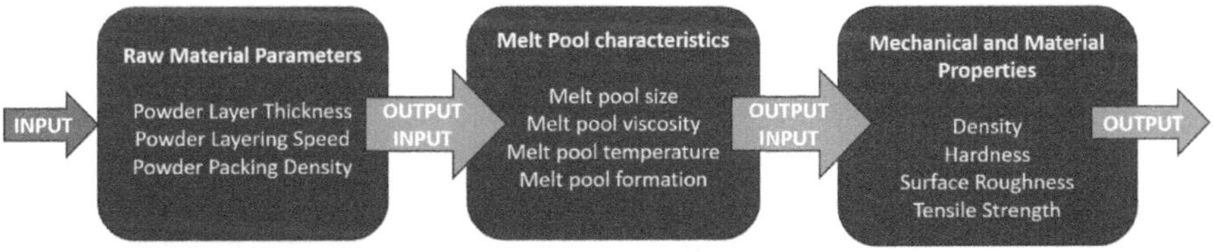

Fig. 18.2 Input and output parameters in L-PBF when raw material, L-PBF process itself and AM part are considered Courtesy of [20]

the interaction of laser beam with powder material. In situ process signal detection is used at this stage of the L-PBF process involving setting up equipment like high-speed cameras, thermal imaging devices etc. to gather data in real time. Melt pool characteristics such as size, viscosity, temperature and formation constitute this block. The last building (see Fig. 18.2) block is the manufactured parts and their mechanical and material properties such as density, hardness, surface roughness and tensile strength [21].

18.2 Aim and Purpose of This Study

The research [22–24] on the topic of simulation of the melt pool parameters, laser and material parameters, powder behaviour etc. sets the precedent for the use of simulation for studying the interaction between laser beam and metal material which forms the basis the L-PBF process. The precedent is the complex nature of the L-PBF process because of which it is not possible to conduct a qualitative study of the phenomena occurring in the process. The objective of this study is to present state-of-art situation of simulation in the L-PBF process from different perspectives such as scale of feature, type of simulation and also the different commercial software available. The aim of this study is to become the basis to approach the study of microscale and mesoscale simulation of laser powder bed fusion. This study concentrates in simulation of microscale and mesoscale phenomena of L-PBF from (1) material parameter and (2) process parameter point of view. Simulation of mechanical properties of AM part is so wide area that it was decided to exclude form this study.

Figure 18.3 presents the number of publications (based on Scopus) done in 2010–2022 in the field of simulation when different scale of simulation is considered.

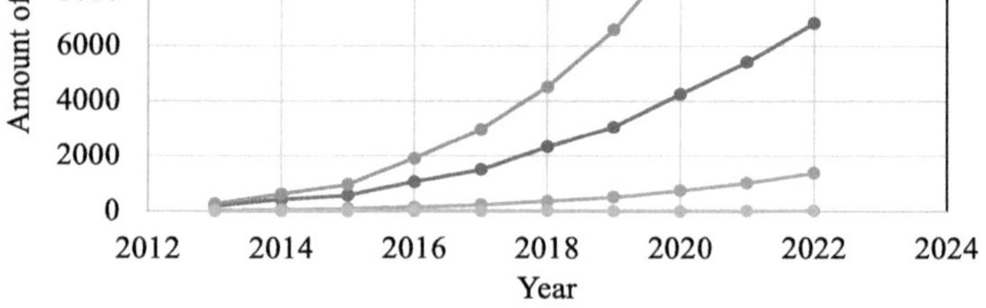

Fig. 18.3 Articles published on simulations in Additive Manufacturing in 2013–2022 (Scopus)

As it can be seen from the Fig. 18.3, research in simulation in L-PBF in the last decade has increased in terms of the articles published, A clear gap can be noticed in the research between the simulation in AM and the simulation in AM at microscale and mesoscale level which is identified by the search term 'DEM-FEM' (Discrete Element Method-Finite Element Method). It can be observed that there is a 3000% increase between the publications in simulation in L-PBF and micro and mesoscale simulation in L-PBF. This is due to emphasis placed by researchers on improving the process itself and also the non-availability of the right kind of simulation software when it comes to the L-PBF process or AM in general. A growth of publications AM in the aerospace industry can also be seen which supports the increase of use of AM in this industry. However, rapid increase of publications in simulations in AM between years 2018–2022 since the turn of the decade can be based on new and developing software in the AM industry. The appearance of new software along with the development of the L-PBF process has been identified as focus of this article, and this is why microscale and mesoscale simulation in the L-PBF is reviewed.

Simulation in the aerospace industry has been used since the beginning stages of evolution of CFD software. Simulation in AM however has been only used since 2016 on a part scale level where mechanical analysis would be performed on the part designs via simulation after manufacturing. Ever since the development of process simulation software, aerospace industry has been utilizing software in AM at the beginning of R&D level [25].

18.3 Simulation in L-PBF

18.3.1 Use of Simulation in L-PBF Process

Simulation in L-PBF can mean a number of things. In the scope of this study, simulation in L-PBF will mean the simulation concerned with meso and microscale level as the focus is on the laser beam and material interaction. Mesoscale in this study means the dimensional scale of the order of 10^{-4} m and Microscale is defined as dimensional scale of the order of 10^{-6} m [26]. Numerical models such as Discrete Element Method (DEM), Finite Element Method (FEM), Computational Fluid Dynamics (CFD) etc. are based on the thermal and mechanical aspects of L-PBF process and used for understanding laser beam and material interaction in both scales. Figure 18.4 is a description the use of different simulation methods depending on the stage at which the L-PBF process is [27].

It can be understood from Fig. 18.4a that DEM is applied to the simulation as the majority of the powder bed consists of individual particles that have different motion and temperature. Figure 18.4b consists of the melt pool which is a fluid consisting of both liquid and gaseous phases and hence CFD method is more suitable and applied. The DEM and CFD models have also been used to study the melt pool characteristics as a function of the thermal history of the L-PBF process [5, 28]. Bayat et al. has

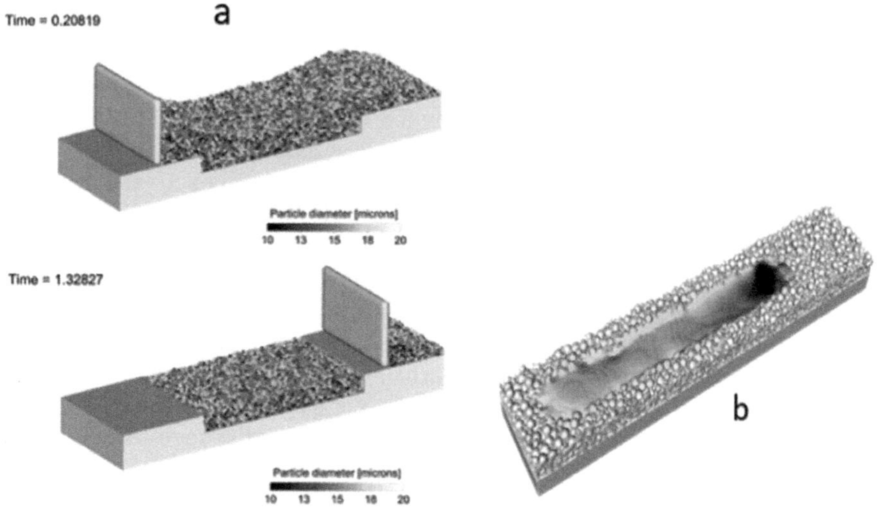

Fig. 18.4 **a** DEM simulation of a powder bed in L-PBF using Flow3D software for SS316L **b** CFD simulation of laser beam track in powder bed in L-PBF using Flow3D software for SS316L, courtesy of [27]

listed the physical phenomena during the laser beam and material interaction as evaporation, melting, solidification, thermal radiation, laser beam absorption, liquid metal flow, recoil pressure etc. [27] Bayer et al. concluded that melt pool of L-PBF can be studied as a function of the phenomena like evaporation, melting, solidification, liquid metal flow etc. In-situ process signal detection of these physical phenomena like the melt pool formation, solidification etc. in L-PBF in real time includes the use of complicated setup of sensors and processing large amounts of data which is seen in the work of Bugatti et al. In-situ process signal detection can be defined as monitoring of the physical phenomena occurring during the laser beam and material interaction within the L-PBF process by the use of cyber-physical equipment to detect signals emitted by the process [29]. The data can be collected in the form of images, videos, digital and analog data etc. Hence, researchers [30, 31] adopted the concept of thermal and mechanical based simulation to study the melt pool related phenomena occurring during the laser beam and material interaction.

Studying the melt pool characteristics in the short time frame is challenging and hence numerical models like thermo-mechanical models like FEM, DEM etc. have been suggested according to this research [32]. Song et al. concluded that the use simulation in L-PBF removes the prospect of trial-and-error methodology to develop the L-PBF process by repeating a designed set of experiments [33]. The idea of removal of trial-and-error methodology establishes a strong background for the use of simulation in L-PBF process.

The scale at which simulation in L-PBF is often performed is at macroscale level or part scale level, as the software in the past was only limited to analyzing the L-PBF process and to estimate the mechanical properties of the fabricated components such

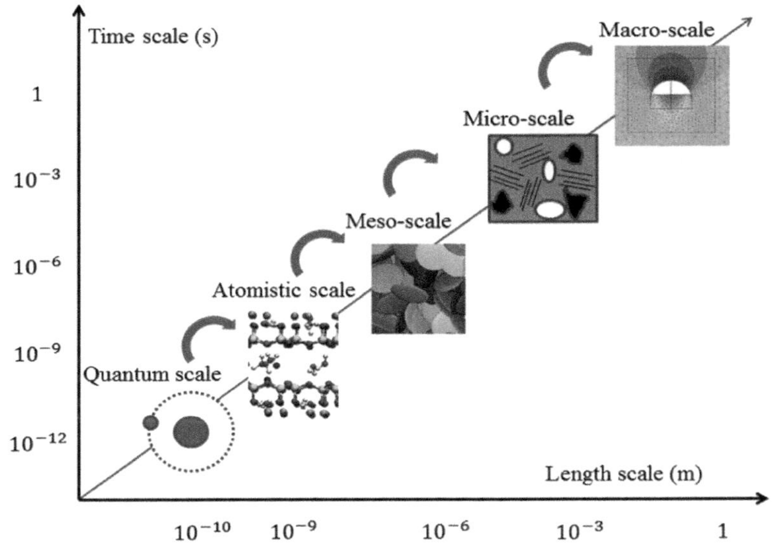

Fig. 18.5 Length scales at which simulation occurs in L-PBF, courtesy of [34]

as tensile strength in real time conditions. While macroscale simulation continues to dominate the aspect of simulation in L-PBF, Fig. 18.5 indicates different dimensional scales at which simulation is performed in L-PBF [34].

The two length scales from Fig. 18.5 that are in the focus of this review article are microscale and mesoscale. Liu et al. and Wang et al. have concluded that control over the L-PBF process at these scales allows the control of microstructure for Ti6Al4V. Wang et al. have taken a multi-track scan approach to understand the evolution of titanium alloy powder on a microscale level. The motivation behind choosing multi track approach was to understand the nature of repeatability of the simulation. This study concluded that simulating multiple tracks of printing in conduction mode not only results in the formation of a long columnar microstructure, but also allows the control in formation of it. This finding alone shows the importance of studying the L-PBF process at these scale levels [35].

Mesoscale simulation of L-PBF has been used in various studies to understand the behavior of metal powder after the laser beam interacts with it [36–38]. This has been closely studied in the research of Mede et al. where a single linear track of melt pool formation for Ti6Al4V has been simulated. This study concluded that melt pool surface profile, i.e., the length with respect to the scan direction largely depended on the size of the powder particles and the packing density of the powder bed. The powder particles studied were of the order of 40 μm which is on a mesoscale level and an individual mesh size of the order of 3 μm was used in this model to solve the continuum conservation equations related to the powder particles. Figure 18.6 shows the simulation of the melt pool formation when the powder bed model was simulated to be interacted with the laser beam [36].

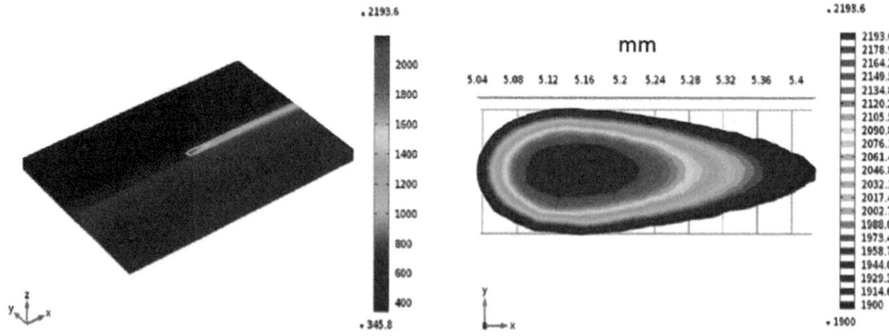

Fig. 18.6 Temperature profile of the melt pool for a single-track scan, courtesy of [36]

It can be seen from Fig. 18.6 that the uniform particle size distribution resulted in an even temperature gradient in this given time step. This is enabled because of a high packing density in the powder bed that allows uniform melting of the powder particles once the laser scans the surface. Microscale simulation in L-PBF is used to understand the formation of the microstructure of the parts being manufactured. By controlling the formation of the microstructure, the mechanical properties like hardness, strength and toughness can be controlled as well. This is done by controlling the grain growth and grain structure [39]. The microscale simulation model is often paired with a macroscale thermal simulation model, as seen in the research of Pei et al. The phase formation during the laser beam and material interaction was studied and this was coupled with an FEM model where four layers of powder was deposited and L-PBF was performed [40]. It was concluded from this approach that the thermal history and solidification phenomenon of the L-PBF process is key to understand the microstructure of the printed parts.

Microscale in L-PBF is often associated with microstructure characteristics of the parts being printed [41]. Babu et al. have combined microscale models with thermal models to study the formation of microstructure. This study used the thermal history from the laser beam and material interaction simulation model and developed the phase field model for simulating microstructure [42]. They have concluded that the part properties like tensile strength, hardness and ductility are dependent on the different phases formed during microstructure evolution. Figure 18.7 is the explanation of the relationship established between the different physical phenomena such as heat transfer, phase change and mechanical analysis. The thermal data gathered from this loop was utilized to generate a phase field model to predict the formation of the microstructure.

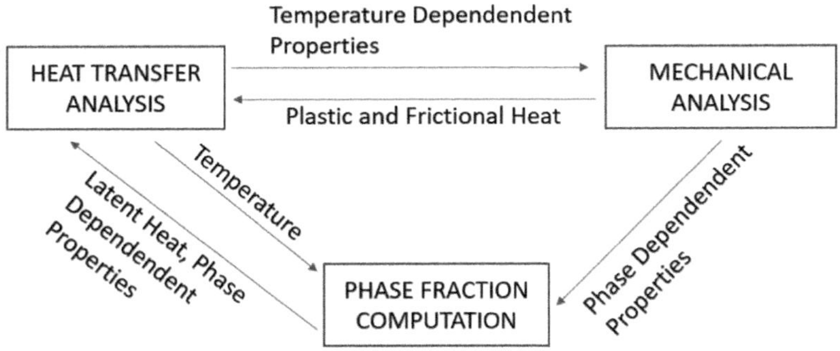

Fig. 18.7 Building blocks of a phase field model for microstructure generation [42]

18.3.2 Methods Used to Simulate the Laser Beam and Material Interaction in Microscale and Mesoscale

The simulation of the laser beam and material interaction in L-PBF consists of developing a thermo-mechanical model containing a powder substrate and a heat source, as seen across different studies [43]. A thermo-mechanical model was defined in the study of Heigl et al. as a model created to analyze the thermal and mechanical phenomena in a given process [44]. In L-PBF, heat source is the laser beam which interacts with the powder substrate. Powder bed is defined in the research of Chen et al. as the layer of the powder that is spread on the build plate where the powder is melted and solidified to form a component according to the required design. Figure 18.8 is an illustration of the powder substrate that is designed for the simulation of laser beam and material interaction in L-PBF process [44].

The method used to carry out powder substrate modelling in this research is DEM (Discrete Element method) because DEM is a suitable method when dealing with

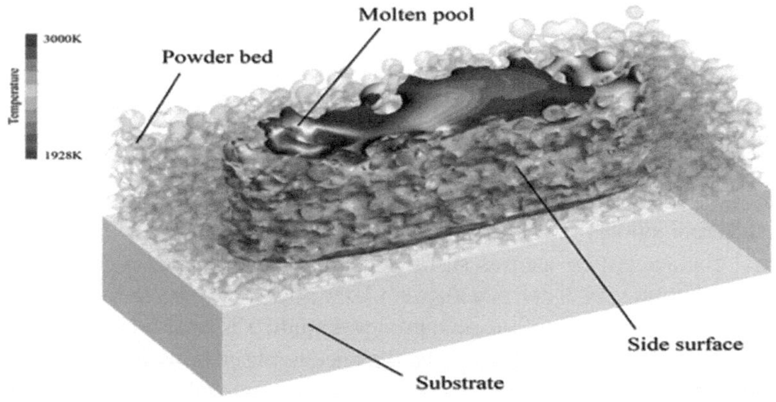

Fig. 18.8 Generic model of powder bed and melt pool for L-PBF process, courtesy of [44]

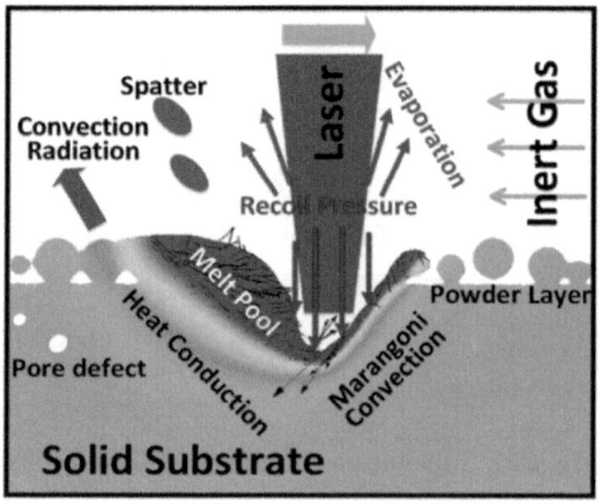

Fig. 18.9 Illustration of Marangoni effect in the melt pool for L-PBF process, courtesy of [48]

powder particles for L-PBF. This is based on the ability of DEM to calculate the kinetics and thermal history of the individual particles. The powder bed in L-PBF consists of a large number of powder particles which are differentiated by size, morphology, packing etc. Haeri et al. explains the motivation behind choosing DEM as a suitable method when dealing with granular particles to be due to the ability of DEM to include the effect of material properties, size distribution and morphology on the simulation which is conducted [45].

The simulation of melting of the powder and melt pool formation is performed by different methods such as Finite Element Method (FEM), Finite Volume Method (FVM) and a combination of DEM-FVM/FEM [46]. FEM and FVM are defined as mathematical methods which are applied within certain boundaries in a given volume. The formation of the melt pool is driven by capillary forces and the Marangoni effect [47], which means that fluid dynamics concerning the melt pool and heat transfer phenomena come into play. The melt pool consists of molten metal, plume, spatter etc. The temperature gradient of the melt pool is really high as the solidification happens in the order of 0.2 s. Figure 18.9 illustrates the Marangoni effect, which has a profound effect on melt pool characteristics.

It can be seen from Fig. 18.9 that the Marangoni effect plays a key role in the movement of the melt pool after its formation as it guides the change in surface tension as a function of the temperature. This supports the use of FVM or FEM, which is used to solve the complex equations governing these phenomena, as the melt pool is assumed as a confined space with a given amount of volume [49]. The use of DEM-FVM method to simulate the melt pool formation is supported by the different studies of Xia et al. and Wang et al. as FVM was used to analyze melt pool characteristics such as surface tension and velocity, in correlation with laser parameters such as hatch spacing and laser power [50]. FVM was applied here because the surface tension of the melt pool was studied as a parameter to understand the effect of hatch spacing, which is the distance between consecutive laser beams, on the melt pool. It was found that surface tension at the edges of the molten pool was lower when the hatch spacing was high. This data was in good agreement with the data from the FVM simulation.

After the formation of the melt pool, solidification of molten metal drives the formation of the microstructure of the component being manufactured since the shape and size of the grains in the microstructure depends on the solidification behavior [51]. The mechanical properties of the material such as hardness, tensile strength, toughness, ductility etc. are influenced by the grain sizes and shapes within the microstructure [51]. To simulate the solidification phenomenon and predict the microstructure, phase field (PF) model can be utilized and the data required for coupling a thermo-mechanical model to the PF model can be seen in Fig. 18.10 [52].

This model solves the problems related to solidification dynamics, thereby predicting the microstructure of the material [52]. PF model is used by combining the thermal data of the L-PBF process as seen from Fig. 18.10, with the PF simulation which predicts the microstructure by estimating the grain boundaries and grain shapes.

Table 18.2 summarizes the methods of simulation used in the process of powder spreading and powder melting in L-PBF process.

As it can be seen from Table 18.2, different simulation methods are listed with the suitable application areas within the L-PBF process. The various physical phenomena

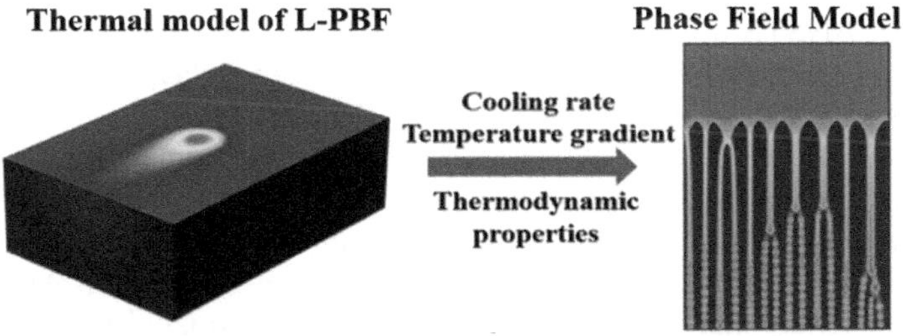

Fig. 18.10 Relationship between a thermo-mechanical model and PF model for L-PBF, courtesy of [52]

Table 18.2 Applications of simulation approaches [53]

Simulation approach	Application area	Physical phenomenon	L-PBF phenomena	References
Finite volume method (FVM)	Powder melting	Thermal fluid behavior	Melt pool behaviour	[54, 55]
Finite element method (FEM)	Powder melting	Thermal fluid behavior	Melt pool behaviour	[56, 57]
Discrete element method (DEM)	Powder spreading	Particle dynamics	Powder bed behaviour	[45, 58]
DEM-FVM and DEM-FEM	Powder melting	Particle dynamics and Thermal fluid behavior	Powder bed and melt pool behaviour	[59, 60]
Phase field (PF)	Melt pool solidification	Microstructure evolution	Microstructure study	[61, 62]

occurring in the L-PBF process along with the specific areas of application for the simulation methods have been listed.

18.4 Simulation Software

Simulation software have been developed by various organizations to understand the laser beam and material interaction at microscale and mesoscale. Among many different software, Flow 3D from Flow Science has been developed to understand the powder spreading and melt pool formation by utilizing a combination of DEM and CFD methodology. Flow3D uses DEM to simulate the spreading of the powder particles on the powder bed and CFD to simulate the melt pool formation. Figure 18.11 describes the different blocks of Flow3D software in the sequence in which the simulation takes place [63–65].

As Fig. 18.11 suggests, different simulation methods are applied by the Flow3D software at different stages of the L-PBF process. This is done by the different modules of the same software. ANSYS Additive provides a multiscale simulation approach for the user to understand the L-PBF process from microscale to macroscale. ANSYS utilizes multiple modules like CFD, FEA and FEM to simulate different aspects of the L-PBF like powder spreading, laser beam and material interaction, residual stress and distortion [66]. Digimat from Hexagon is a software which is under development that focuses on material development from an atomistic scale to microscale based on integrated computational materials engineering [67].

Additive Lab has an unnamed software in development which uses thermo-mechanical CFD models to study the melt pool at mesoscale [68]. Amphyon is a mesoscale software that provides an estimation of the thermal history of the L-PBF at mesoscale dimensions [69]. Figure 18.5 summarizes the different software in use for mesoscale and microscale simulation of L-PBF.

As Fig. 18.12 shows, various softwares are presented with respect to dimension scale, physical phenomena and their impact on research in melt pool study of L-PBF. A comparison of different upcoming software is illustrated along with the description of the frequently used software in the area of laser beam and material interaction for L-PBF.

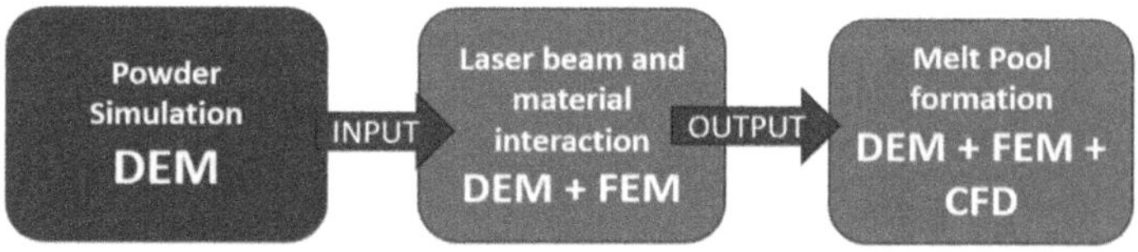

Fig. 18.11 Building blocks of Flow3D software simulation sequence, courtesy of [63— 65]

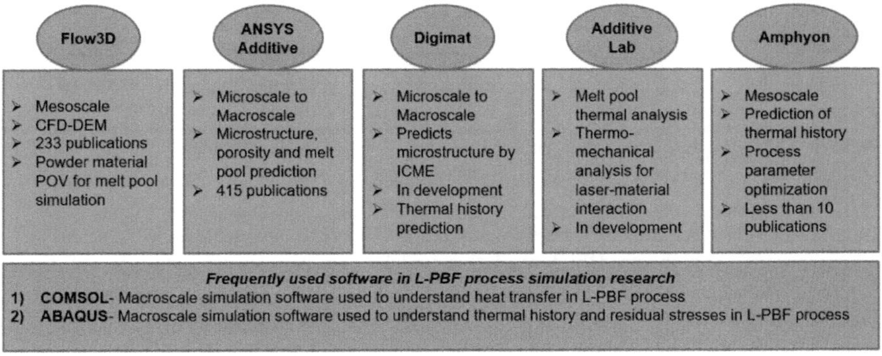

Fig. 18.12 Different simulation software used in L-PBF [5, 21, 27, 48, 70]

18.5 Challenges with Microscale and Mesoscale Simulation

The scope of simulation software in the recent years has expanded to microscale and mesoscale level from macroscale level [71]. This is especially valid for aerospace industry because improving the L-PBF process and eliminating the need for extensive experiments and reducing material wastage have been the key requirements in this industry [72]. Simulation of physical phenomena occurring in L-PBF on microscale and mesoscale allows a larger control over the L-PBF process. This means that the data obtained from the simulation allows parameter optimization and process control and improves the process. Simulation data can be used for testing the L-PBF process instead of conducting an experiment with a given set of parameters. The entire L-PBF process is simulated which allows removal of design of experiments which is used for improving the efficiency of the process. The challenge of accurately implementing simulation in L-PBF process arises from the side of software development. While there is continuous development of different software, there are still some areas where the software are lacking. Table 18.3 is a collection of the advantages and disadvantages of different software for L-PBF at mesoscale and microscale level.

It can be understood from the Table 18.3 that every software has different types of disadvantages when it reaches the end user. This makes the implementation of simulation of L-PBF not straightforward. Simulation of the L-PBF process, irrespective of the scale at which it is performed, requires validation before it can be used directly for research. Validation of a simulation is a method where the data gathered from simulation is compared with the results from the experiment. This is done via mechanical testing and in-situ process signal detection. Multiple iterations of simulation need to be carried out before the results agree with the experimental data, as seen in the approaches of different researchers [76, 77]. This is a limitation when using simulation for the optimization of L-PBF process, which suggests that simulation is not at a level where it can completely remove the experimental aspect of research and development of L-PBF process. The need for experimental validation of simulation and vice versa in order to optimize the L-PBF process indicates a

Table 18.3 Pros and cons of different software for L-PBF

Software	Pros	Cons	References
Flow3D by flow science	Ease of use, accurate, high-speed simulations	Limited material library	[5, 64]
Digimat AM by eXstream for metals	Accurate prediction of material property effect on manufactured parts	In-development	[67, 73]
Simufact additive	Accurate non-linear simulations, high-speed, high-resolution models	Not available on a microscopic scale yet	[21, 74]
ANSYS additive suite	Includes different work benches increasing the ease of usage	Unavailability of select few work benches	[70, 75]

gap that exists between simulation and real time iterations of the process. The gap suggests that until the simulation software is at an optimum level, the application of simulation for L-PBF will always need to include experimental validation of the simulation results.

18.6 Future Trends in Simulation

Additive manufacturing, also known as 3D printing, is a rapidly growing field with a wide range of applications in industries such as healthcare, aerospace, and automotive. As the technology continues to advance, it is likely that we will see a number of significant trends emerge in the simulation of additive manufacturing processes.

One of the key trends in the simulation of additive manufacturing is the increasing use of machine learning and artificial intelligence (AI) to improve the accuracy and efficiency of simulations. Machine learning algorithms can be trained on large datasets to identify patterns and trends that can be used to optimize the simulation process. For example, machine learning can be used to identify the most efficient printing path for a given design, or to predict the likely outcome of a printing process based on a range of input variables. Another trend in the simulation of additive manufacturing is the development of more sophisticated simulation tools that can take into account the complex interactions between the various materials and processes involved in 3D printing. These tools will be able to simulate the behavior of materials at the microscopic level, allowing manufacturers to better predict the performance and durability of their printed products. This will be particularly important for applications such as aerospace, where the performance of an additively manufactured component can have a significant impact on the safety and reliability of the overall system.

In the aerospace industry, simulation is likely to be used at every stage of research and development in the future. The mesoscale and microscale simulation is key to the industry as the application of simulation at this stage improves the speed, efficiency and functionality of the L-PBF parts and the process. The specific requirement of AM in the aerospace industry in related to the fluid flow, especially in the parts which have a unique design with complex structures for cooling channels. In addition to the design, simulation in AM may be used in the aerospace industry at micro and mesoscale to improve the mechanical properties of the parts produced by improving the powder properties and microstructure tailoring.

In addition to these technical developments, there is also likely to be an increased focus on the use of simulation to optimize the overall manufacturing process. This could involve using simulation to identify bottlenecks in the production line and find ways to improve efficiency, or to develop new manufacturing processes that can reduce waste and improve sustainability. As the use of additive manufacturing continues to grow, the ability to simulate and optimize these processes will become increasingly important. Overall, the future of simulation in the field of additive manufacturing looks bright, with a range of exciting developments on the horizon. As the technology continues to evolve, we can expect to see ever more sophisticated simulation tools that will help manufacturers to optimize their processes and improve the performance of their 3D printed products.

18.7 Conclusions

The aim of this study was to identify the importance of microscale and mesoscale simulation in L-PBF process. Simulation in L-PBF is usually performed at macroscale or part scale level due to abundance of software at this length scale. The purpose of conducting this study is to understand the possibility to improve the understanding of the laser beam and material interaction at microscale and mesoscale level by the use of simulation, for which there is not as much software available when compared to macroscale level.

Literature review was conducted in the field of simulation in L-PBF for the laser beam and material interaction. From this review, it was found out research regarding the simulation of laser beam and material interaction at microscale and mesoscale level is lacking when compared to the simulation at a part scale level. A background study was conducted on the existing research on simulation at microscale and mesoscale level and the use of simulation at this scale was justified. An insight was provided into the simulation methods at microscale and meso scale level with suitable methods suggested for different physical phenomena occurring in the L-PBF process. This was then followed by an analysis of existing software and their impact on research in this field was studied.

The findings of this study indicate the procedure that needs to be followed to conduct simulation at microscale and mesoscale level. An FEM-DEM or an FVM-DEM combination of simulation provides qualitative results for studying the physical phenomena occurring during the L-PBF process. The thermal history of the process combined with the PF model is the method followed by different researchers to simulate microstructure at meso scale level. It was also concluded that among the existing software, COMSOL and ABAQUS have been used frequently followed by newly developed software like Flow3D and Amphyon. The main criteria used to compare different software here are the number of existing publications where the software have been used. A brief description of challenges while using simulation has been presented where experimental validation of the simulation has been identified as a key challenge.

As further studies based on this study, it is concluded that there is need for developing a methodology for using simulation at microscale and mesoscale level to understand the laser beam and material interaction in a qualitative manner and improve the L-PBF process by using the data from this.

Acknowledgements Authors of this article acknowledge project of DREAMS (Database for Radically Enhancing Additive Manufacturing and Standardization). This project was funded by Business Finland and the duration of project is from April 2022 to February 2024. The participants of project were the Finnish Additive Manufacturing Ecosystem (FAME). Aim of the project was to establish standards for AM and develop a powder material database for AM. Authors would like to thank partners of this project for their contribution to this article.

References

1. Kunovjanek M, Wankmüller C (2020) An analysis of the global additive manufacturing response to the COVID-19 pandemic. J Manuf Technol Manag 32(9):75–100. https://doi.org/10.1108/JMTM-07-2020-0263/FULL/PDF
2. Advincula RC, Dizon JRC, Chen Q, Niu I, Chung J, Kilpatrick L, Newman R (2020) Additive manufacturing for COVID-19: devices, materials, prospects, and challenges. MRS Commun 10(3):413–427. https://doi.org/10.1557/MRC.2020.57/FIGURES/4
3. Blakey-Milner B, Gradl P, Snedden G, Brooks M, Pitot J, Lopez E, Leary M, Berto F, du Plessis A (2021) Metal additive manufacturing in aerospace: a review. Mater Des 209:110008. https://doi.org/10.1016/J.MATDES.2021.110008
4. Frazier WE (n.d) Metal additive manufacturing: a review. https://doi.org/10.1007/s11665-014-0958-z
5. Khairallah SA, Anderson AT, Rubenchik A, King WE (2016) Laser powder-bed fusion additive manufacturing: physics of complex melt flow and formation mechanisms of pores, spatter, and denudation zones. Acta Mater 108:36–45. https://doi.org/10.1016/J.ACTAMAT.2016.02.014
6. Gamon A, Arrieta E, Gradl PR, Katsarelis C, Murr LE, Wicker RB, Medina F (2021) Microstructure and hardness comparison of as-built inconel 625 alloy following various additive manufacturing processes. Results Mater 12:100239. https://doi.org/10.1016/J.RINMA.2021.100239
7. Fischer FG, Birk N, Rooney L, Jauer L, Schleifenbaum JH (2021) Optical process monitoring in laser powder bed fusion using a recoater-based line camera. Addit Manuf 47:102218. https://doi.org/10.1016/J.ADDMA.2021.102218
8. Soltani-Tehrani A, Shrestha R, Phan N, Seifi M, Shamsaei N (2021) Establishing specimen property to part performance relationships for laser beam powder bed fusion additive manufacturing. Int J Fatigue 151:106384. https://doi.org/10.1016/J.IJFATIGUE.2021.106384
9. Seifi M, Gorelik M, Waller J, Hrabe N, Shamsaei N, Daniewicz S, Lewandowski JJ (2017) Progress towards metal additive manufacturing standardization to support qualification and certification. JOM 69(3):439–455. https://doi.org/10.1007/S11837-017-2265-2/FIGURES/7

10. Liu R, Liu S, Zhang, · Xiaoli (nd) A physics-informed machine learning model for porosity analysis in laser powder bed fusion additive manufacturing. https://doi.org/10.1007/s00170-021-06640-3
11. Brika SE, Letenneur M, Dion CA, Brailovski V (2020) Influence of particle morphology and size distribution on the powder flowability and laser powder bed fusion manufacturability of Ti-6Al-4V alloy. Addit Manuf 31:100929. https://doi.org/10.1016/J.ADDMA.2019.100929
12. Sutton AT, Kriewall CS, Leu MC, Newkirk JW (2017) Virtual and physical prototyping powder characterisation techniques and effects of powder characteristics on part properties in powder-bed fusion processes) powder characterisation techniques and effects of powder characteristics on part properties in powder-bed fusion processes. Virtual and Physical Prototyping Powder characterisation techniques and effects of powder characteristics on part properties in powder-bed fusion processes. 12:3–29. https://doi.org/10.1080/17452759.2016.1250605
13. Oliveira JP, LaLonde AD, Ma J (2020) Processing parameters in laser powder bed fusion metal additive manufacturing. Mater Des 193:108762. https://doi.org/10.1016/J.MATDES.2020.108762
14. Shen H, Rometsch P, Wu X, Huang A (2020) Influence of gas flow speed on laser plume attenuation and powder bed particle pickup in laser powder bed fusion. JOM 72(3):1039–1051. https://doi.org/10.1007/S11837-020-04020-Y/FIGURES/10
15. Sing SL, Kuo CN, Shih CT, Ho CC, Chua CK (2021) Perspectives of using machine learning in laser powder bed fusion for metal additive manufacturing, vol 16, no 3, pp 372–386. https://doi.org/10.1080/17452759.2021.1944229. https://doi.org/10.1080/17452759.2021.1944229
16. How process parameters drive successful metal AM part production (n.d) https://www.metal-am.com/articles/70927-2/. Retrieved 10 Feb 2023
17. Martin AA, Calta NP, Hammons JA, Khairallah SA, Nielsen MH, Shuttlesworth RM, Sinclair N, Matthews MJ, Jeffries JR, Willey TM, Lee JRI (2019) Ultrafast dynamics of laser-metal interactions in additive manufacturing alloys captured by in situ X-ray imaging. Mater Today Adv 1:100002. https://doi.org/10.1016/J.MTADV.2019.01.001
18. Kruth JP, Froyen L, van Vaerenbergh J, Mercelis P, Rombouts M, Lauwers B (2004) Selective laser melting of iron-based powder. J Mater Process Technol 149(1–3):616–622. https://doi.org/10.1016/J.JMATPROTEC.2003.11.051
19. Wang P, Yang Y, Moghaddam NS (2022) Process modeling in laser powder bed fusion towards defect detection and quality control via machine learning: the state-of-the-art and research challenges. J Manuf Process 73:961–984. https://doi.org/10.1016/J.JMAPRO.2021.11.037
20. Megahed M, Mindt HW, N'Dri N, Duan H, Desmaison O (2016) Metal additive-manufacturing process and residual stress modeling. Int Mater Manuf Innov 5(1):61–93. https://doi.org/10.1186/S40192-016-0047-2/FIGURES/21
21. Jawade SA, Joshi RS, Desai SB (2021) Comparative study of mechanical properties of additively manufactured aluminum alloy. Mater Today Proceed 46:9270–9274. https://doi.org/10.1016/J.MATPR.2020.02.096
22. Bennett J, Glerum J, Cao J (2021) Relating additively manufactured part tensile properties to thermal metrics. CIRP Ann 70(1):187–190. https://doi.org/10.1016/J.CIRP.2021.04.053
23. (PDF) Melt Pool Evolution Study in Selective Laser Melting (n.d) https://www.researchgate.net/publication/283017877_Melt_Pool_Evolution_Study_in_Selective_Laser_Melting. Retrieved 21 Jan 2022
24. Wischeropp TM, Emmelmann C, Brandt M, Pateras A (2019) Measurement of actual powder layer height and packing density in a single layer in selective laser melting. Addit Manuf 28:176–183. https://doi.org/10.1016/J.ADDMA.2019.04.019
25. Wiederkehr P, Siebrecht T (2016) Virtual machining: capabilities and challenges of process simulations in the aerospace industry. Proced Manuf 6:80–87. https://doi.org/10.1016/J.PROMFG.2016.11.011
26. Ibrahimbegovic A, Delaplace A (2003) Microscale and mesoscale discrete models for dynamic fracture of structures built of brittle material. Comput Struct 81(12):1255–1265. https://doi.org/10.1016/S0045-7949(03)00040-3

27. Bayat M, Thanki A, Mohanty S, Witvrouw A, Yang S, Thorborg J, Tiedje NS, Hattel JH (2019) Keyhole-induced porosities in laser-based powder bed fusion (L-PBF) of Ti6Al4V: high-fidelity modelling and experimental validation. Addit Manuf 30:100835. https://doi.org/10.1016/J.ADDMA.2019.100835
28. Zhang Z, Tan ZJ, Yao XX, Hu CP, Ge P, Wan ZY, Li JY, Wu Q (2019) Numerical methods for microstructural evolutions in laser additive manufacturing. Comput Math Appl 78(7):2296–2307. https://doi.org/10.1016/J.CAMWA.2018.07.011
29. Bugatti M, Bianca, Colosimo M (2022) Towards real-time in-situ monitoring of hot-spot defects in L-PBF: a new classification-based method for fast video-imaging data analysis. J Intell Manuf 33:293–309.https://doi.org/10.1007/s10845-021-01787-y
30. Köhnen P, Létang M, Voshage M, Schleifenbaum JH, Haase C (2019) Understanding the process-microstructure correlations for tailoring the mechanical properties of L-PBF produced austenitic advanced high strength steel. Addit Manuf 30:100914. https://doi.org/10.1016/J.ADDMA.2019.100914
31. Thermal and melting track simulations of laser powder bed fusion (L-PBF) (n.d) https://doi.org/10.1088/1757-899X/526/1/012030
32. Chouhan A, Aggarwal A, Kumar A (2021) Role of melt flow dynamics on track surface morphology in the L-PBF additive manufacturing process. Int J Heat Mass Transf 178:121602. https://doi.org/10.1016/J.IJHEATMASSTRANSFER.2021.121602
33. Song X, Feih S, Zhai W, Sun CN, Li F, Maiti R, Wei J, Yang Y, Oancea V, Romano Brandt L, Korsunsky AM (2020) Advances in additive manufacturing process simulation: residual stresses and distortion predictions in complex metallic components. Mater Des 193:108779. https://doi.org/10.1016/J.MATDES.2020.108779
34. Multiscale Modeling of Clays | Andrew J. Whittle's Research Group. (n.d) https://ajw-group.mit.edu/multiscale-modeling-clays. Retrieved 11 Feb 2023
35. Liu D, Wang Y (2019) Mesoscale multi-physics simulation of rapid solidification of Ti-6Al-4V alloy. Addit Manuf 25:551–562. https://doi.org/10.1016/J.ADDMA.2018.12.005
36. Mede T, Kocjan A, Paulin I, Godec M (2020) Numerical mesoscale modelling of microstructure evolution during selective laser melting. Metals 10(6):800. https://doi.org/10.3390/MET10060800
37. Strantza M, Vrancken B, Prime MB, Truman CE, Rombouts M, Brown DW, Guillaume P, van Hemelrijck D (2019) Directional and oscillating residual stress on the mesoscale in additively manufactured Ti-6Al-4V. Acta Mater 168:299–308. https://doi.org/10.1016/J.ACTAMAT.2019.01.050
38. Tang Y (2017) A multiscale and multifunctional design method for mesoscale lattice structures fabricated by additive manufacturing processes
39. Behera D, Chizari S, Shaw LA, Porter M, Hensleigh R, Xu Z, Zheng X, Connolly LG, Roy NK, Panas RM, Saha SK, Zheng X (Rayne), Hopkins JB, Chen SC, Cullinan MA (2021) Current challenges and potential directions towards precision microscale additive manufacturing–part IV: future perspectives. Precis Eng 68:197–205.https://doi.org/10.1016/J.PRECISIONENG.2020.12.014
40. Vastola G, Zhang G, Pei QX, Zhang Y-W (2016) Modeling the microstructure evolution during additive manufacturing of Ti6Al4V: a comparison between electron beam melting and selective laser melting. JOM 68(5). https://doi.org/10.1007/s11837-016-1890-5
41. Nie P, Ojo OA, Li Z (2014) Numerical modeling of microstructure evolution during laser additive manufacturing of a nickel-based superalloy. Acta Mater 77:85–95. https://doi.org/10.1016/J.ACTAMAT.2014.05.039
42. Babu B, Lundbäck A, Lindgren L-E (2019) Simulation of Ti-6Al-4V additive manufacturing using coupled physically based flow stress and metallurgical model. Materials 12(23):3844. https://doi.org/10.3390/MA12233844
43. Gu D, Ma C, Xia M, Dai D, Shi Q (2017) A multiscale understanding of the thermodynamic and kinetic mechanisms of laser additive manufacturing. Engineering 3(5):675–684. https://doi.org/10.1016/J.ENG.2017.05.011

44. Heigel JC, Michaleris P, Reutzel EW (2015) Thermo-mechanical model development and validation of directed energy deposition additive manufacturing of Ti–6Al–4V. Addit Manuf 5:9–19. https://doi.org/10.1016/J.ADDMA.2014.10.003
45. Haeri S (2017) Optimisation of blade type spreaders for powder bed preparation in additive manufacturing using DEM simulations. Powder Technol 321:94–104. https://doi.org/10.1016/J.POWTEC.2017.08.011
46. Pal D, Patil N, Zeng K, Stucker B (2014) An integrated approach to additive manufacturing simulations using physics based, coupled multiscale process modeling. J Manuf Sci Eng Trans ASME 136(6). https://doi.org/10.1115/1.4028580/377646
47. Shuja SZ, Yilbas BS (2011) Laser produced melt pool: influence of laser intensity parameter on flow field in melt pool. Opt Laser Technol 43(4):767–775. https://doi.org/10.1016/J.OPTLASTEC.2010.12.003
48. Cheng B, Loeber L, Willeck H, Hartel U, Tuffile C (2019) Computational investigation of melt pool process dynamics and pore formation in laser powder bed fusion. J Mater Eng Perform 28(11):6565–6578. https://doi.org/10.1007/S11665-019-04435-Y/FIGURES/18
49. Plotkowski A, Kirka MM, Babu SS (2017) Verification and validation of a rapid heat transfer calculation methodology for transient melt pool solidification conditions in powder bed metal additive manufacturing. Addit Manuf 18:256–268. https://doi.org/10.1016/J.ADDMA.2017.10.017
50. Xia M, Gu D, Yu G, Dai D, Chen H, Shi Q (2016) Influence of hatch spacing on heat and mass transfer, thermodynamics and laser processability during additive manufacturing of Inconel 718 alloy. Int J Mach Tools Manuf 109:147–157. https://doi.org/10.1016/J.IJMACHTOOLS.2016.07.010
51. Bajaj P, Hariharan A, Kini A, Kürnsteiner P, Raabe D, Jägle EA (2020) Steels in additive manufacturing: a review of their microstructure and properties. Mater Sci Eng A 772:138633. https://doi.org/10.1016/J.MSEA.2019.138633
52. Steinbach I (2013) MR43CH04-Steinbach ARI 1 June 2013 12:18 phase-field model for microstructure evolution at the mesoscopic scale keynote topic. https://doi.org/10.1146/annurev-matsci-071312-121703
53. Gouge M, Michaleris P, Denlinger E, Irwin J (2018) The finite element method for the thermo-mechanical modeling of additive manufacturing processes. Thermo-Mech Model Addit Manuf 19–38.https://doi.org/10.1016/B978-0-12-811820-7.00003-3
54. Lian Y, Gan Z, Yu C, Kats D, Liu WK, Wagner GJ (2019) A cellular automaton finite volume method for microstructure evolution during additive manufacturing. Mater Des 169:107672. https://doi.org/10.1016/J.MATDES.2019.107672
55. Roussel N, Spangenberg J, Wallevik J, Wolfs R (2020) Numerical simulations of concrete processing: from standard formative casting to additive manufacturing. Cem Concr Res 135:106075. https://doi.org/10.1016/J.CEMCONRES.2020.106075
56. Montevecchi F, Venturini G, Scippa A, Campatelli G (2016) Finite element modelling of wire-arc-additive-manufacturing process. Proced CIRP 55:109–114. https://doi.org/10.1016/J.PROCIR.2016.08.024
57. Luo Z, Zhao Y (2018) A survey of finite element analysis of temperature and thermal stress fields in powder bed fusion Additive Manufacturing. Addit Manuf 21:318–332. https://doi.org/10.1016/J.ADDMA.2018.03.022
58. Desai PS, Mehta A, Dougherty PSM, Higgs FC (2019) A rheometry based calibration of a first-order DEM model to generate virtual avatars of metal additive manufacturing (AM) powders. Powder Technol 342:441–456. https://doi.org/10.1016/J.POWTEC.2018.09.047
59. Li MJ, Chen J, Lian Y, Xiong F, Fang D (2023) An efficient and high-fidelity local multi-mesh finite volume method for heat transfer and fluid flow problems in metal additive manufacturing. Comput Methods Appl Mech Eng 404:115828. https://doi.org/10.1016/J.CMA.2022.115828
60. Chen H, Yan W (2020) Spattering and denudation in laser powder bed fusion process: multi-phase flow modelling. Acta Mater 196:154–167. https://doi.org/10.1016/J.ACTAMAT.2020.06.033

61. Lu LX, Sridhar N, Zhang YW (2018) Phase field simulation of powder bed-based additive manufacturing. Acta Mater 144:801–809. https://doi.org/10.1016/J.ACTAMAT.2017.11.033
62. Ji Y, Chen L, Chen LQ (2018) Understanding microstructure evolution during additive manufacturing of metallic alloys using phase-field modeling. Thermo-Mech Model Addit Manuf 93–116.https://doi.org/10.1016/B978-0-12-811820-7.00008-2
63. Qu M, Guo Q, Escano LI, Clark SJ, Fezzaa K, Chen L (2022) Mitigating keyhole pore formation by nanoparticles during laser powder bed fusion additive manufacturing. Addit Manuf Lett 3:100068. https://doi.org/10.1016/J.ADDLET.2022.100068
64. Sukhotskiy V, Vishnoi P, Karampelas IH, Vader S, Vader Z, Furlani EP (n.d) Magnetohydrodynamic drop-on-demand liquid metal additive manufacturing: system overview and modelling. https://doi.org/10.11159/ffhmt18.155
65. Wang H, Zou Y (2019) Microscale interaction between laser and metal powder in powder-bed additive manufacturing: conduction mode versus keyhole mode. Int J Heat Mass Transf 142:118473. https://doi.org/10.1016/J.IJHEATMASSTRANSFER.2019.118473
66. ben Slama M, Chatti S (2022) Numerical prediction of melt pool size and defect appearing during selective laser melting process of AlSi10Mg at various scan speeds. Lect Notes Mech Eng 226–234.https://doi.org/10.1007/978-3-030-86446-0_30/COVER/
67. Hexagon's e-Xstream engineering presents improved Digimat software | CompositesWorld (n.d) https://www.compositesworld.com/products/hexagons-e-xstream-engineering-presents-improved-digimat-software. Retrieved 11 July 2022
68. Melt pool simulation with AdditiveLab (n.d) https://www.additive-lab.com/post/melt-pool-simulation-with-additivelab Retrieved 11 July 2022
69. Stief P, Dantan J-Y, Etienne A, Siadat A (2018). ScienceDirect ScienceDirect a new methodology to analyze the functional and physical architecture of existing products for an assembly oriented product family identification. https://doi.org/10.1016/j.procir.2018.08.058
70. Mayer T, Brändle G, Schönenberger A, Eberlein R (2020) Simulation and validation of residual deformations in additive manufacturing of metal parts. Heliyon 6(5):e03987. https://doi.org/10.1016/J.HELIYON.2020.E03987
71. Afazov S, Roberts A, Wright L, Jadhav P, Holloway A, Basoalto H, Milne K, Brierley N (2022) Metal powder bed fusion process chains: an overview of modelling techniques. Prog Addit Manuf 7(2):289–314. https://doi.org/10.1007/S40964-021-00230-1
72. Najmon JC, Raeisi S, Tovar A (2019) Review of additive manufacturing technologies and applications in the aerospace industry. Addit Manuf Aerosp Indus 7–31.https://doi.org/10.1016/B978-0-12-814062-8.00002-9
73. Brenken B, Barocio E, Favaloro A, Kunc V, Pipes RB (2019) Development and validation of extrusion deposition additive manufacturing process simulations. Addit Manuf 25:218–226. https://doi.org/10.1016/J.ADDMA.2018.10.041
74. Bandyopadhyay A, Traxel KD (2018) Invited review article: metal-additive manufacturing—modeling strategies for application-optimized designs. Addit Manuf 22:758–774. https://doi.org/10.1016/J.ADDMA.2018.06.024
75. Gibson I, Rosen D, Stucker B, Khorasani M (2021) Design for additive manufacturing. Addit Manuf Technol 555–607.https://doi.org/10.1007/978-3-030-56127-7_19
76. Ansari P, Salamci MU (2022) On the selective laser melting based additive manufacturing of AlSi10Mg: the process parameter investigation through multiphysics simulation and experimental validation. J Alloy Compd 890:161873. https://doi.org/10.1016/J.JALLCOM.2021.161873
77. Lee YS, Nandwana P, Zhang W (n.d) Dynamic simulation of powder packing structure for powder bed additive manufacturing. https://doi.org/10.1007/s00170-018-1697-3

GPSR Compliance

The European Union's (EU) General Product Safety Regulation (GPSR) is a set of rules that requires consumer products to be safe and our obligations to ensure this.

If you have any concerns about our products, you can contact us on ProductSafety@springernature.com

In case Publisher is established outside the EU, the EU authorized representative is:

Springer Nature Customer Service Center GmbH
Europaplatz 3
69115 Heidelberg, Germany

Batch number: 10755227

Printed by Printforce, the Netherlands